ENDORSED BY

# Complete Chemistry
## for Cambridge IGCSE®

Second Edition

RoseMarie Gallagher
Paul Ingram

Oxford and Cambridge
leading education together

**OXFORD**
UNIVERSITY PRESS

# OXFORD
**UNIVERSITY PRESS**

Great Clarendon Street, Oxford OX2 6DP

Oxford University Press is a department of the University of Oxford.
It furthers the University's objective of excellence in research, scholarship,
and education by publishing worldwide in

Oxford  New York

Auckland  Cape Town  Dar es Salaam  Hong Kong  Karachi
Kuala Lumpur  Madrid  Melbourne  Mexico City  Nairobi
New Delhi  Shanghai  Taipei  Toronto

With offices in

Argentina  Austria  Brazil  Chile  Czech Republic  France  Greece
Guatemala  Hungary  Italy  Japan  Poland  Portugal  Singapore
South Korea  Switzerland  Thailand  Turkey  Ukraine  Vietnam

© RoseMarie Gallagher and Paul Ingram 2011

The moral rights of the authors have been asserted

Database right Oxford University Press (maker)

First published as *Complete Chemistry* (ISBN 9780199147991)

This edition first published in 2007

All rights reserved. No part of this publication may be reproduced, stored in a
retrieval system, or transmitted, in any form or by any means, without the prior
permission in writing of Oxford University Press, or as expressly permitted by law,
or under terms agreed with the appropriate reprographics rights organization.
Enquiries concerning reproduction outside the scope of the above should be sent to
the Rights Department, Oxford University Press, at the address above.

You must not circulate this book in any other binding or cover and you must
impose this same condition on any acquirer

British Library Cataloguing in Publication Data

Data available

ISBN 978-0-19-913878-4

20 19 18 17 16 15 14 13 12 11 10

Printed in Singapore by KHL Printing Co. Pte Ltd

Paper used in the production of this book is a natural, recyclable product made
from wood grown in sustainable forests. The manufacturing process conforms to
the environmental regulations of the country of origin.

**Acknowledgments**

®IGCSE is the registered trademark of Cambridge International Examinations.

The publisher would like to thank Cambridge International Examinations for their
kind permission to reproduce past paper questions.

Cambridge International Examinations bears no responsibility for the example
answers to questions taken from its past question papers which are contained in
this publication.

The acknowledgments for the photographs are on page 320.

# Introduction

If you are taking IGCSE chemistry, using the Cambridge International Examinations syllabus 0620, then this book is for you. It covers the syllabus fully, and has been endorsed by the exam board.

**Finding your way around the book**
The contents list on the next page shows how the book is organised. Take a look. Note the extra material at the back of the book too: for example the questions from past exam papers, and the glossary.

**Finding your way around the chapters**
Each chapter is divided into two-page units. Some colour coding is used within the units, to help you use them properly. Look at these notes:

| Core curriculum | Extended curriculum |
|---|---|
| If you are following the Core curriculum, you can ignore any material with a red line beside it. | For this, you need *all* the material on the white pages, including the material marked with a red line. |

| Extra material | Chapter checkups |
|---|---|
| Pages of this colour contain extra material for some topics. We hope that you will find it interesting – but it is not needed for the exam. | There is a revision checklist at the end of each chapter, and also a set of exam-level questions about the chapter, on a coloured background. |

**Making the most of the book and CD**
We want you to understand chemistry, and do well in your exams. This book, and the CD, can help you. So make the most of them!

**Work through the units**   The two-page units will help you build up your knowledge and understanding of the chemistry on your syllabus.

**Use the glossary**   If you come across a chemical term that you do not understand, try the glossary. You can also use the glossary to test yourself.

**Answer the questions**   It is a great way to get to grips with a topic. This book has lots of questions: at the end of each unit and each chapter, and questions from past exam papers at the end of the book.

Answers to the numerical questions are given at the back of the book. Your teacher can provide the answers for all the others.

**Use the CD**   The CD has an interactive test for each chapter, advice on revision, sample exam papers, and more.

**And finally, enjoy!**   Chemistry is an important and exciting subject. We hope this book will help you to enjoy it, and succeed in your course.

*RoseMarie Gallagher*
*Paul Ingram*

# Contents

## 1 States of matter
| | | |
|---|---|---|
| 1.1 | Everything is made of particles | 6 |
| 1.2 | Solids, liquids, and gases | 8 |
| 1.3 | The particles in solids, liquids, and gases | 10 |
| 1.4 | A closer look at gases | 12 |
| | Checkup on Chapter 1 | 14 |

## 2 Separating substances
| | | |
|---|---|---|
| 2.1 | Mixtures, solutions, and solvents | 16 |
| 2.2 | Pure substances and impurities | 18 |
| 2.3 | Separation methods (part I) | 20 |
| 2.4 | Separation methods (part II) | 22 |
| 2.5 | More about paper chromatography | 24 |
| | The chromatography detectives | 26 |
| | Checkup on Chapter 2 | 28 |

## 3 Atoms and elements
| | | |
|---|---|---|
| 3.1 | Atoms and elements | 30 |
| 3.2 | More about atoms | 32 |
| 3.3 | Isotopes and radioactivity | 34 |
| 3.4 | How electrons are arranged | 36 |
| | How our model of the atom developed | 38 |
| | The atom: the inside story | 40 |
| 3.5 | The metals and non-metals | 42 |
| | Checkup on Chapter 3 | 44 |

## 4 Atoms combining
| | | |
|---|---|---|
| 4.1 | Compounds, mixtures, and chemical change | 46 |
| 4.2 | Why do atoms form bonds? | 48 |
| 4.3 | The ionic bond | 50 |
| 4.4 | More about ions | 52 |
| 4.5 | The covalent bond | 54 |
| 4.6 | Covalent compounds | 56 |
| 4.7 | Comparing ionic and covalent compounds | 58 |
| 4.8 | Giant covalent structures | 60 |
| 4.9 | The bonding in metals | 62 |
| | Checkup on Chapter 4 | 64 |

## 5 Reacting masses, and chemical equations
| | | |
|---|---|---|
| 5.1 | The names and formuale of compounds | 66 |
| 5.2 | Equations for chemical reactions | 68 |
| 5.3 | The masses of atoms, molecules. and ions | 70 |
| 5.4 | Some calculations about masses and % | 72 |
| | Checkup on Chapter 5 | 74 |

## 6 Using moles
| | | |
|---|---|---|
| 6.1 | The mole | 76 |
| 6.2 | Calculations from equations, using the mole | 78 |
| 6.3 | Reactions involving gases | 80 |
| 6.4 | The concentration of a solution | 82 |
| 6.5 | Finding the empirical formula | 84 |
| 6.6 | From empirical to final formula | 86 |
| 6.7 | Finding % yield and % purity | 88 |
| | Checkup on Chapter 6 | 90 |

## 7 Redox reactions
| | | |
|---|---|---|
| 7.1 | Oxidation and reduction | 92 |
| 7.2 | Redox and electron transfer | 94 |
| 7.3 | Redox and changes in oxidation state | 96 |
| 7.4 | Oxidising and reducing agents | 98 |
| | Checkup on Chapter 7 | 100 |

## 8 Electricity and chemical change
| | | |
|---|---|---|
| 8.1 | Conductors and insulators | 102 |
| 8.2 | The principles of electrolysis | 104 |
| 8.3 | The reactions at the electrodes | 106 |
| 8.4 | The electrolysis of brine | 108 |
| 8.5 | Two more uses of electrolysis | 110 |
| | Checkup on Chapter 8 | 112 |

## 9 Energy changes, and reversible reactions
| | | |
|---|---|---|
| 9.1 | Energy changes in reactions | 114 |
| 9.2 | Explaining energy changes | 116 |
| 9.3 | Energy from fuels | 118 |
| 9.4 | Giving out energy as electricity | 120 |
| | The batteries in your life | 122 |
| 9.5 | Reversible reactions | 124 |
| 9.6 | Shifting the equilibrium | 126 |
| | Checkup on Chapter 9 | 128 |

## 10 The speed of a reaction
| | | |
|---|---|---|
| 10.1 | Rates of reaction | 130 |
| 10.2 | Measuring the rate of a reaction | 132 |
| 10.3 | Changing the rate of a reaction (part I) | 134 |
| 10.4 | Changing the rate of a reaction (part II) | 136 |
| 10.5 | Explaining rates | 138 |
| 10.6 | Catalysts | 140 |
| | More about enzymes | 142 |
| 10.7 | Photochemical reactions | 144 |
| | Checkup on Chapter 10 | 146 |

## 11 Acids and bases

| | | |
|---|---|---|
| 11.1 | Acids and alkalis | 148 |
| 11.2 | A closer look at acids and alkalis | 150 |
| 11.3 | The reactions of acids and bases | 152 |
| 11.4 | A closer look at neutralisation | 154 |
| 11.5 | Oxides | 156 |
| 11.6 | Making salts | 158 |
| 11.7 | Making insoluble salts by precipitation | 160 |
| 11.8 | Finding concentrations by titration | 162 |
| | Checkup on Chapter 11 | 164 |

## 12 The Periodic Table

| | | |
|---|---|---|
| 12.1 | An overview of the Periodic Table | 166 |
| 12.2 | Group I: the alkali metals | 168 |
| 12.3 | Group VII: the halogens | 170 |
| 12.4 | Group 0: the noble gases | 172 |
| 12.5 | The transition elements | 174 |
| 12.6 | Across the Periodic Table | 176 |
| | How the Periodic Table developed | 178 |
| | Checkup on Chapter 12 | 180 |

## 13 The behaviour of metals

| | | |
|---|---|---|
| 13.1 | Metals: a review | 182 |
| 13.2 | Comparing metals for reactivity | 184 |
| 13.3 | Metals in competition | 186 |
| 13.4 | The reactivity series | 188 |
| 13.5 | Making use of the reactivity series | 190 |
| | Checkup on Chapter 13 | 192 |

## 14 Making use of metals

| | | |
|---|---|---|
| 14.1 | Metals in the Earth's crust | 194 |
| 14.2 | Extracting metals from their ores | 196 |
| 14.3 | Extracting iron | 198 |
| 14.4 | Extracting aluminium | 200 |
| 14.5 | Making use of metals and alloys | 202 |
| 14.6 | Steels and steel-making | 204 |
| | Metals, civilisation, and you | 206 |
| | Checkup on Chapter 14 | 208 |

## 15 Air and water

| | | |
|---|---|---|
| 15.1 | What is air? | 210 |
| 15.2 | Making use of air | 212 |
| 15.3 | Pollution alert! | 214 |
| 15.4 | The rusting problem | 216 |
| 15.5 | Water supply | 218 |
| | Living in space | 220 |
| | Checkup on Chapter 15 | 222 |

## 16 Some non-metals and their compounds

| | | |
|---|---|---|
| 16.1 | Hydrogen, nitrogen, and ammonia | 224 |
| 16.2 | Making ammonia in industry | 226 |
| 16.3 | Fertilisers | 228 |
| 16.4 | Sulfur and sulfur dioxide | 230 |
| 16.5 | Sulfuric acid | 232 |
| 16.6 | Carbon and the carbon cycle | 234 |
| 16.7 | Some carbon compounds | 236 |
| 16.8 | Greenhouse gases, and global warming | 238 |
| 16.9 | Limestone | 240 |
| | Checkup on Chapter 16 | 242 |

## 17 Organic chemistry

| | | |
|---|---|---|
| 17.1 | Petroleum: a fossil fuel | 244 |
| 17.2 | Refining petroleum | 246 |
| 17.3 | Cracking hydrocarbons | 248 |
| 17.4 | Families of organic compounds | 250 |
| 17.5 | The alkanes | 252 |
| 17.6 | The alkenes | 254 |
| 17.7 | The alcohols | 256 |
| 17.8 | The carboxylic acids | 258 |
| | Checkup on Chapter 17 | 260 |

## 18 Polymers

| | | |
|---|---|---|
| 18.1 | Introducing polymers | 262 |
| 18.2 | Addition polymerisation | 264 |
| 18.3 | Condensation polymerisation | 266 |
| 18.4 | Making use of synthetic polymers | 268 |
| 18.5 | Plastics: here to stay? | 270 |
| 18.6 | The macromolecules in food (part I) | 272 |
| 18.7 | The macromolecules in food (part II) | 274 |
| 18.8 | Breaking down the macromolecules | 276 |
| | Checkup on Chapter 18 | 278 |

## 19 In the lab

| | | |
|---|---|---|
| 19.1 | Chemistry: a practical subject | 280 |
| 19.2 | Example of an experiment | 282 |
| 19.3 | Working with gases in the lab | 284 |
| 19.4 | Testing for ions in the lab | 286 |
| | Checkup on Chapter 19 | 288 |

Answers to the numerical questions in this book    290

### Your Cambridge IGCSE chemistry exam

| | |
|---|---|
| About the Cambridge IGCSE chemistry exam | 291 |
| Exam questions from Paper 2 | 292 |
| Exam questions from Paper 3 | 298 |
| Exam questions from Paper 6 | 304 |

### Reference

| | |
|---|---|
| Glossary | 310 |
| The Periodic Table and atomic masses | 314 |
| Index | 316 |

# STATES OF MATTER

## 1.1 Everything is made of particles

### Made of particles

Rock, air, and water look very different. But they have one big thing in common: they are all made of very tiny pieces, far too small to see. For the moment, we will call these pieces **particles**.

In fact everything around you is made of particles – and so are you!

### Particles on the move

In rock and other solids, the particles are not free to move around. But in liquids and gases, they move freely. As they move they collide with each other, and bounce off in all directions.

So the path of one particle, in a liquid or gas, could look like this:

The particle moves in a random way, changing direction every time it hits another particle. We call this **random motion**.

▲ All made of particles!

### Some evidence for particles

There is evidence all around you that things are made of particles, and that they move around in liquids and gases. Look at these examples.

**Evidence outside the lab**

1 Cooking smells can spread out into the street. This is because 'smells' are caused by gas particles mixing with, and moving through, the air. They dissolve in moisture in the lining of your nose.

2 You often see dust and smoke dancing in the air, in bright sunlight. The dust and smoke are clusters of particles. They dance around because they are being bombarded by tiny particles in the air.

# STATES OF MATTER

### Evidence in the lab

**1** Place a crystal of potassium manganate(VII) in a beaker of water. The colour spreads through the water. Why? First, particles leave the crystal – it **dissolves**. Then they mix among the water particles.

**2** Place an open gas jar of air upside down on an open gas jar containing a few drops of red-brown bromine. The colour spreads upwards because particles of bromine vapour mix among the particles of air.

## Diffusion

In all those examples, particles mix by colliding with each other and bouncing off in all directions. This mixing process is called **diffusion**.

The overall result is the flow of particles from where they are more concentrated to where they are less concentrated, until they are evenly spread out.

## So what are these particles?

The very smallest particles, that we cannot break down further by chemical means, are called **atoms**.

- In some substances, the particles are just single atoms. For example argon, a gas found in air, is made up of single argon atoms.
- In many substances, the particles consist of two or more atoms joined together. These particles are called **molecules**. Water, bromine, and the gases nitrogen and oxygen in air, are made up of molecules.
- In other substances the particles consist of atoms or groups of atoms that carry a charge. These particles are called **ions**. Potassium manganate(VII) is made of ions.

You'll find out more about all these particles in Chapters 2 and 3.

## 'Seeing' particles

We are now able to 'see' the particles in some solids, using very powerful microscopes. For example the image on the right shows palladium atoms sitting on carbon atoms. In this image, the atoms appear over 70 million times larger than they really are!

▲ This image was taken using a tunneling electron microscope. The white blobs are palladium atoms, the blue ones are carbon. (The colour was added to help us see them.)

---

**Q**

**1** The particles in liquids and gases show *random motion*. What does that mean, and why does it occur?

**2** Why does the purple colour spread when a crystal of potassium manganate(VII) is placed in water?

**3** Bromine vapour is heavier than air. Even so, it spreads upwards in the experiment above. Why?

**4 a** What is *diffusion*?  **b** Use the idea of diffusion to explain how the smell of perfume travels.

## STATES OF MATTER

# 1.2 Solids, liquids, and gases

### What's the difference?
It is easy to tell the difference between a solid, a liquid and a gas:

A solid has a fixed shape and a fixed volume. It does not flow. Think of all the solid things around you: their shapes and volumes do not change.

A liquid flows easily. It has a fixed volume, but its shape changes. It takes the shape of the container you pour it into.

A gas does not have a fixed volume or shape. It spreads out to fill its container. It is much lighter than the same volume of solid or liquid.

### Water: solid, liquid and gas
Water can be a solid (ice), a liquid (water), and a gas (water vapour or steam). Its state can be changed by heating or cooling:

1  **Ice** slowly changes to **water**, when it is put in a warm place. This change is called **melting**. The thermometer shows 0 °C until all the ice has melted. So 0 °C is called its **melting point**.

2  When the water is heated its temperature rises, and some of it changes to **water vapour**. This change is called **evaporation**. The hotter the water gets, the more quickly it evaporates.

3  Soon bubbles appear in the water. It is **boiling**. The water vapour shows up as steam. The thermometer stays at 100 °C while the water boils off. 100 °C is the **boiling point** of water.

And when steam is cooled, the opposite changes take place:

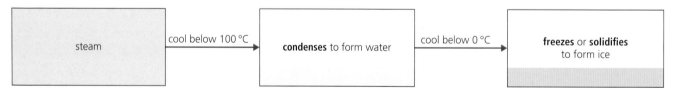

You can see that:
- condensing is the opposite of evaporating
- freezing is the opposite of melting
- the freezing point of water is the same as the melting point of ice, 0 °C.

## Other things can change state too

It's not just water! Nearly all substances can exist as solid, liquid and gas. Even iron and diamond can melt and boil! Some melting and boiling points are given below. Look how different they are.

| Substance | Melting point/°C | Boiling point/°C |
|---|---|---|
| oxygen | −219 | −183 |
| ethanol | −15 | 78 |
| sodium | 98 | 890 |
| sulfur | 119 | 445 |
| iron | 1540 | 2900 |
| diamond | 3550 | 4832 |

▲ Molten iron being poured out at an iron works. Hot – over 1540°C!

## Showing changes of state on a graph

Look at this graph. It shows how the temperature changes as a block of ice is steadily heated. First the ice melts to water. Then the water gets warmer and warmer, and eventually turns to steam:

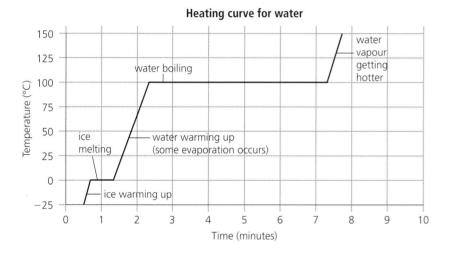

A graph like this is called a **heating curve**.

Look at the step where the ice is melting. Once melting starts, the temperature stays at 0°C until *all* the ice has melted. When the water starts to boil, the temperature stays at 100°C until *all* the water has turned to steam. So the melting and boiling points are clear and sharp.

▲ Evaporation in the sunshine …

**Q**

1 Write down two properties of a solid, two of a liquid, and two of a gas.
2 Which word means the opposite of:
 a boiling?  b melting?
3 Which has a lower freezing point, oxygen or ethanol?
4 Which has a higher boiling point, oxygen or ethanol?
5 Look at the heating curve above.
 a About how long did it take for the ice to melt, once melting started?
 b How long did boiling take to complete, once it started?
 c Try to think of a reason for the difference in **a** and **b**.
6 See if you can sketch a heating curve for sodium.

STATES OF MATTER

## 1.3 The particles in solids, liquids, and gases

### How the particles are arranged

Water can change from solid to liquid to gas. Its *particles* do not change. They are the same in each state. But their *arrangement* changes. The same is true for all substances.

| State | How the particles are arranged | Diagram of particles |
|---|---|---|
| **Solid** | The particles in a solid are arranged in a fixed pattern or **lattice**. Strong forces hold them together. So they cannot leave their positions. The only movements they make are tiny vibrations to and fro. | |
| **Liquid** | The particles in a liquid can move about and slide past each other. They are still close together, but not in a lattice. The forces that hold them together are weaker than in a solid. | |
| **Gas** | The particles in a gas are far apart, and they move about very quickly. There are almost no forces holding them together. They collide with each other and bounce off in all directions. | |

### Changing state

**Melting** When a solid is heated, its particles get more energy and vibrate more. This makes the solid **expand**. At the melting point, the particles vibrate so much that they break away from their positions. The solid turns liquid.

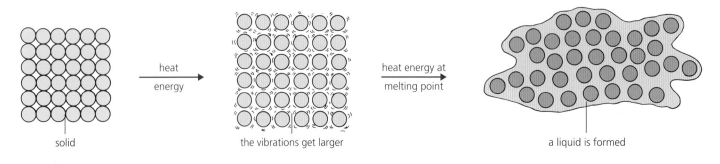

solid → heat energy → the vibrations get larger → heat energy at melting point → a liquid is formed

STATES OF MATTER

**Boiling** When a liquid is heated, its particles get more energy and move faster. They bump into each other more often, and bounce further apart. This makes the liquid expand. At the boiling point, the particles get enough energy to overcome the forces between them. They break away to form a gas:

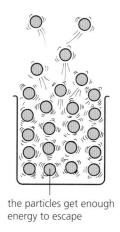

**Evaporating** Some particles in a liquid have more energy than others. Even well below the boiling point, some have enough energy to escape and form a gas. This is called **evaporation**. It is why puddles of rain dry up in the sun.

## How much heat is needed?
The amount of heat needed to melt or boil a substance is different for every substance. That's because the particles in each substance are different, with different forces between them.

The stronger the forces, the more heat energy is needed to overcome them. So the higher the melting and boiling points will be.

## Reversing the changes
You can reverse those changes again by cooling. As a gas cools, its particles lose energy and move more slowly. When they collide, they do not have enough energy to bounce away. So they stay close, and form a liquid. On further cooling, the liquid turns to a solid.

Look at this diagram for water:

**The kinetic particle theory**
Look at the key ideas you have met:
- A substance can be a solid, a liquid, or a gas, and change from one state to another.
- It has different characteristics in each state. (For example, solids do not flow.)
- The differences are due to the way its particles are arranged, and move, in each state.

Together, these ideas make up the **kinetic particle theory**.
(*Kinetic* means *about motion*.)

---

**Q**

1 Using the idea of particles, explain why:
   **a** you can pour liquids   **b** solids expand on heating
2 Draw a diagram to show what happens to the particles, when a liquid cools to a solid.
3 Oxygen is the gas we breathe in. It can be separated from the air. It boils at −219 °C and freezes at −183 °C.
   **a** In which state is oxygen, at: **i** 0 °C? **ii** −200 °C?
   **b** How would you turn oxygen gas into solid oxygen?

STATES OF MATTER

## 1.4 A closer look at gases

### What is gas pressure?
When you blow up a balloon, you fill it with air particles. They collide with each other. They also hit the sides of the balloon, and exert **pressure** on it. This pressure keeps the balloon inflated.

In the same way, *all* gases exert a pressure. The pressure depends on the **temperature** of the gas and the **volume** it takes up, as you'll see below.

### When you heat a gas

 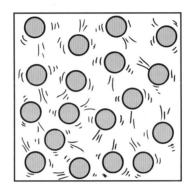

The particles in this gas are moving fast. They hit the walls of the container and exert pressure on them. If you now heat the gas …

… the particles take in heat energy and move even faster. They hit the walls more often, and with more force. So the gas pressure increases.

The same happens with all gases:
**When you heat a gas in a closed container, its pressure increases.**
That is why the pressure gets very high inside a pressure cooker.

▲ The harder you blow, the greater the pressure inside the balloon.

▲ In a pressure cooker, water vapour (gas) is heated to well over 100 °C. So it is at high pressure. You must let a pressure cooker cool before you open it!

### When you squeeze a gas into a smaller space

There is a lot of space between the particles in a gas. You can **compress** the gas, or force its particles closer, by pushing in the plunger …

… like this. Now the particles are in a smaller space – so they hit the walls more often. So the gas pressure increases.

The same thing is true for all gases:
**When a gas is compressed into a smaller space, its pressure increases.**

*All* gases can be compressed. If enough force is applied, the particles can be pushed so close that the gas turns into a liquid. But liquids and solids cannot be compressed, because their particles are already very close together.

▲ When you blow up a bicycle tyre, you compress air into the inner tube.

# The rate of diffusion of gases

On page 7 you saw that gases **diffuse** because the particles collide with other particles, and bounce off in all directions. But gases do not all diffuse at the same rate, every time. It depends on these two factors:

## 1 The mass of the particles

The particles in hydrogen chloride gas are twice as heavy as those in ammonia gas. So which gas do you think will diffuse faster? Let's see:

- Cotton wool soaked in ammonia solution is put into one end of a long tube (at A below). It gives off ammonia gas.
- *At the same time*, cotton wool soaked in hydrochloric acid is put into the other end of the tube (at B). It gives off hydrogen chloride gas.
- The gases diffuse along the tube. White smoke forms where they meet:

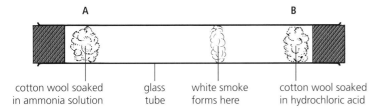

cotton wool soaked in ammonia solution | glass tube | white smoke forms here | cotton wool soaked in hydrochloric acid

The white smoke forms closer to B. So the ammonia particles have travelled further than the hydrogen chloride particles – which means they have travelled *faster*.

**The lower the mass of its particles, the faster a gas will diffuse.**

That makes sense when you think about it. When particles collide and bounce away, the lighter particles will bounce further.

The particles in the two gases above are molecules. The mass of a molecule is called its **relative molecular mass**. So we can also say:
**The lower its relative molecular mass, the faster a gas will diffuse.**

## 2 The temperature

When a gas is heated, its particles take in heat energy, and move faster. They collide with more energy, and bounce further away. So the gas diffuses faster. **The higher the temperature, the faster a gas will diffuse.**

▲ The scent of flowers travels faster in a warm room. Can you explain why?

▲ The faster a particle is moving when it hits another, the faster and further it will bounce away. Just like snooker balls!

---

**Q**

1. What causes the *pressure* in a gas?
2. Why does a balloon burst if you keep on blowing?
3. A gas is in a sealed container. How do you think the pressure will change if the container is cooled? Explain your answer.
4. A gas flows from one container into a larger one. What do you think will happen to its pressure? Draw diagrams to explain.
5. **a** Why does the scent of perfume spread?
   **b** Why does the scent of perfume wear off faster in warm weather than in cold?
6. Of all gases, hydrogen diffuses fastest at any given temperature. What can you tell from this?
7. Look at the glass tube above. Suppose it was warmed a little in an oven, before the experiment. Do you think that would change the result? If so, how?

# STATES OF MATTER

## Checkup on Chapter 1

### Revision checklist

#### Core curriculum
Make sure you can …
- [ ] give two examples of evidence, from the lab, that matter is made of particles
- [ ] explain what *diffusion* is, and how it happens
- [ ] name the three states of matter, and give their physical properties (hard, fixed shape, and so on)
- [ ] describe, and sketch, the particle arrangement in each state
- [ ] describe how a substance changes state when you heat it, and explain this using the idea of particles
- [ ] explain, and use, these terms:
   melt    boil    evaporate    condense
   melting point    boiling point    freezing point
- [ ] sketch, and label, a heating curve
- [ ] explain why a gas exerts a pressure
- [ ] explain why the pressure increases when you:
   – heat a gas
   – push it into a smaller space

#### Extended curriculum
Make sure you can also …
- [ ] describe an experiment to show that a gas will diffuse faster than another gas that has heavier particles
- [ ] say how, and why, the temperature affects the rate at which a gas diffuses

### Questions

#### Core curriculum

1 A large crystal of potassium manganate(VII) was placed in the bottom of a beaker of cold water, and left for several hours.

   a Describe what would be seen:
      i after five minutes    ii after several hours
   b Explain your answers using the idea of particles.
   c Name the two processes that took place during the experiment.

2 Use the idea of particles to explain why:
   a solids have a definite shape
   b liquids fill the bottom of a container
   c you can't store gases in open containers
   d you can't squeeze a sealed plastic syringe that is completely full of water
   e a balloon expands as you blow into it.

3 Below is a heating curve for a pure substance. It shows how the temperature rises over time, when the substance is heated until it melts, then boils.

   a What is the melting point of the substance?
   b What happens to the temperature while the substance changes state?
   c The graph shows that the substance takes longer to boil than to melt. Suggest a reason for this.
   d How can you tell that the substance is not water?
   f Sketch a rough heating curve for pure water.

## STATES OF MATTER

**4** A **cooling curve** is the opposite of a heating curve. It shows how the temperature of a substance changes with time, as it is cooled from a gas to a solid. Here is the cooling curve for one substance:

  a What is the state of the substance at room temperature (20 °C)?
  b Use the list of melting and boiling points on page 9 to identify the substance.
  c Sketch a cooling curve for pure water.

**5** Using the idea of particles explain why:
  a the smell of burnt food travels through the house
  b when two solids are placed on top of each other, they do not mix
  c pumping up your bike tyres gives a smooth ride
  d smokers can cause lung damage in other people
  e heating a gas in a closed container will increase its pressure
  f a liquid is used in a car's breaking system, to transfer the pressure from the brake pedal
  g poisonous gases from a factory chimney can affect a large area.

**6** a Which of these are examples of diffusion?
   i a helium-filled balloon rising in air
   ii a hydrogen-filled balloon deflating, due to gas passing through the skin
   iii the smell of perfume from a person standing on the other side of a room
   iv sucking a drink from a bottle, using a straw
   v an ice lolly turning liquid when it is left out of the freezer
   vi the tea in the cup changing colour when you add milk, without stirring
   vii a light, coloured gas, spreading down through a gas jar
   viii a blue crystal forming a blue solution, when it is left sitting in a glass of water
   ix spraying paint from a spray can.
  b For *one* of the examples of diffusion, draw a diagram showing the particles before and after diffusion has taken place.

### Extended curriculum

**7** You can measure the rate of diffusion of a gas using this apparatus. The gas enters through the thin tube:

The measuring tube is sealed at the top with a plug of porous plaster. Air and other gases can diffuse in and out through the tiny holes in the plug.

The water rises in the measuring tube if the chosen gas diffuses out through the plug faster than air diffuses in. Air is mainly nitrogen and oxygen.

  a When you use hydrogen gas, the water rises in the measuring tube. Why?
  b What does this tell you about the rate of diffusion of hydrogen, compared to the gases in air?
  c Explain your answer to **b**. Use the term *mass*!
  d The molecules in carbon dioxide are heavier than those in nitrogen and oxygen. So what do you think will happen to the water in the measuring tube, when you use carbon dioxide? Explain your answer.

**8**

| Gas | Formula | Relative atomic or molecular mass |
|---|---|---|
| methane | $CH_4$ | 16 |
| helium | He | 4 |
| oxygen | $O_2$ | 32 |
| nitrogen | $N_2$ | 28 |
| chlorine | $Cl_2$ | 71 |

Look at the table above.
  a Which two gases will mix fastest? Explain.
  b Which gas will take least time to escape from a gas syringe?
  c Would you expect chlorine to diffuse more slowly than the gases in air? Explain.
  d An unknown gas diffuses faster than nitrogen, but more slowly than methane. What you can say about its relative molecular mass?

SEPARATING SUBSTANCES

## 2.1 Mixtures, solutions, and solvents

### Mixtures

A **mixture** contains more than one substance. The substances are just mixed together, and not chemically combined. For example:

- air is a mixture of nitrogen, oxygen, and small amounts of other gases
- shampoo is a mixture of several chemicals and water.

### Solutions

When you mix sugar with water, the sugar seems to disappear. That is because its particles spread all through the water particles, like this:

The sugar has **dissolved** in the water, giving a mixture called a **solution**. Sugar is the **solute**, and water is the **solvent**:

**solute + solvent = solution**

You can't get the sugar out again by filtering.

### Not everything dissolves so easily

Now think about chalk. If you mix chalk powder with water, most of the powder eventually sinks to the bottom. You can get it out again by filtering.

Why is it so different for sugar and chalk? Because their particles are very different! How easily a substance dissolves depends on the particles in it. Look at the examples in this table:

| Compound | Mass (g) dissolving in 100 g of water at 25 °C |
|---|---|
| silver nitrate | 241.3 |
| calcium nitrate | 102.1 |
| sugar (glucose) | 91.0 |
| potassium nitrate | 37.9 |
| potassium sulfate | 12.0 |
| calcium hydroxide | 0.113 |
| calcium carbonate (chalk) | 0.0013 |
| silver chloride | 0.0002 |

decreasing solubility

So silver nitrate is much more soluble than sugar – but potassium nitrate is a lot less soluble than sugar. It all depends on the particles.

Look at calcium hydroxide. It is only very slightly or **sparingly soluble** compared with the compounds above it. Its solution is called **limewater**.

Now look at the last two substances in the table. They are usually called **insoluble** since so very little dissolves.

▲ A mixture of sugar and water. This mixture is a solution.

▲ A mixture of chalk powder and water. This is not a solution. The tiny chalk particles do not separate and spread through the water particles. They stay in clusters big enough to see. In time, most sink to the bottom.

**What's soluble, what's not?**
- The solubility of every substance is different.
- But there are some overall patterns. For example *all* sodium compounds are soluble.
- Find out more on page 160.

# Helping a solute dissolve

Sugar dissolves quite slowly in water at room temperature. If you stir the liquid, that helps. But if you keep on adding sugar …

… eventually no more of it will dissolve, no matter how hard you stir. The extra sinks to the bottom. The solution is now **saturated**.

But look what happens if you heat the solution. The extra sugar dissolves. Add more sugar and it will dissolve too, as the temperature rises.

So sugar is **more soluble** in hot water than in cold water.

**A soluble solid usually gets more soluble as the temperature rises.**

**A solution is called *saturated* when it can dissolve no more solute, at that temperature.**

## Water is not the only solvent

Water is the world's most common solvent. A solution in water is called an **aqueous solution** (from *aqua*, the Latin word for water).

But many other solvents are used in industry and about the house, to dissolve substances that are insoluble in water. For example:

| Solvent | It dissolves |
| --- | --- |
| white spirit | gloss paint |
| propanone (acetone) | grease, nail polish |
| ethanol | glues, printing inks, the scented substances that are used in perfumes and aftershaves |

All three of these solvents evaporate easily at room temperature – they are **volatile**. This means that glues and paints dry easily. Aftershave feels cool because ethanol cools the skin when it evaporates.

▲ Nail polish is insoluble in water. It can be removed later by dissolving it in propanone.

> **About volatile liquids**
> - A **volatile** liquid is one that evaporates easily.
> - This is a sign that the forces between its particles are weak.
> - So volatile liquids have low boiling points too. (Propanone boils at 56.5 °C.)

**Q**

1. Explain each term in your own words:
   a  soluble    b  insoluble    c  aqueous solution
2. Look at the table on page 16.
   a  Which substance in it is the most soluble?
   b  About how many times more soluble is this substance than potassium sulfate, at 25 °C?
   c  The substance in **a** gives a colourless solution. What will you see if you add 300 g of it to 100 g of water at 25 °C?
   d  What will you see if you heat up the mixture in **c**?
3. Now turn to the table at the top of page 160.
   a  Name two metals that have *no* insoluble salts.
   b  Name one other group of salts that are *always* soluble.
4. See if you can give three examples of:
   a  solids you dissolve in water, at home
   b  insoluble solids you use at home.
5. Name two solvents other than water that are used in the home. What are they used for?
6. Many gases dissolve in water. Try to give some examples.

SEPARATING SUBSTANCES

## 2.2 Pure substances and impurities

### What is a pure substance?

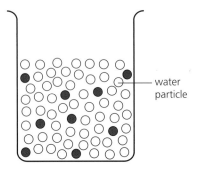

This is water. It has only water particles in it, and nothing else. So it is 100% **pure**.

This water has particles of other substances mixed with it. So it is not pure.

This water has particles of a harmful substance in it. So it is not pure – and could make you ill.

**A pure substance has no particles of any other substance mixed with it.**

In real life, very few substances are 100% pure. For example tap water contains small amounts of many different particles (such as calcium ions and chloride ions). The particles in it are not usually harmful – and some are even good for you.

Distilled water is much purer than tap water, but still not 100% pure. For example it may contain particles of gases, dissolved from the air.

### Does purity matter?

Often it does not matter if a substance is not pure. We wash in tap water, without thinking too much about what is in it. But sometimes purity is very important. If you are making a new medical drug, or a flavouring for food, you must make sure it contains nothing that could harm people.

An unwanted substance, mixed with the substance you want, is called an **impurity**.

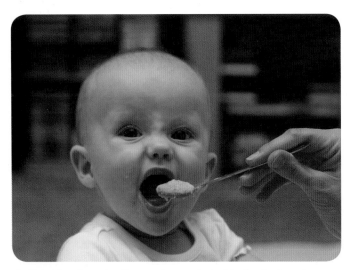

▲ Baby foods and milk powder are tested in the factory, to make sure they contain no harmful impurities.

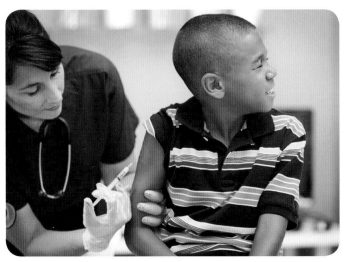

▲ Getting ready for a jab. Vaccines and medicines must be safe, and free of harmful impurities. So they are tested heavily.

# SEPARATING SUBSTANCES

## How can you tell if a substance is pure?

Chemists use some complex methods to check purity. But there is one simple method *you* can use in the lab: **you can check melting and boiling points**.

- A pure substance has a definite, sharp, melting point and boiling point. These are different for each substance. You can look them up in tables.
- When a substance contains an impurity:
  - its melting point falls and its boiling point rises
  - it melts and boils over a range of temperatures, not sharply.
- The more impurity there is:
  - the bigger the change in melting and boiling points
  - the wider the temperature range over which melting and boiling occur.

**ID check!**
- Every substance has a unique pair of melting and boiling points.
- So you can also use melting and boiling points to **identify** a substance.
- First, measure them. Then look up data tables to find out what the substance is.

For example:

| Substance | sulfur | water |
|---|---|---|
| Melts at (°C) | 119 | 0 |
| Boils at (°C) | 445 | 100 |

These are the melting and boiling points for two pure substances: sulfur and water.

This sulfur sample melts sharply at 119 °C and boils at 445 °C. So it must be pure.

This water freezes around −0.5 °C and boils around 101 °C. So it is not pure.

## Separation: the first step in obtaining a pure substance

When you carry out a reaction, you usually end up with a *mixture* of substances. Then you have to separate the one you want.

The table below shows some separation methods. These can give quite pure substances. For example when you filter off a solid, and rinse it well with distilled water, you remove a lot of impurity. But it is just not possible to remove every tiny particle of impurity, in the school lab.

| Method of separation | Used to separate... |
|---|---|
| filtration | a solid from a liquid |
| crystallisation | a solute from its solution |
| evaporation | a solute from its solution |
| simple distillation | a solvent from a solution |
| fractional distillation | liquids from each other |
| paper chromatography | different substances from a solution |

There is more about these methods in the next three units.

▲ At the end of this reaction, the beaker may contain several products, plus reactants that have not reacted. Separating them can be a challenge!

**Q**

1. What does *a pure substance* mean?
2. You mix instant coffee with water, to make a cup of coffee. Is the coffee an *impurity*? Explain.
3. Explain why melting and boiling points can be used as a way to check purity.
4. Could there be impurities in a gas? Explain.

SEPARATING SUBSTANCES

## 2.3 Separation methods (part I)

### Separating a solid from a liquid
Which method should you use? It depends on whether the solid is dissolved, and how its solubility changes with temperature.

**1 By filtering**

For example, chalk is insoluble in water. So it is easy to separate by filtering. The chalk is trapped in the filter paper, while the water passes through. The trapped solid is called the **residue**. The water is the **filtrate**.

▲ Filtering in the kitchen ...

> **Saturated solutions**
> - Remember, most solutes get *more* soluble as the temperature rises – so *less* soluble as it falls!
> - A saturated solution can hold no more solute, at that temperature.

**2 By crystallisation**

You can obtain many solids from their solutions by letting crystals form. The process is called **crystallisation**. It works because soluble solids tend to be *less soluble at lower temperatures*. For example:

**1** This is a solution of copper(II) sulfate in water. You want to obtain solid copper(II) sulfate from it.

**2** So you heat the solution to evaporate some of the water. It becomes more concentrated.

**3** Eventually the solution becomes **saturated**. If you cool it now, crystals will start to form.

**4** Check that it is ready by placing a drop on a microscope slide. Crystals will form quickly on the cool glass.

**5** Leave the solution to cool. Crystals start to form in it, as the temperature falls.

**6** Remove the crystals by filtering. Then rinse them with distilled water and dry them with filter paper.

# SEPARATING SUBSTANCES

Making a living from crystallisation. Seawater is led into shallow ponds. The water evaporates in the sun. He collects the sea salt, and sells it.

## 3 By evaporating all the solvent

For some substances, the solubility changes very little as the temperature falls. So crystallisation does not work for these. Salt is an example.

To obtain salt from an aqueous solution, you need to keep heating the solution, to evaporate the water.

When there is only a little water left, the salt will start to appear. Heat carefully until it is dry.

▲ Evaporating the water from a solution of salt in water.

## Separating a mixture of two solids

To separate two solids, you could choose a solvent that will dissolve just one of them.

For example, water dissolves salt but not sand. So you could separate a mixture of salt and sand like this:

1 Add water to the mixture, and stir. The salt dissolves.
2 Filter the mixture. The sand is trapped in the filter paper, but the salt solution passes through.
3 Rinse the sand with water, and dry it in an oven.
4 Evaporate the water from the salt solution, to give dry salt.

Water could *not* be used to separate salt and sugar, because it dissolves both. But you could use ethanol, which dissolves sugar but not salt. Ethanol is flammable, so should be evaporated over a water bath, as shown here.

▲ Evaporating the ethanol from a solution of sugar in ethanol, over a water bath.

## Q

1 What does this term mean? Give an example.
   a filtrate
   b residue
2 You have a solution of sugar in water. You want to obtain the sugar from it.
   a Explain why filtering will not work.
   b Which method will you use instead?
3 Describe how you would crystallise potassium nitrate from its aqueous solution.
4 How would you separate salt and sugar? Mention any special safety precaution you would take.
5 Now see if you can think of a way to get clean sand from a mixture of sand and little bits of iron wire.

SEPARATING SUBSTANCES

## 2.4 Separation methods (part II)

### Simple distillation

This is a way to obtain the *solvent* from a solution. The apparatus is shown on the right. It could be used to obtain water from salt water, for example. Like this:

1. Heat the solution in the flask. As it boils, water vapour rises into the condenser, leaving salt behind.
2. The condenser is cold, so the vapour condenses to water in it.
3. The water drips into the beaker. It is called **distilled water**. It is almost pure.

You could get drinking water from seawater, in this way. Many countries in the Middle East obtain drinking water by distilling seawater in giant distillation plants.

### Fractional distillation

This is used to separate *a mixture of liquids* from each other. It makes use of their different boiling points. You could use it to separate a mixture of ethanol and water, for example. The apparatus is shown on the right.

These are the steps:

1. Heat the mixture in the flask. At about 78 °C, the ethanol begins to boil. Some water evaporates too. So a mixture of ethanol and water vapours rises up the column.
2. The vapours condense on the glass beads in the column, making them hot.
3. When the beads reach about 78 °C, ethanol vapour no longer condenses on them. Only the water vapour does. So water drips back into the flask. The ethanol vapour goes into the condenser.
4. There it condenses. Pure liquid ethanol drips into the beaker.
5. Eventually, the thermometer reading rises above 78 °C – a sign that all the ethanol has gone. So you can stop heating.

### Fractional distillation in industry

Fractional distillation is very important in industry. It is used:

- in the petroleum industry, to **refine** crude oil into petrol and other groups of compounds. The oil is heated and the vapours rise to different heights, up a tall steel fractionating column. See page 247.
- in producing **ethanol**. The ethanol is made by fermentation, using sugar cane or other plant material. It is separated from the fermented mixture by fractional distillation. Ethanol is used as a solvent, and as car fuel. See page 256.
- to separate the gases in **air**. The air is cooled until it is liquid, then warmed up. The gases boil off one by one. See page 212.

▲ A petroleum refinery. It produces petrol and many other useful substances, with the help of fractional distillation.

SEPARATING SUBSTANCES

## Paper chromatography

This method can be used to separate a mixture of substances. For example, you could use it to find out how many different dyes there are in black ink:

**1** Place a drop of black ink in the centre of some filter paper. Let it dry. Then add three or four more drops on the same spot, in the same way.

**2** Now drip water onto the ink spot, one drop at a time. The ink slowly spreads out and separates into rings of different colours.

**3** Suppose there are three rings: yellow, red and blue. This shows that the ink contains three dyes, coloured yellow, red and blue.

The dyes in the ink have different solubilities in water. So they travel across the paper at different rates. (The most soluble one travels fastest.) That is why they separate into rings. The filter paper with the coloured rings is called a **chromatogram**. (*Chroma* means *colour*.)

Paper chromatography can also be used to **identify** substances. For example, mixture **X** is thought to contain substances **A**, **B**, **C**, and **D**, which are all soluble in propanone. You could check the mixture like this:

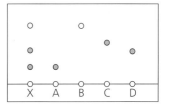

**1** Prepare concentrated solutions of **X**, **A**, **B**, **C**, and **D**, in propanone. Place a spot of each along a line, on chromatography paper. Label them.

**2** Stand the paper in a little propanone, in a covered glass tank. The solvent rises up the paper. When it's near the top, remove the paper.

**3** **X** has separated into three spots. Two are at the same height as **A** and **B**, so **X** must contain substances **A** and **B**. Does it also contain **C** and **D**?

Note that you must use a pencil to draw the line on the chromatography paper. If you use a biro or felt-tipped pen, the ink will run.

---

**Q**

1. How would you obtain pure water from seawater? Draw the apparatus, and explain how the method works.
2. Why are *condensers* called that? What is the cold water for?
3. You would not use *exactly* the same apparatus you described in **1**, to separate ethanol and water. Why not?
4. Explain how fractional distillation works.
5. In the last chromatogram above, how can you tell that **X** does *not* contain substance **C**?
6. Look at the first chromatogram above. Can you think of a way to separate the coloured substances from the paper?

SEPARATING SUBSTANCES

## 2.5 More about paper chromatography

### How paper chromatography works
Paper chromatography depends on how the substances in a mixture interact with the chromatography paper and the solvent.

chromatography paper

**1** These coloured dots represent a mixture of two substances. The mixture is dissolved in a suitable solvent.

**2** The two substances travel over the paper at different speeds, because of their different solubilities in the solvent, and attraction to the paper.

**3** Eventually they get completely separated from each other. Now you can identify the substances – and even collect them if you wish.

**The more soluble a substance is in the solvent, the further it will travel up the chromatography paper.**

### Making use of paper chromatography
You can use paper chromatography to:
- identify a substance
- separate mixtures of substances
- purify a substance, by separating it from its impurities.

### Example: Identify substances in a colourless mixture
On page 23, paper chromatography was used to identify *coloured* substances. Now for a bigger challenge!

Test-tubes **A–E** on the right below contain five *colourless* solutions of amino acids, dissolved in water. The solution in **A** contains several amino acids. The other solutions contain just one each.

Your task is to identify *all* the amino acids in **A–E**.

1 Place a spot of each solution along a line drawn in pencil on slotted chromatography paper, as shown below. (The purpose of the slots is to keep the samples separate.) Label each spot in pencil at the *top* of the paper.

▲ Amino acids coming up! When you digest food, the proteins in it are broken down to amino acids. Your body needs 20 different amino acids to stay healthy.

▲ The five mystery solutions.

SEPARATING SUBSTANCES

2. Place a suitable solvent in the bottom of a beaker. (For amino acids, a mixture of water, ethanoic acid and butanol is suitable.)
3. Roll the chromatography paper into a cylinder and place it in the beaker. Cover the beaker.
4. The solvent rises up the paper. When it has almost reached the top, remove the paper.
5. Mark a line in pencil on it, to show where the solvent reached. (You can't tell where the amino acids are, because they are colourless.)
6. Put the paper in an oven to dry out.
7. Next spray it with a **locating agent** to make the amino acids show up. **Ninhydrin** is a good choice. (Use it in a fume cupboard!) After spraying, heat the paper in the oven for 10 minutes. The spots turn purple. So now you have a proper chromatogram.
8. Mark a pencil dot at the centre of each spot. Measure from the base line to each dot, and to the line showing the final solvent level.

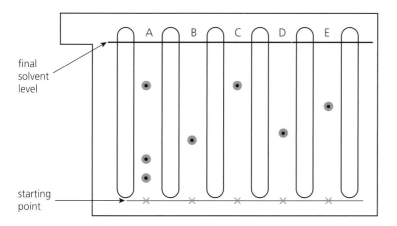

9. Now work out the $R_f$ value for each amino acid. Like this:

$$R_f \text{ value} = \frac{\text{distance moved by amino acid}}{\text{distance moved by solvent}}$$

10. Finally, look up $R_f$ tables to **identify** the amino acids. Part of an $R_f$ table for the solvent you used is shown on the right. The method works because: **the $R_f$ value of a compound is always the same for a given solvent, under the same conditions**.

**$R_f$ values for amino acids**
(for water/butanol/ethanoic acid as solvent)

| amino acid | $R_f$ value |
|---|---|
| cysteine | 0.08 |
| lysine | 0.14 |
| glycine | 0.26 |
| serine | 0.27 |
| alanine | 0.38 |
| proline | 0.43 |
| valine | 0.60 |
| leucine | 0.73 |

---

**Q**

1. Explain in your own words how paper chromatography works.
2. a  What do you think a *locating agent* is?
   b  Why would you need one, in an experiment to separate amino acids by chromatography?
3. What makes $R_f$ values so useful?
4. For the chromatogram above:
   a  Were any of the amino acids in **B–E** also present in **A**? How can you tell at a glance?
   b  Using a ruler, work out the $R_f$ values for the amino acids in **A–E**.
   c  Now use the $R_f$ table above to name them.

SEPARATING SUBSTANCES

# The chromatography detectives

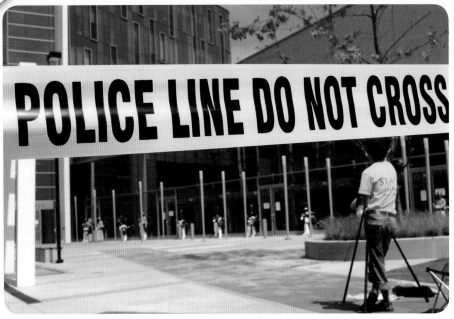

▲ After a crime, the forensic detectives move in, looking for fingerprints and other samples they can use in evidence.

## The key ideas in chromatography.

Much of chromatography is detective work. You have already met paper chromatography. There are many other kinds too. But the key ideas are always the same.

- You need two phases:
  - a non-moving or **stationary** phase, such as filter paper
  - a moving or **mobile** phase. This consists of the mixture you want to separate, dissolved in a solvent.
- The substances in the mixture separate because each has different levels of attraction to the solvent and the stationary phase. Look at the diagram on the right.
- You can then identify each separated substance. Depending on the technique you use, you can also collect them.

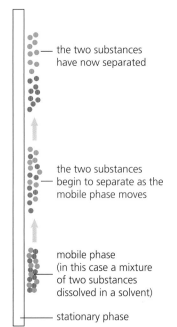

▲ How chromatography works.

## Ringing the changes

Although those key ideas are always the same, the techniques used for chromatography can be quite different. For example:

| The stationary phase could be ... | The mobile phase could be ... | To analyse the substances, you could ... |
|---|---|---|
| • paper, as in paper chromatography<br>• a thin coat of an adsorbent substance on a glass plate, or inside a tube<br>• plastic beads packed into a tube | • a mixture of substances dissolved in a liquid, as in paper chromatography<br>• a mixture of gases, carried in an **inert** (unreactive) gas; this is called **gas chromatography** | • study the coloured spots on the chromatogram, as in paper chromatography<br>• pass them through a machine that will help you analyse them |

SEPARATING SUBSTANCES

## Chromatography and crime detection

Chromatography is widely used in crime detection. For example it is used to analyse samples of fibre from crime scenes, check people's blood for traces of illegal drugs, and examine clothing for traces of explosives.

This shows how a blood sample could be analysed, for traces of illegal drugs, or a poison, using gas chromatography:

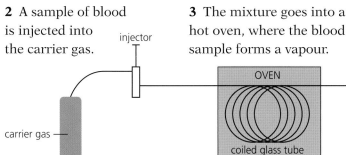

**2** A sample of blood is injected into the carrier gas.

**3** The mixture goes into a hot oven, where the blood sample forms a vapour.

**5** The separated substances pass into a mass spectrometer, where they are analysed.

**6** The data is fed into a recorder. The police study it. They might make an arrest …

**1** The carrier gas is fed in. It could be helium or nitrogen, for example.

**4** The vapour moves over the stationary phase: an adsorbent substance lining a coiled glass tube.

## Other uses

Chromatography can be used on a small scale in the lab, or on a very large scale in industry. For example it is used on a small scale to:

- identify substances (such as amino acids, on page 277)
- check the purity of substances
- help in crime detection (as above)
- identify pollutants in air, or in samples of river water.

It is used on a large scale to:

- separate pure substances (for example for making medical drugs or food flavourings) from tanks of reaction mixtures, in factories
- separate individual compounds from the groups of compounds (fractions) obtained in refining petroleum.

So chromatography is a really powerful and versatile tool.

▲ Injecting a sample into the carrier gas, at the start of gas chromatography.

◄ Collecting water samples, to analyse for pollutants. The factories that produce them could then be identified – and fined.

SEPARATING SUBSTANCES

# Checkup on Chapter 2

## Revision checklist

### Core curriculum
Make sure you can ...
- define and use these terms:
  *mixture*   *solute*   *solvent*
  *solution*   *aqueous solution*
- give at least three examples of solvents
- state that most solids become more soluble as the temperature of the solvent rises
- explain what these terms mean:
  *pure substance*   *impurity*
- give examples of where purity is very important
- say how melting and boiling points change, when an impurity is present
- decide whether a substance is pure, from melting and boiling point data
- describe these methods for separating mixtures, and sketch and label the apparatus:
  *filtration*
  *crystallisation*
  *evaporation to dryness*
  *simple distillation*
  *fractional distillation*
  *paper chromatography*
- explain why each of those separation methods works
- say which method you would choose for a given mixture, and why
- identify the coloured substances present in a mixture, using chromatography

### Extended curriculum
Make sure you can also ...
- explain what a *locating agent* is
- describe how to carry out chromatography, to identify colourless substances
- define $R_f$ *value*
- identify the substances in a mixture, given a chromatogram and a table of $R_f$ values.

## Questions

### Core curriculum

1. This question is about ways to separate and purify substances. Match each term on the left with the correct description on the right.

   A evaporation — i a solid appears as the solution cools
   B condensing — ii used to separate a mixture of two liquids
   C filtering — iii the solvent is removed as a gas
   D crystallising — iv this method allows you to recycle a solvent
   E distillation — v a gas changes to a liquid, on cooling
   F fractional distillation — vi separates an insoluble substance from a liquid

2. This apparatus can be used to obtain pure water from salt water.

   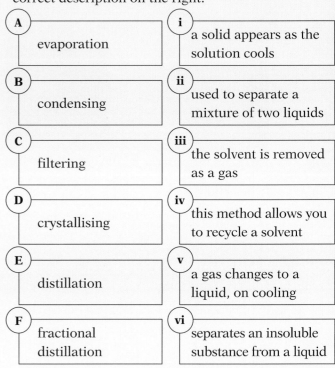

   a What is the purpose of the ice-cold water?
   b The glass arm must reach far down into the second test-tube. Why?
   c Where in the apparatus does this take place?
      i evaporation
      ii condensation
   d What is this separation method called?
   e What will remain in the first test-tube, at the end of the experiment?

**3** Seawater can be purified using this apparatus:

**a i** What is the maximum temperature recorded on the thermometer, during the distillation?
**ii** How does this compare to the boiling point of the seawater?
**b** In which piece of apparatus does evaporation take place? Give its name.
**c i** Which is the condenser, A, B, or C?
**ii** Where does the supply of cold water enter?
**d** Distillation is used rather than filtration, to purify seawater for drinking. Why?

**4** Gypsum is insoluble in water. You are asked to purify a sample of gypsum that is contaminated with a soluble salt.
**a** Which of these pieces of apparatus will you use?
Bunsen burner     filter funnel     tripod
distillation flask    conical flask     pipette
thermometer        condenser        gauze
stirring rod          filter paper      beaker
**b** Write step-by-step instructions for the procedure.

**5** Argon, oxygen, and nitrogen are obtained from air by fractional distillation. Liquid air, at −250 °C, is warmed up, and the gases are collected one by one.
**a** Is liquid air a mixture, or a pure substance?
**b** Explain why fractional distillation is used, rather than simple distillation.
**c** During the distillation, nitrogen gas is obtained first, then argon and oxygen. What can you say about the boiling points of these three gases?

**6** A mixture of salt and sugar has to be separated, using the solvent ethanol.
**a** Draw a diagram to show how you will separate the salt.
**b** How could you obtain sugar crystals from the sugar solution, *without* losing the ethanol?
**c** Draw a diagram of the apparatus for **b**.

**7** In a chromatography experiment, eight coloured substances were spotted onto a piece of filter paper. Three were the basic colours red, blue, and yellow. The others were unknown substances, labelled A–E. This shows the resulting chromatogram:

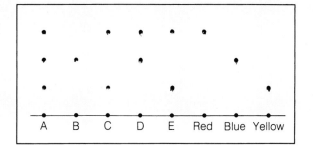

**a** Which one of substances A–E contains only one basic colour?
**b** Which contains all three basic colours?
**c** The solvent was propanone. Which of the three basic colours is the most soluble in propanone?

**Extended curriculum**

**8** The diagram below shows a chromatogram for a mixture of amino acids.

The solvent was a mixture of water, butanol, and ethanoic acid.
**a** Using the table of $R_f$ values on page 25, identify the two amino acids.
**b** Which of them is less soluble in the solvent?
**c** How will the $R_f$ values change if the solvent travels only 6 cm?

**9** You have three colourless solutions. Each contains an amino acid you must identify.
Explain how to do this using chromatography.
Use the terms $R_f$ and *locating agent* in your answer, and show that you understand what they mean.

ATOMS AND ELEMENTS

## 3.1 Atoms and elements

### Atoms

Sodium is made of tiny particles called sodium atoms.

Diamond is made of carbon atoms – different from sodium atoms.

Mercury is made of mercury atoms – different again!

**Atoms are the smallest particles of matter, that we cannot break down further by chemical means.**

Single atoms are far too small to see. Perhaps a million sodium atoms could fit in a line across this full stop. So you can see sodium only if there are enough sodium atoms together in one place!

In fact atoms are mostly empty space. Each consists of a **nucleus** and a cloud of particles called **electrons** that whizz around it. This drawing shows how a sodium atom might look, magnified many millions of times.

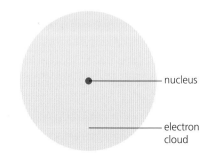

### The elements

Sodium is made of sodium atoms only, so it is an **element**.
**An element contains only one kind of atom.**

Around 90 elements have been found in the Earth and atmosphere. Scientists have made nearly 30 others in the lab. Many of the 'artificial' elements are very unstable, and last just a few seconds before breaking down into other elements. (That is why they are not found in nature.)

### Symbols for the elements

To make life easy, each element has a symbol. For example the symbol for carbon is C. The symbol for potassium is K, from its Latin name *kalium*. Some elements are named after the people who discovered them.

▲ This painting shows Hennig Brand, who discovered the element phosphorus, in 1669. It glows in the dark!

◀ Collecting the element sulfur from a volcano crater in Indonesia. It is used as an ingredient in many cosmetics.

ATOMS AND ELEMENTS

# The Periodic Table

| Group | | | | | | | | | | | | | | | | | Group | | | | | | 0 |
|---|---|---|---|---|---|---|---|---|---|---|---|---|---|---|---|---|---|---|---|---|---|---|---|
| I | II | | | | | | | | | | | | | | | | III | IV | V | VI | VII | | 4/2 He helium |
| | | 1 | 1/1 H hydrogen | | | | | | | | | | | | | | | | | | | | |
| 2 | 7/3 Li lithium | 9/4 Be beryllium | | | | | | | | | | | | | | | 11/5 B boron | 12/6 C carbon | 14/7 N nitrogen | 16/8 O oxygen | 19/9 F fluorine | | 20/10 Ne neon |
| 3 | 23/11 Na sodium | 24/12 Mg magnesium | | The transition elements | | | | | | | | | | | | | 27/13 Al aluminium | 28/14 Si silicon | 31/15 P phosphorus | 32/16 S sulfur | 35.5/17 Cl chlorine | | 40/18 Ar argon |
| 4 | 39/19 K potassium | 40/20 Ca calcium | 45/21 Sc scandium | 48/22 Ti titanium | 51/23 V vanadium | 52/24 Cr chromium | 55/25 Mn manganese | 56/26 Fe iron | 59/27 Co cobalt | 59/28 Ni nickel | 64/29 Cu copper | 65/30 Zn zinc | 70/31 Ga gallium | 73/32 Ge germanium | 75/33 As arsenic | 79/34 Se selenium | 80/35 Br bromine | 84/36 Kr krypton |
| 5 | 85/37 Rb rubidium | 88/38 Sr strontium | 89/39 Y yurium | 91/40 Zr zirconium | 93/41 Nb niobium | 96/42 Mo molybdenum | 99/43 Tc technetium | 101/44 Ru ruthenium | 103/45 Rh rhodium | 106/46 Pd palladium | 108/47 Ag silver | 112/48 Cd cadmium | 115/49 In indium | 119/50 Sn tin | 122/51 Sb antimony | 128/52 Te tellurium | 127/53 I iodine | 131/54 Xe xenon |
| 6 | 133/55 Cs caesium | 137/56 Ba barium | 139/57 La lanthanium | 178.5/72 Hf hafnium | 181/73 Ta tantalum | 184/74 W tungsten | 186/75 Re rhenium | 190/76 Os osmium | 192/77 Ir iridium | 195/78 Pt platinum | 197/79 Au gold | 201/80 Hg mercury | 204/81 Tl thallium | 207/82 Pb lead | 209/83 Bi bismuth | 210/84 Po polonium | 210/85 At astatine | 222/86 Rn radon |
| 7 | 223/87 Fr francium | 226/88 Ra radium | 227/89 Ac actinuim | | | | | | | | | | | | | | | |

Lanthanides: 140/58 Ce cerium, 141/59 Pr praseodymium, 144/60 Nd neodymium, 147/61 Pm promethium, 150/62 Sm samarium, 152/63 Eu europium, 157/64 Gd gadolinium, 159/65 Tb terbium, 162/66 Dy dysprosium, 165/67 Ho holmium, 167/68 Er erbium, 169/69 Tm thutium, 173/70 Yb ytterbium, 175/71 Lu lutetium

Actinides: 232/90 Th thorium, 231/91 Pa protactinium, 238/92 U uranium, 237/93 Np neptunium, 244/94 Pu plutonium, 243/95 Am americium, 247/96 Cm curium, 247/97 Bk berkelium, 251/98 Cf californium, 252/99 Es einsteinium, 257/100 Fm fermium, 258/101 Md mendelevium, 259/102 No nobelium, 262/103 Lw lawrencium

The table above is called the **Periodic Table**.

- It gives the names and symbols for the elements.
- The column and row an element is in gives us lots of clues about it. For example, look at the columns numbered I, II, III …
  The elements in these form families or **groups**, with similar properties. So if you know how one element in Group I behaves, for example, you can make a good guess about the others.
- The rows are called **periods**.
- Look at the zig-zag line. It separates metals from non-metals, with the non-metals on the right of the line, except for hydrogen. So there is a change from metal to non-metal, as you go across a period.

Now look at the small numbers beside each symbol. These tell us a lot about the atoms of the element, as you will soon see.

▲ The element chlorine is a poisonous gas. It was used as a weapon in World War I. This soldier was prepared.

**Q**

1 What is: **a** an atom? **b** an element?
2 If you could look inside an atom, what would you see?
3 The symbols for some elements come from their Latin names. See if you can identify the element whose Latin name is:
  **a** natrium  **b** ferrum  **c** plumbum  **d** argentum
4 Which element has this symbol? **a** Ca  **b** Mg  **c** N
5 See if you can pick out an element named after the famous scientist Albert Einstein.
6 From the Periodic Table, name
  **a** three metals  **a** three non-metals
  that you expect to behave in a similar way.

ATOMS AND ELEMENTS

## 3.2 More about atoms

### Protons, neutrons, and electrons

Atoms consist of a **nucleus** and a cloud of **electrons** that move around the nucleus. The nucleus is itself a cluster of two kinds of particles, **protons** and **neutrons**.

All the particles in an atom are very light. So their mass is measured in **atomic mass units**, rather than grams. Protons and electrons also have an **electric charge**:

| Particle in atom | Mass | Charge |
|---|---|---|
| proton ● | 1 unit | positive charge (1+) |
| neutron ● | 1 unit | none |
| electron · | almost nothing | negative charge (1−) |

Since electrons are so light, their mass is usually taken as zero.

### How the particles are arranged

The sodium atom is a good one to start with. It has **11** protons, **11** electrons, and **12** neutrons. They are arranged like this:

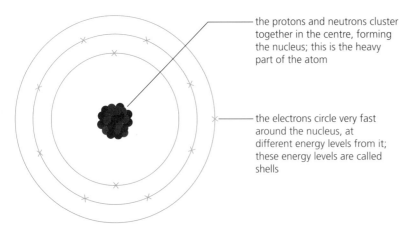

- the protons and neutrons cluster together in the centre, forming the nucleus; this is the heavy part of the atom
- the electrons circle very fast around the nucleus, at different energy levels from it; these energy levels are called shells

▲ The nucleus is very tiny compared with the rest of the atom. If the atom were the size of a football stadium, the nucleus would be the size of a pea!

**Note**
Since they make up the atom, protons, neutrons and electrons are often called **sub-atomic particles**.

### Proton number

A sodium atom has 11 protons. This can be used to identify it, since *only* a sodium atom has 11 protons. Every other atom has a different number. **You can identify an atom by the number of protons in it**.

The number of protons in an atom is called its **proton number**. The proton number of sodium is 11.

### How many electrons?

The sodium atom also has 11 electrons. So it has an equal number of protons and electrons. The same is true for every sort of atom:
**Every atom has an equal number of protons and electrons. So atoms have no overall charge.**

Look at the box on the right. It shows that the positive and negative charges cancel each other, for the sodium atom.

**The charge on a sodium atom:**
●●●●    11 protons
●●●●    Each has a charge of 1+
●●●     Total charge 11+
××××   11 electrons
××××   Each has a charge of 1−
×××    Total charge 11−
Adding the charges: 11+

The answer is zero.
The atom has no overall charge.

32

## Nucleon number

Protons and neutrons form the nucleus, so are called **nucleons.**
**The total number of protons and neutrons in an atom is called its nucleon number.**
The nucleon number for the sodium atom is 23. (11 + 12 = 23)

So sodium can be described in a short way like this: $^{23}_{11}$Na.

The lower number is *always* the proton number. The other number is the nucleon number. So you can tell straight away that sodium atoms have 12 neutrons. (23 − 11 = 12)

> **Try it yourself!**
> You can describe any element in a short way like this:
>
> nucleon number
> proton number   symbol
>
> For example: $^{16}_{8}$O

## The atoms of the first 20 elements

In the Periodic Table, the elements are arranged in order of increasing proton number. Here are the first 20 elements, shown as a list:

| Element | Symbol | Proton number | Electrons | Neutrons | Nucleon number (protons + neutrons) |
|---|---|---|---|---|---|
| hydrogen | H | 1 | 1 | 0 | 1 |
| helium | He | 2 | 2 | 2 | 4 |
| lithium | Li | 3 | 3 | 4 | 7 |
| beryllium | Be | 4 | 4 | 5 | 9 |
| boron | B | 5 | 5 | 6 | 11 |
| carbon | C | 6 | 6 | 6 | 12 |
| nitrogen | N | 7 | 7 | 7 | 14 |
| oxygen | O | 8 | 8 | 8 | 16 |
| fluorine | F | 9 | 9 | 10 | 19 |
| neon | Ne | 10 | 10 | 10 | 20 |
| sodium | Na | 11 | 11 | 12 | 23 |
| magnesium | Mg | 12 | 12 | 12 | 24 |
| aluminium | Al | 13 | 13 | 14 | 27 |
| silicon | Si | 14 | 14 | 14 | 28 |
| phosphorus | P | 15 | 15 | 16 | 31 |
| sulfur | S | 16 | 16 | 16 | 32 |
| chlorine | Cl | 17 | 17 | 18 | 35 |
| argon | Ar | 18 | 18 | 22 | 40 |
| potassium | K | 19 | 19 | 20 | 39 |
| calcium | Ca | 20 | 20 | 20 | 40 |

So the numbers of protons and electrons increase by 1 at a time – and are always equal. What do you notice about the number of neutrons?

1. Name the particles that make up the atom.
2. Which particle has:
   a  a positive charge?   b  no charge?   c  almost no mass?
3. An atom has 9 protons. Which element is it?
4. Why do atoms have no overall charge?
5. What does this term mean?
   a  proton number   b  nucleon number
6. Name each of these atoms, and say how many protons, electrons, and neutrons it has:
   $^{12}_{6}$C   $^{16}_{8}$O   $^{24}_{12}$Mg   $^{27}_{13}$Al   $^{64}_{29}$Cu

ATOMS AND ELEMENTS

## 3.3 Isotopes and radioactivity

### How to identify an atom: a reminder
Only sodium atoms have 11 protons.
**You can identify an atom by the number of protons in it.**

### Isotopes
All carbon atoms have 6 protons. But not all carbon atoms are identical. Some have more *neutrons* than others.

 6 protons
6 electrons
6 neutrons

Most carbon atoms are like this, with **6** neutrons. That makes **12** nucleons (protons + neutrons) in total, so it is called **carbon-12**.

 6 protons
6 electrons
7 neutrons

But about one in every hundred carbon atoms is like this, with **7** neutrons. It has **13** nucleons in total, so is called **carbon-13**.

 6 protons
6 electrons
8 neutrons

And a very tiny number of carbon atoms are like this, with **8** neutrons. It has **14** nucleons in total, so is called **carbon-14**.

The three atoms above are called **isotopes** of carbon.
**Isotopes are atoms of the same element, with different numbers of neutrons.**

Most elements have isotopes. For example calcium has six, magnesium has three, iron has four, and chlorine has two.

### Some isotopes are radioactive
A carbon-14 atom behaves in a strange way. It is **radioactive**. That means its nucleus is unstable. Sooner or later the atom breaks down naturally or **decays**, giving out **radiation** in the form of rays and particles, plus a large amount of energy.

Like carbon, a number of other elements have radioactive isotopes – or **radioisotopes** – that occur naturally, and eventually decay.

But the other two isotopes of carbon (like most natural isotopes) are **non-radioactive**.

> **Radiation may contain …**
> - **alpha particles** – made up of 2 protons and 2 electrons
> - **beta particles** – electrons moving at high speed
> - **neutrons**
> - **gamma rays** – high energy rays

> **Decay is a random process**
> We can't tell whether a given atom of carbon-14 will decay in the next few seconds, or in a thousand years. But we *do* know how long it takes for *half* the radioisotopes in a sample to decay. This is called the **half-life**.
>
> The half-life for carbon-14 is 5730 years. So if you have a hundred atoms of carbon-14, fifty of them will have decayed 5730 years from now.
>
> Half-lives vary a lot. For example:
> - for radon-220        55.5 seconds
> - for cobalt-60        5.26 years
> - for potassium-40     1300 million years

▲ Radioisotopes are dangerous. This scientist is using a glove box, for safety.

# ATOMS AND ELEMENTS

## Radiation can harm you

If the radiation from radioisotopes gets into your body, it will kill body cells. A large dose causes **radiation sickness**. Victims vomit a lot, and feel really tired. Their hair falls out, their gums bleed, and they die within weeks. Even small doses of radiation, over a long period, will cause cancer.

## Making use of radioisotopes

Radioisotopes are dangerous – but they are also useful. For example:

**To check for leaks** Engineers can check oil and gas pipes for leaks by adding radioisotopes to the oil or gas. If a **Geiger counter** detects radiation outside the pipe, it means there is a leak. Radioisotopes used in this way are called **tracers**.

▲ Checking for radiation using a Geiger counter. The meter gives a reading, and you may also hear beeps.

**To treat cancer** Radioisotopes can cause cancer. But they are also used in **radiotherapy** to *cure* cancer – because the gamma rays in radiation kill cancer cells more readily than healthy cells. Cobalt-60 is usually used for this. The beam of gamma rays is aimed carefully at the site of the cancer in the body.

**To kill germs and bacteria** Gamma rays kill germs too. So they are used to sterilise syringes and other disposable medical equipment. They also kill the bacteria that cause food to decay. So in many countries, foods like vegetables, fruit, spices, and meat, are treated with a low dose of radiation. Cobalt-60 and cesium-137 are used for this.

▲ Radioisotopes are used as fuel in nuclear power stations, because they give out so much energy when they break down. See page 119 for more.

▲ Another use for radiation: carbon-dating. Our bodies contain some carbon-14, taken in in food. When we die, we take no more in. But the carbon-14 atoms continue to decay. So scientists can tell the age of ancient remains by measuring the radioactivity from them. This mummy was found to be around 5300 years old.

## Q

1. **a** What are *isotopes*?
   **b** Name the three isotopes of carbon, and write symbols for them.
2. Carbon-14 is *radioactive*. What does that mean?
3. What is a *radioisotope*? Give two examples.
4. **a** Radiation can kill us. Why?
   **b** So why are radioisotopes used to treat cancer?
5. Radioisotopes can be used to check pipes for leaks.
   **a** Explain how this works.
   **b** How could you tell that a pipe had no leak?
6. Spices are shipped all over the world, and are often stored for long periods.
   **a** They are usually treated with radiation. Why?
   **b** Name two radioisotopes used for this.

ATOMS AND ELEMENTS

## 3.4 How electrons are arranged

### Electron shells

Electrons are arranged in shells around the nucleus.
The first shell, closest to the nucleus, is the lowest energy level.
The further a shell is from the nucleus, the higher the energy level.

Each shell can hold only a certain number of electrons. These are the rules:

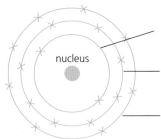

The first shell can hold only 2 electrons. It fills first.

The second shell can hold 8 electrons. It fills next.

The third shell can hold 18 electrons. But it fills up to 8. The next 2 go into the fourth shell (not shown). Then the rest of the third shell fills.

▲ The Danish scientist Niels Bohr (1885 – 1962) was the first person to put forward the idea of electron shells.

The distribution of electrons in the atom above is written in a short way as **288**. (Or sometimes as **2,8,8** or **2.8.8**.)

### The electron shells for the first 20 elements

Below are the electron shells for the first 20 elements of the Periodic Table.

The number of electrons increases by 1 each time. (It is the same as the proton number.) The shells fill according to the rules above.

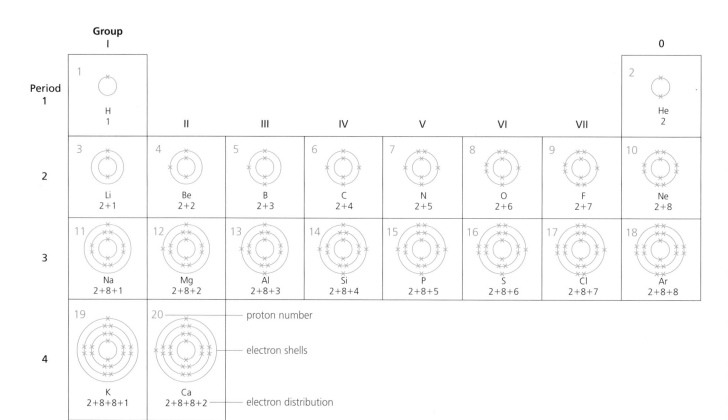

## Patterns in the Periodic Table

Note these patterns for the table of the first 20 elements, on page 36:

- The **period number** tells you how many shells there are.
- All the elements in a group have the same number of electrons in their outer shells. So Group I elements have 1, Group II have 2, and so on. These outer-shell electrons are also called the **valency electrons**.
- The **group number** is the same as the number of outer-shell electrons, *except for* Group 0.
- The valency electrons dictate how an element reacts. So the elements in Group I all have similar reactions, for example.

## Group 0, a special group

The elements in Group 0 have a very stable arrangement of electrons. Their atoms all have **8 outer-shell electrons**, *except for* helium, which has **2**. (It has only one shell.)

▲ Sodium reacts with water to give an alkaline solution. The other Group I metals react in a similar way – because their atoms all have one outer electron.

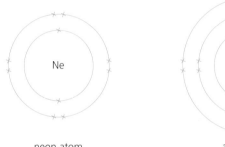

helium atom
full outer shell of 2 electrons
**stable**

neon atom
full outer shell of 8 electrons
**stable**

argon atom
outer shell of 8 electrons
**stable**

This stable arrangement of electrons has a very important result: it makes the Group 0 elements **unreactive**.

## The elements after calcium

After the 20th element, calcium, the electron shells fill in a more complex order. But you should be able to answer questions about electron distribution for later elements, if you remember the points above.

*Example* The element rubidium, Rb, is the 37th element in the Periodic Table. It is in Group I, Period 5. Its proton number is 37. What is its electron distribution?

*Group I* tells you there is 1 electron in the outer shell.
*Period 5* tells you there are five shells.
The proton number is 37, so there are also 37 electrons.
The third shell holds 18 electrons, when full.

So the electron distribution for rubidium is: **2+8+18+8+1**.

> **They all mean the same …**
> The terms
>   electron arrangement
>   electron distribution
>   electronic configuration
> all mean the same thing: how the electrons are arranged in shells.

**Q**

1. One element has atoms with 13 electrons.
   a Draw a diagram to show the electron distribution.
   b Write the electron distribution in this form: 2+ …
   c Name the element.
2. The electron distribution for boron is 2+3. What is it for:
   a lithium?   b magnesium?   c hydrogen?
3. An element has 5 valency electrons. Which group is it in?
4. How many electron shells do atoms of Period 3 have?
5. The element krypton, Kr, is in Group 0, Period 4. Its proton number is 36.
   a Write down the electronic configuration for krypton.
   b What can you say about the reactivity of krypton?

ATOMS AND ELEMENTS

# How our model of the atom developed

## The two big ideas
All chemistry depends on these two big ideas:

- everything is made of particles, and …
- atoms are the simplest particles of an element, that cannot be broken down in a chemical reaction.

But how did chemists find out about atoms? It's a long story.

## It began with the Ancient Greeks
In Ancient Greece (around 750 BC – 150 BC), the philosophers thought hard about the world around them. Is water continuous matter, or lots of separate bits? Is air just empty space? If you crush a stone to dust, then crush the dust, will you end up with bits that will not break up further?

The philosopher Democritus came up with an answer: everything is made of tiny particles that cannot be divided. He called them **atoms**. He said they came in four colours: white, black, red, and green. And in different shapes and sizes: large round atoms that taste sweet, and small sharp ones that taste sour. White atoms are smooth, and black ones jagged.

He said everything is made up of these atoms, mixed in different amounts.

Other philosophers thought this was nonsense. Aristotle (384–270 BC) believed that everything was made of four elements – earth, air, fire, and water – mixed in different amounts. A stone has a lot of earth but not much water. No matter how much you crush it, each tiny bit will still have the properties of stone.

## On to the alchemists
The Greek philosophers did a lot of heavy thinking – but no experiments. The **alchemists** were different. They experimented day and night, mixing this with that. Their main quests were to find the **elixir of life** (to keep us young), and turn common metals into **gold**.

From about 600 AD, the practice of alchemy spread to many countries, including Persia (Iran), India, China, Greece, France, and Britain.

▲ The Greek philosopher Democritus (around 460–370 BC), shown here on a Greek bank note. A lot of thinking – but no experiments!

▲ The alchemists developed many secret recipes.

◄ The Persian alchemist Geber (around 721–815 AD) is often called 'the father of chemistry'.

The alchemists did not succeed in making gold. But they made many substances look like gold, by using secret recipes to coat them with other substances. They also developed many of the techniques we use in the lab today, such as distillation and crystallisation.

## Make way for us chemists

Some alchemists got a reputation as cheats, who swindled 'grants' from rich men with the promise of gold. In the end, by around 1600 AD, the alchemists gave way to a new breed of **chemists**.

By now the idea of atoms was almost forgotten. But in 1661 the scientist Robert Boyle showed that a gas can be compressed into a smaller space. He deduced that gas is made of particles with empty space between them.

In 1799, over 130 years later, the French chemist Joseph Louis Proust showed that copper(II) carbonate always contained the same proportions by mass of copper, carbon, and oxygen, no matter how it was made: 5.3 parts of copper to 1 of carbon to 4 of oxygen. This suggested that copper, carbon, and oxygen were made of particles, and these always combined in the same ratios.

▲ Robert Boyle (1627–91). He was born in Ireland but did most of his work in England. He put forward Boyle's Law for gases. And yes, it is a wig.

## Dalton's dilemma

The English chemist John Dalton puzzled over these discoveries. In 1803 he concluded that if elements *really were made of indivisible particles* then everything made sense. He called the particles **atoms**, as a tribute to the Greek philosophers. He suggested that atoms of one element could combine with atoms of another element only in a fixed ratio.

This time the idea of atoms caught on really fast, because it fitted with the results from so many experiments.

## Jiggling pollen grains

There was still one problem. No one could *prove* that matter was made of separate particles, since they were too small to see. But in 1827, a Scottish botanist called Robert Brown was studying some pollen grains in water, under a microscope. He saw them jiggling around. He deduced that they were being struck by water particles. That meant tiny separate particles really did exist. They were not just theory.

## And then …

In 1955 Erwin Müller, an American, developed a machine called a **field-ion microscope**. It could 'picture' the tip of a needle, magnified 5 million times! The atoms in the needle showed up as dots.

Today, microscopes are much more powerful. The **scanning tunneling microscope** gives us images of individual atoms, magnified by up to 100 million times. (See page 7 for an example.)

Meanwhile, for many decades, scientists wondered what was *inside* atoms. And that is another story.

▶ Getting ready to use the scanning tunneling miscroscope.

ATOMS AND ELEMENTS

# The atom: the inside story

### Bring on the physicists
By 200 years ago, chemists had accepted that everything was indeed made of tiny indivisible particles: atoms. But now we know they are not quite indivisible! In the last 120 years or so, we have learned a great deal about the particles inside atoms, thanks to physicists.

### First, the electrons
In 1897, the English physicist J.J. Thomson was investigating **cathode rays**. These mystery rays glowed inside an empty glass tube, when it was plugged into an electric circuit.

He deduced that these rays were streams of charged particles, *much smaller than atoms*. In fact they were bits from atoms. He called them *corpuscles*, but soon the name got changed to **electrons**.

It was a shock to find that atoms were not the smallest particle after all! Thomson imagined that electrons were stuck on the atoms like raisins on a bun. The rest of the atom (the bun) had a positive charge.

### More strange rays
A year earlier, a French physicist called Becquerel had been working with crystals of a uranium salt. He found that they glowed in the dark. By accident, he left some in a drawer, wrapped in thick paper, on top of a photographic plate. To his surprise, he found an image of the crystals on the plate. They had given out rays of some kind, that could pass through paper! He had discovered **radioactivity**.

Later, the English physicist Ernest Rutherford found that radiation could be separated into **alpha particles, beta particles,** and **gamma rays.** Alpha particles were found to be 7000 times heavier than electrons, with a positive charge. You could speed them up and shoot them like tiny bullets! (We know now they consist of two protons and two neutrons.)

Dalton's atom

Thomson's atom

▲ Becquerel's plate, showing the image of the crystals.

◀ The Polish scientist Marie Curie (1867–1934) heard about Becquerel's discovery. She began to look for other radioactive substances – and discovered the elements polonium and radium. Marie Curie spent years searching for ways to use radiation in medicine. Sadly, she herself died from cancer caused by exposure to radium.

## The nucleus and protons

In 1911, in England, Ernest Rutherford was experimenting with alpha particles. He shot a stream of them at some gold foil. Most went right through it. But some bounced back!

Rutherford deduced that an atom is mostly empty space, which the alpha particles can pass through. But there is something small and dense at the centre of the atom – and if an alpha particle hits this it will bounce back. He had discovered the **nucleus**. He assumed it was made up of particles of positive charge, and called them **protons**.

## Those electron shells

If the nucleus is positive, why don't the negative electrons rush straight into it? In 1913 Niels Bohr came up with the theory of 'electron shells'. It fitted all the experiments.

## At last, the neutrons

In 1930, two German physicists, Bothe and Becker, shot alpha particles at beryllium – and knocked a stream of new particles from it. In 1932 the English physicist James Chadwick found that these particles had the same mass as protons, but no charge. He named them **neutrons**.

So finally, 129 years after Dalton proposed the atom, the chemist's model of it was complete.

## The whole truth

But now for the whole truth. The model of the atom that we use works well for chemists. It explains how the elements behave. But it is only a **model** – a simplified picture.

In fact atoms are far more complex than our model suggests. Physicists have discovered around 50 different **elementary particles** within atoms. They include the **up** and **anti-up**, the **charm** and **anti-charm**, and the **strange** and **anti-strange**. There may be even more to discover.

So those tiny atoms, far far too small to see, are each a throbbing universe of particles. And you are made up of atoms. Think about that!

Rutherford's atom

Bohr's atom

Chadwick's atom

▲ Sir James Chadwick, who gave the neutron its name.

◄ The Large Hadron Collider. Scientists hope this machine will tell them more about the particles inside atoms. It lies in a huge circular tunnel, 27 km across, on the border between France and Switzerland. Protons are accelerated through the pipes to enormous speeds, and allowed to collide.

ATOMS AND ELEMENTS

# 3.5 The metals and non-metals

## Two groups of elements

Look again at the Periodic Table on page 31. The zig-zag line separates the elements into two groups: **metals** and **non-metals**. The non-metals lie to the right of the line, except for hydrogen.

As you can see, there are many more metals than non-metals. In fact over 80% of the known elements are metals.

## What is the difference between them?

The metals and non-metals have very different general properties. Look at this table:

| General properties of metals | General properties of non-metals |
|---|---|
| • good conductors of electricity and heat | • do not conduct electricity or heat |
| • high melting and boiling points – which means they are solid at room temperature | • lower melting and boiling points – many are gases at room temperature |
| • hard, strong, do not shatter if you hammer them | • solid non-metals break up easily – they are **brittle** |
| • can be hammered into different shapes (they are **malleable**) and drawn out to make wires (they are **ductile**) | • solid non-metals are not malleable or ductile – they are brittle |
| • look shiny when they are polished | • look dull, in the solid state |
| • make a ringing noise when struck – they are **sonorous** | • solid non-metals break up when you strike them |
| • have high density – they feel 'heavy' | • solid non-metals have low density |
| • form positive ions when they react. For example sodium forms sodium ions ($Na^+$). You will learn about ions in Chapter 4. | • often form negative ions when they react. For example oxygen forms oxide ions ($O^{2-}$). |
| • react with oxygen to form oxides that are **bases**. (In other words, the oxides can neutralise acids.) | • react with oxygen to form oxides that are **acidic**. (Their aqueous solutions will turn litmus red.) |

The properties in the last two rows above are called **chemical properties**, since they are about chemical change. The others are **physical properties.** You will find out more about many of those properties later.

## Exceptions to those properties

The properties above are *general* properties of metals and non-metals. But there are exceptions. For example:

- not all metals are hard solids. You can cut sodium and potassium with a knife, and mercury is a liquid at room temperature.
- hydrogen is a non-metal, but forms positive ions ($H^+$) like metals do.
- carbon is a non-metal, but one form of it (graphite) is a good conductor; another form (diamond) is very hard, with a very high melting point.

▲ Gold: malleable, ductile, attractive, unreactive, scarce – and expensive.

# ATOMS AND ELEMENTS

▲ Think of two reasons why metals are used to make drums...

▲ ...and three reasons why they are used for saucepans.

## Making use of the metals

Because metals are generally hard and strong, and good conductors, we make great use of them. For example:

- Iron is the most-used metal in the world. It is used in buildings, bridges, cars, tin cans (coated with tin), needles, and nails.
- Copper is used for electrical wiring in homes.
- Aluminium is strong but light. So it is used in planes and space rockets.

## Non-metals are everywhere

There are far fewer non-metals than metals. But they are all around us – and inside us.

- Air is almost 80% nitrogen, and about 20% oxygen.
- Water is a compound of hydrogen and oxygen.
- Our bodies are mostly water, plus hundreds of carbon compounds. Many of these contain atoms of other non-metals too, such as nitrogen, phosphorus, and iodine. (Plus metals such as calcium and iron.)
- Sand is mainly the compound silicon dioxide, formed from silicon and oxygen.

▲ Sea, sand, sky, palms – made almost all of non-metals.

**Q**

1. Without looking at the Periodic Table, see if you can quickly list 30 elements, and give their symbols. Then underline the metals.
2. Explain what these terms mean. (The glossary may help.)
   conductor   ductile   malleable
   brittle     sonorous  density
3. Aluminium is used for outdoor electricity cables. See if you can suggest three reasons why.
4. Write down what you think are the three *main* general properties that distinguish metals from non-metals.
5. Give one example of a physical property, and one of a chemical property, for non-metals.

ATOMS AND ELEMENTS

# Checkup on Chapter 3

## Revision checklist
### Core curriculum
Make sure you can …
- define these terms:
  *atom*    *element*    *compound*
- say where in the atom the nucleus is, and which particles it contains
- define *proton number* and *nucleon number*
- state the number of protons, neutrons and electrons in an atom, from a short description like this: $^{23}_{11}Na$
- explain what a *radioisotope* is
- give one medical and one industrial use, for radioisotopes
- sketch the structure of an atom, showing the nucleus and electron shells
- state the order in which electrons fill the electron shells
- name the first 20 elements of the Periodic Table, in order of proton number, and give their symbols
- sketch the electron distribution for any of the first 20 elements of the Periodic Table, when you are given the proton number
- show electron distribution in this form: 2 + 8 + …
- define the term *valency electron*
- state the connection between the number of valency electrons and the group number in the Periodic Table
- state the connection between the number of electron shells and the period number in the Periodic Table
- work out the electron distribution for an element, given its period and group numbers
- say how many outer-shell electrons there are in the atoms of Group 0 elements
- explain why the Group 0 elements are unreactive
- point out where the metals and non-metals are, in the Periodic Table
- give at least five key differences between metals and non-metals
- name and give the symbols for the common metals and non-metals (including metals from the transition block of the Periodic Table)

## Questions
### Core curriculum

**1**

| Particle | Electrons | Protons | Neutrons |
|---|---|---|---|
| A | 12 | 12 | 12 |
| B | 12 | 12 | 14 |
| C | 10 | 12 | 12 |
| D | 10 | 8 | 8 |
| E | 9 | 9 | 10 |

The table above describes some particles.
a Which three particles are neutral atoms?
b Which particle is a negative ion? What is the charge on this ion?
c Which particle is a positive ion? What is the charge on this ion?
d Which two particles are isotopes?
e Use the table on page 33 to identify A to E.

**2** The following statements are about the particles that make up the atom. For each statement write:
   **p** if it describes the **proton**
   **e** if it describes the **electron**
   **n** if it describes the **neutron**
A  the positively-charged particle
B  found with the proton, in the nucleus
C  the particle that can occur in different numbers, in atoms of the same element
D  held in shells around the nucleus
E  the negatively-charged particle
F  the particle with negligible mass
G  the number of these particles is found by subtracting the proton number from the nucleon number
H  the particle with no charge
I  the particle with the same mass as a neutron
J  the particle that dictates the position of the element in the Periodic Table

**3** The atoms of an element can be represented by a set of three letters, as shown on the right. $^{y}_{z}X$
a What does this letter stand for?
   i X    ii y    iii z
b How many neutrons are there in these atoms?
   i $^{107}_{47}Ag$    ii $^{63}_{29}Cu$    iii $^{1}_{1}H$    iv $^{20}_{10}Ne$    v $^{238}_{92}U$
c Bromine atoms have 36 neutrons. Describe a bromine atom, using the method in **b**.

4 For each of the six elements aluminium (Al), boron (B), nitrogen (N), oxygen (O), phosphorus (P), and sulfur (S), write down:
  a  i   which period of the Periodic Table it belongs to
     ii  its group number in the Periodic Table
     iii its proton number
     iv  the number of electrons in its atoms
     v   its electronic configuration
     vi  the number of outer electrons in its atoms
  b  The outer electrons are also called the _____ electrons. What is the missing word? (7 letters!)
  c  Which of the above elements would you expect to have similar properties? Why?

5 Boron has two types of atom, shown below.

atom A        atom B

○ proton
⊗ neutron
× electron

  a  What is different about these two atoms?
  b  What name is given to atoms like these?
  c  Describe each atom in shorthand form, as in 3.
  d  What is the nucleon number of atom A?
  e  Is atom B heavier, or lighter, than atom A?
  f  i   Give the electronic configuration for A and B.
     ii  Comment on your answer for i.

6 The two metals sodium (proton number 11) and magnesium (proton number 12) are found next to each other in the Periodic Table.
  a  Say whether this is the same, or different, for their atoms:
     i   the number of electron shells
     ii  the number of outer (valency) electrons
  The relative atomic mass of sodium is 23.0.
  The relative atomic mass of magnesium is 24.3.
  b  Which of the two elements may exist naturally as a single isotope? Explain your answer.

7 Strontium, proton number 38, is in the fifth period of the Periodic Table. It belongs to Group II. Copy and complete the following.
  An atom of strontium has:
  a  ............... electrons
  b  ............... shells of electrons
  c  ............... electrons in its outer shell

8 This diagram represents the electronic arrangement in an atom of an element.

  a  i   Give the electron distribution for the atom.
     ii  What is special about this arrangement?
  b  Which group of the Periodic Table does the element belong to?
  c  Name another element with the same number of outer-shell electrons in its atoms.

9 Gallium exists naturally as a mixture of two non-radioactive isotopes, gallium-69 and gallium-71. The proton number of gallium is 31.
  a  i   How many neutrons are there in gallium-69?
     ii  How many neutrons are there in gallium-71?
  Gallium also has a radioactive isotope, gallium-67. As gallium-67 decays, it gives out rays called gamma rays.
  b  How does the radioactive isotope differ from the non-radioactive isotope?
  c  Name two possible uses, one medical and one non-medical, for gallium-67.

10 Read this passage about metals.
  Elements are divided into metals and non-metals. All metals are <u>electrical conductors</u>. Many of them have a high <u>density</u> and they are usually <u>ductile</u> and <u>malleable</u>. All these properties influence the way the metals are used. Some metals are <u>sonorous</u> and this leads to special uses for them.
  a  Explain the underlined terms.
  b  Copper is ductile. How is this property useful in everyday life?
  c  Aluminium is hammered and bent to make large structures for use in ships and planes. What property allows it to be shaped like this?
  d  Name one metal that has a *low* density.
  e  Some metals are cast into bells. What property must the chosen metals have?
  f  Give the missing word: *Metals are good conductors of .......... and electricity.*
  g  Choose another physical property of metals, and give two examples of how it is useful.
  h  Phosphorus is a solid non-metal at room temperature. What other physical properties would you expect it to have?
  i  Explain how the chemical properties of metals and non-metals can be used to tell them apart.

ATOMS COMBINING

# 4.1 Compounds, mixtures, and chemical change

### Elements: a reminder
An **element** contains only one kind of atom. For example the element sodium contains only sodium atoms.

### Compounds
**A compound is made of atoms of different elements, bonded together.**

The compound is described by a **formula**, made from the symbols of the atoms in it. (The plural of formula is **formulae**.)

There are millions of compounds. This table shows three common ones.

| Name of compound | Elements in it | How the atoms are joined | Formula of compound |
|---|---|---|---|
| water | hydrogen and oxygen | | $H_2O$ |
| carbon dioxide | carbon and oxygen | | $CO_2$ |
| ethanol | carbon, hydrogen, and oxygen | | $C_2H_5OH$ |

Water has two hydrogen atoms joined or bonded to an oxygen atom. So its formula is $H_2O$. Note where the 2 is written. Now check the formulae for carbon dioxide and ethanol. Are they correct?

### Compounds and mixtures: the difference
**A mixture contains different substances that are *not* bonded together.**
So you can usually separate the substances quite easily, using methods like those you met in Chapter 2. For example:

This is a **mixture** of iron powder and sulfur. You could separate them by dissolving the sulfur in methylbenzene (a solvent), and filtering the iron off.

But if you heat the end of a metal rod in a Bunsen burner, and push it into the mixture, the mixture starts to glow brightly. A **chemical change** is taking place.

The result is a black **compound** called iron(II) sulfide. It is made of iron and sulfur atoms bonded together. Its formula is FeS. It will not dissolve in methylbenzene.

# ATOMS COMBINING

## The signs of a chemical change

When you heat a mixture of iron and sulfur, a chemical change takes place. The iron and sulfur atoms bond together to form a compound.

You can tell when a chemical change has taken place, by these three signs:

1 **One or more new chemical substances are formed.**
   You can describe the change by a **word equation** like this:

   iron + sulfur $\longrightarrow$ iron(II) sulfide

   The + means *reacts with*, and the $\longrightarrow$ means *to form*.

   The new substances usually look different from the starting substances. For example sulfur is yellow, but iron(II) sulfide is black.

2 **Energy is taken in or given out, during the reaction.**
   Energy was needed to start off the reaction between iron and sulfur, in the form of heat from the hot metal rod. But the reaction gave out heat once it began – the mixture glowed brightly.

3 **The change is usually difficult to reverse.**
   You would need to carry out several reactions to get the iron and sulfur back from iron sulfide. (But it can be done!)

**A chemical change is usually called a chemical reaction.**

▲ Burning gas, to fry eggs. Are chemical changes taking place?

## It is different from physical change

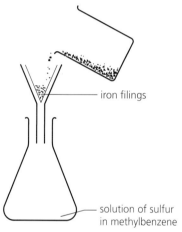

When you mix iron powder with sulfur, that is a **physical change**. No new substance has formed. If you then dissolve the sulfur …

… in methylbenzene, that is also a physical change. The solvent could be removed again by distilling it. (Danger! It is highly flammable.)

Now separate the iron by filtering. That is a physical change. You can reverse it by putting the iron back into the filtrate again.

No new chemical substances are formed in these changes.
**If no new chemical substance is formed, a change is a physical change.**
Unlike chemical changes, a physical change is usually easy to reverse.

---

**Q**

1 Explain the difference between a *mixture* of iron and sulfur and the *compound* iron sulfide.
2 When you light a piece of magnesium ribbon, it burns with a dazzling white light. A white ash forms. What signs are there that a chemical change has taken place?
3 Is it a chemical change or a physical change? Give reasons.
   a a glass bottle breaking
   b butter and sugar being made into toffee
   c cotton being woven to make sheets
   d coal burning in air

ATOMS COMBINING

## 4.2 Why do atoms form bonds?

### The reaction between sodium and chlorine

Sodium and chlorine are both **elements**. When sodium is heated and placed in a jar of chlorine, it burns with a bright flame.

The result is a white solid that has to be scraped from the sides of the jar. It looks completely different from the sodium and chlorine.

So a chemical reaction has taken place. The white solid is **sodium chloride**. Atoms of sodium and chlorine have bonded (joined together) to form a compound. The word equation for the reaction is:

sodium + chlorine ⟶ sodium chloride

### Why do atoms form bonds?

Like sodium and chlorine, the atoms of most elements form bonds.

Why? We get a clue by looking at the elements of Group 0, **the noble gases**. Their atoms *do not* form bonds.

This is because the atoms have a very stable arrangement of electrons in the outer shell. This makes the noble gases **unreactive**.

helium atom:
full outer shell of 2 electrons – *stable*

neon atom:
full outer shell of 8 electrons – *stable*

argon atom:
outer shell of 8 electrons – *stable*

2    2 + 8    2 + 8 + 8

And that gives us the answer to our question:

**Atoms bond with each other in order to gain a stable arrangement of outer-shell electrons, like the atoms of Group 0.**

**In other words, they bond in order to gain 8 electrons in their outer shell (or 2, if they have only one shell).**

▲ Neon: the unreactive gas used in light tubes for advertising.

▲ Welding is often carried out in an atmosphere of argon, which will not react with hot metals (unlike oxygen).

# ATOMS COMBINING

## How sodium atoms gain a stable outer shell

A sodium atom has just 1 electron in its outer shell. To obtain a stable outer shell of 8 electrons, it loses this electron to another atom. It becomes a **sodium ion**:

The sodium ion has 11 protons but only 10 electrons, so it has a charge of 1+, as you can see from the panel on the right.

The symbol for sodium is Na, so the symbol for the sodium ion is **Na$^+$**.

The + means *1 positive charge*. Na$^+$ is a **positive ion**.

| The charge on a sodium ion | |
|---|---|
| charge on 11 protons | 11+ |
| charge on 10 electrons | 10− |
| total charge | 1+ |

## How chlorine atoms gain a stable outer shell

A chlorine atom has 7 electrons in its outer shell. It can reach 8 electrons by accepting 1 electron from another atom. It becomes a chloride ion:

The chloride ion has a charge of 1−, so it is a **negative ion**. Its symbol is **Cl$^-$**.

## Ions

An atom becomes an ion when it loses or gains electrons.
**An ion is a charged particle. It is charged because it has an unequal number of protons and electrons.**

| The charge on a chloride ion | |
|---|---|
| charge on 17 protons | 17+ |
| charge on 18 electrons | 18− |
| total charge | 1− |

---

**Q**

1. Why are the atoms of the Group 0 elements unreactive?
2. Explain why all other atoms are reactive.
3. Draw a diagram to show how this atom gains a stable outer shell of 8 electrons:
   a. a sodium atom  b. a chlorine atom
4. Explain why
   a. a sodium ion has a charge of 1+
   b. a chloride ion has a charge of 1−.
5. Explain what an *ion* is, in your own words.
6. Atoms of Group 0 elements do *not* form ions. Why not?

ATOMS COMBINING

## 4.3 The ionic bond

### How sodium and chlorine atoms bond together

As you saw on page 49, a sodium atom must lose one electron, and a chlorine atom must gain one, to obtain stable outer shells of 8 electrons.

So when a sodium atom and a chlorine atom react together, the sodium atom loses its electron *to the chlorine atom*, and two ions are formed.

Here, sodium electrons are shown as • and chlorine electrons as ×:

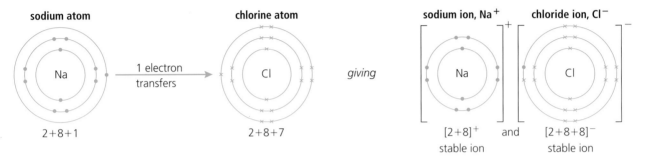

The two ions have opposite charges, so they attract each other. The force of attraction between them is strong. It is called an **ionic bond**.

**The ionic bond is the bond that forms between ions of opposite charge.**

### How solid sodium chloride is formed

When sodium reacts with chlorine, billions of sodium and chloride ions form. But they do not stay in pairs. They form a regular pattern or **lattice** of alternating positive and negative ions, as shown below. The ions are held together by strong ionic bonds.

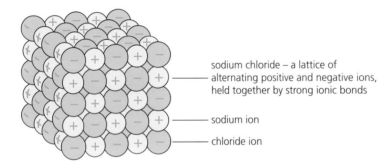

sodium chloride – a lattice of alternating positive and negative ions, held together by strong ionic bonds

sodium ion

chloride ion

The lattice grows to form a giant 3-D structure. It is called 'giant' because it contains a very large number of ions. This giant structure is the compound **sodium chloride**, or **common salt**.

Since it is made of ions, sodium chloride is called an **ionic compound**. It contains one $Na^+$ ion for each $Cl^-$ ion, so its formula is **NaCl**.

The charges in the structure add up to zero:
    the charge on each sodium ion is    1+
    the charge on each chloride ion is    1−
                 total charge    0

So the compound has no overall charge.

> **Bonding diagrams**
> To show the bonding clearly:
> - use dots and crosses (o, •, and ×) for electrons from atoms of different elements
> - write the symbol for the element in the centre of each atom.

▲ These polystyrene balls were given opposite charges. So they are attracted to each other, and cling together. The same happens with ions of opposite charge.

# ATOMS COMBINING

## Other ionic compounds
Sodium is a metal. Chlorine is a non-metal. They react together to form an ionic compound. Other metals and non-metals follow the same pattern.

**A metal reacts with a non-metal to form an ionic compound. The metal atoms lose electrons. The non-metal atoms gain them. The ions form a lattice. The compound has no overall charge.**

Below are two more examples.

## Magnesium oxide
A magnesium atom has 2 outer electrons and an oxygen atom has 6. When magnesium burns in oxygen, each magnesium atom loses its 2 outer electrons to an oxygen atom. Magnesium and oxide ions are formed:

The ions attract each other because of their opposite charges. Like the sodium and chloride ions, they group to form a lattice.

The resulting compound is called **magnesium oxide**. It has one magnesium ion for each oxide ion, so its formula is **MgO**. It has no overall charge.

> **The charge on magnesium oxide**
> charge on a magnesium ion    2+
> charge on an oxide ion       2−
> total charge                 0

## Magnesium chloride
When magnesium burns in chlorine, each magnesium atom reacts with *two* chlorine atoms, to form **magnesium chloride**. Each ion has 8 outer electrons:

The ions form a lattice with two chloride ions for each magnesium ion. So the formula of the compound is **MgCl$_2$**. It has no overall charge.

---

**Q**

1. Draw a diagram to show what happens to the electrons, when a sodium atom reacts with a chlorine atom.
2. What is an *ionic bond*?
3. Describe in your own words the structure of solid sodium chloride, and explain why its formula is NaCl.
4. Explain why:
   a. a magnesium ion has a charge of 2+
   b. the ions in magnesium oxide stay together
   c. magnesium chloride has no overall charge
   d. the formula of magnesium chloride is MgCl$_2$.

ATOMS COMBINING

## 4.4 More about ions

### Ions of the first twenty elements

Not every element forms ions during reactions. In fact, out of the first twenty elements in the Periodic Table, only twelve easily form ions. These ions are given below, with their names.

| Group I | II | | III | IV | V | VI | VII | 0 |
|---|---|---|---|---|---|---|---|---|
| | | $H^+$ hydrogen | | | | | | none |
| $Li^+$ lithium | $Be^{2+}$ beryllium | | | | | $O^{2-}$ oxide | $F^-$ fluoride | none |
| $Na^+$ sodium | $Mg^{2+}$ magnesium | | $Al^{3+}$ aluminium | | | $S^{2-}$ sulfide | $Cl^-$ chloride | none |
| $K^+$ potassium | $Ca^{2+}$ calcium | transition elements | | | | | | |

Note that:

- Hydrogen and the metals lose electrons and form **positive ions**. The ions have the same names as the atoms.
- Non-metals form **negative ions**, with names ending in **-ide**.
- The elements in Groups IV and V do not usually form ions, because their atoms would have to gain or lose several electrons, and that takes too much energy.
- Group 0 elements do not form ions: their atoms already have stable outer shells, so do not need to gain or lose electrons.

### The names and formulae of ionic compounds

**The names** To name an ionic compound, you just put the names of the ions together, with the positive one first:

| Ions in compound | Name of compound |
|---|---|
| $K^+$ and $F^-$ | potassium fluoride |
| $Ca^{2+}$ and $S^{2-}$ | calcium sulfide |

▲ Bath time. Bath salts contain ionic compounds such as magnesium sulfate (Epsom salts) and sodium hydrogen carbonate (baking soda). Plus scent!

**The formulae** The formulae of ionic compounds can be worked out using these four steps. Look at the examples that follow.

1 Write down the name of the ionic compound.
2 Write down the symbols for its ions.
3 The compound must have no overall charge, so balance the ions until the positive and negative charges add up to zero.
4 Write down the formula without the charges.

*Example 1*
1 Lithium fluoride.
2 The ions are $Li^+$ and $F^-$.
3 One $Li^+$ is needed for every $F^-$, to make the total charge zero.
4 The formula is LiF.

*Example 2*
1 Sodium sulfide.
2 The ions are $Na^+$ and $S^{2-}$.
3 Two $Na^+$ ions are needed for every $S^{2-}$ ion, to make the total charge zero: $Na^+ Na^+ S^{2-}$.
4 The formula is $Na_2S$. (What does the $_2$ show?)

52

## Some metals form more than one type of ion

Look back at the Periodic Table on page 31. Look for the block of **transition elements**. These include many common metals, such as iron and copper.

Some transition elements form only one type of ion:
- silver forms only $Ag^+$ ions
- zinc forms only $Zn^{2+}$ ions.

But most transition elements can form more than one type of ion. For example, copper and iron can each form two:

| Ion | Name | Example of compound |
|---|---|---|
| $Cu^+$ | copper(I) ion | copper(I) oxide, $Cu_2O$ |
| $Cu^{2+}$ | copper(II) ion | copper(II) oxide, $CuO$ |
| $Fe^{2+}$ | iron(II) ion | iron(II) chloride, $FeCl_2$ |
| $Fe^{3+}$ | iron(III) ion | iron(III) chloride, $FeCl_3$ |

The (II) in the name tells you that the ion has a charge of 2+. What do the (I) and (III) show?

▲ The two oxides of copper.

## Compound ions

All the ions you met so far have been formed from single atoms. But ions can also be formed from a **group** of bonded atoms. These are called **compound ions**.

The most common ones are shown on the right. Remember, each is just one ion, even though it contains more than one atom.

The formulae for their compounds can be worked out as before. Some examples are shown below.

$NH_4^+$, the ammonium ion

$OH^-$, the hydroxide ion

$NO_3^-$, the nitrate ion

$SO_4^{2-}$, the sulfate ion

$CO_3^{2-}$, the carbonate ion

$HCO_3^-$, the hydrogen carbonate ion

*Example 3*
1. Sodium carbonate.
2. The ions are $Na^+$ and $CO_3^{2-}$.
3. Two $Na^+$ are needed to balance the charge on one $CO_3^{2-}$.
4. The formula is $Na_2CO_3$.

*Example 4*
1. Calcium nitrate.
2. The ions are $Ca^{2+}$ and $NO_3^-$.
3. Two $NO_3^-$ are needed to balance the charge on one $Ca^{2+}$.
4. The formula is $Ca(NO_3)_2$. Note that brackets are put round the $NO_3$, before the $_2$ is put in.

---

**Q**

1. Explain why a calcium ion has a charge of 2+.
2. Why is the charge on an aluminium ion 3+?
3. Write down the symbols for the ions in:
   a potassium chloride   b calcium sulfide
   c lithium sulfide      d magnesium fluoride
4. Now work out the formula for each compound in **3**.
5. Work out the formula for each compound:
   a copper(II) chloride   b iron(III) oxide
6. Write a name for each compound:
   $CuCl$, $FeS$, $Mg(NO_3)_2$, $NH_4NO_3$, $Ca(HCO_3)_2$
7. Work out the formula for:   a sodium sulfate
   b potassium hydroxide   c silver nitrate

ATOMS COMBINING

# 4.5 The covalent bond

## Why atoms bond: a reminder

As you saw in Unit 4.3, atoms bond in order to gain a stable outer shell of electrons, like the noble gas atoms. So when sodium and chlorine react together, each sodium atom gives up an electron to a chlorine atom.

But that is not the only way. Atoms can also gain stable outer shells by *sharing* electrons with each other.

## Sharing electrons

When two non-metal atoms react together, *both need to gain electrons* to achieve stable outer shells. They manage this by sharing electrons.

We will look at **non-metal elements** in this unit, and at **non-metal compounds** in the next unit. Atoms can share only their outer (valence) electrons, so the diagrams will show only these.

## Hydrogen

A hydrogen atom has only one shell, with one electron. The shell can hold two electrons. When two hydrogen atoms get close enough, their shells overlap and then they can share electrons. Like this:

So each has gained a full shell of two electrons, like helium atoms.

## The bond between the atoms

Each hydrogen atom has a positive nucleus. Both nuclei attract the shared electrons – and this strong force of attraction holds the two atoms together.

This force of attraction is called a **covalent bond**.
**A single covalent bond is formed when atoms share two electrons.**

## Molecules

The two bonded hydrogen atoms above form a **molecule**.
**A molecule is a group of atoms held together by covalent bonds.**

Since it is made up of molecules, hydrogen is a molecular element. Its formula is $H_2$. The $_2$ tells you there are 2 hydrogen atoms in each molecule.

Many other non-metals are also molecular. For example:

iodine, $I_2$    oxygen, $O_2$    nitrogen, $N_2$
chlorine, $Cl_2$    sulfur, $S_8$    phosphorus, $P_4$

Elements made up of molecules containing two atoms are called **diatomic**. So iodine and oxygen are diatomic. Can you give two other examples?

▲ Atoms of non-metals do not *give up* electrons to gain a full shell, because they would have to lose so many. It would take too much energy to overcome the pull of the positive nucleus.

▲ A model of the hydrogen molecule. The molecule can also be shown as H–H. The line represents a single bond.

## Chlorine

A chlorine atom needs a share in one more electron, to obtain a stable outer shell of eight electrons. So two chlorine atoms bond covalently like this:

▲ A model of the chlorine molecule.

Since only one pair of electrons is shared, the bond between the atoms is called a **single covalent bond**, or just a **single bond**. You can show it in a short way by a single line, like this: Cl−Cl.

## Oxygen

An oxygen atom has six outer electrons, so needs a share in *two* more. So two oxygen atoms share two electrons each, giving molecules with the formula $O_2$. Each atom now has a stable outer shell of eight electrons:

▲ A model of the oxygen molecule.

Since the oxygen atoms share two pairs of electrons, the bond between them is called a **double bond**. You can show it like this: O=O.

## Nitrogen

A nitrogen atom has five outer electrons, so needs a share in *three* more. So two nitrogen atoms share three electrons each, giving molecules with the formula $N_2$. Each atom now has a stable outer shell of eight electrons:

▲ A model of the nitrogen molecule.

Since the nitrogen atoms share three pairs of electrons, the bond between them is called a **triple bond**. You can show it like this: N≡N.

---

**1 a** Name the bond between atoms that share electrons.
  **b** What holds the bonded atoms together?
**2** What is a *molecule*?
**3** Give five examples of molecular elements.
**4** Draw a diagram to show the bonding in:
  **a** hydrogen    **b** chlorine
**5** Now explain why the bond in a nitrogen molecule is called a *triple* bond.

ATOMS COMBINING

## 4.6 Covalent compounds

### Covalent compounds

In the last unit you saw that many non-metal elements exist as molecules. A huge number of *compounds* also exist as molecules.

In a molecular compound, atoms of *different* elements share electrons. The compounds are called **covalent compounds**. Here are three examples.

> **Most are molecular ...**
> Most non-metal elements and their compounds exist as molecules.

| Covalent compound | Description | Model of the molecule |
|---|---|---|
| **hydrogen chloride, HCl** 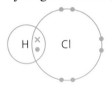 a molecule of hydrogen chloride | The chlorine atom shares one electron with the hydrogen atom. Both now have a stable arrangement of electrons in their outer shells: 2 for hydrogen (like the helium atom) and 8 for chlorine (like the other noble gas atoms). |  |
| **water, $H_2O$** 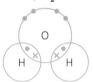 a molecule of water | The oxygen atom shares electrons with the two hydrogen atoms. All now have a stable arrangement of electrons in their outer shells: 2 for hydrogen and 8 for oxygen. |  |
| **methane, $CH_4$** 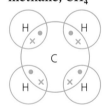 a molecule of methane | The carbon atom shares electrons with four hydrogen atoms. All now have a stable arrangement of electrons in their outer shells: 2 for hydrogen and 8 for carbon. | 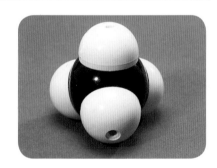 |

### The shapes of the molecules

Look at the models of the methane molecule, above and on the right. The molecule is **tetrahedral** in shape, because the four pairs of electrons around carbon repel each other, and move as far apart as possible.

Now look at the model of the water molecule above. The hydrogen atoms are closer together than in methane. This is because the two non-bonding pairs of atoms repel more strongly than the bonding pairs. So they push these closer together.

The angle between the hydrogen atoms in water is 104.5°.

▲ The methane molecule: the same angle between all the H atoms.

# ATOMS COMBINING

## More examples of covalent compounds

This table shows three more examples of covalent compounds. Each time:
- the atoms share electrons, to gain stable outer shells
- repulsion between pairs of electrons dictates the shape of the molecule.

| Covalent compound | Description | Model of the molecule |
|---|---|---|
| **ammonia, $NH_3$** 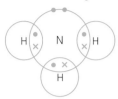 a molecule of ammonia | Each nitrogen atom shares electrons with three hydrogen atoms. So all three atoms now have a stable arrangement of electrons in their outer shells: 2 for hydrogen and 8 for nitrogen. The molecule is shaped like a **pyramid**. |  |
| **methanol, $CH_3OH$**  a molecule of methanol | The carbon atom shares electrons with three hydrogen atoms and one oxygen atom. Look at the shape of the molecule: a little like methane, but changed by the presence of the oxygen atom. |  |
| **carbon dioxide, $CO_2$** 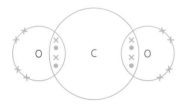 a molecule of carbon dioxide | The carbon atom shares all four of its electrons: two with each oxygen atom. So all three atoms gain stable shells. The two sets of bonding electrons repel each other. They move as far apart as they can, giving a **linear** molecule. All the bonds are double bonds, so we can show the molecule like this: $O=C=O$. |  |
| **ethene, $C_2H_4$** 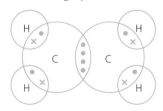 a molecule of ethene | Look how each carbon atom shares its four electrons this time. It shares two with two hydrogen atoms. and two with another carbon atom, giving a carbon-carbon double bond. So the molecule is usually drawn like this:  |  |

**Q**

1. **a** What is a *covalent compound*?
   **b** Give five examples, with their formulae.
2. Draw a diagram to show the bonding in a molecule of:
   **a** methane  **b** water
3. How do the atoms gain stable outer shells, in ammonia?
4. Draw a diagram to show the bonding in carbon dioxide.
5. Why is the carbon dioxide molecule straight, and not bent like the water molecule?.

ATOMS COMBINING

## 4.7 Comparing ionic and covalent compounds

### Remember
Metals and non-metals react together to form **ionic compounds**.
Non-metals react together to form **covalent compounds**.
The covalent compounds you have met so far exist as **molecules**.

### Comparing the structures of the solids
In Chapter 1, you met the idea that solids are a **regular lattice** of particles. In ionic compounds, these particles are **ions**. In the covalent compounds you have met so far, they are **molecules**. Let's compare their lattices.

**A solid ionic compound**    Sodium chloride is a typical ionic compound:

In sodium chloride, the ions are held in a regular lattice like this. They are held by strong ionic bonds.

The lattice grows in all directions, giving a crystal of sodium chloride. This one is magnified 35 times.

The crystals look white and shiny. We add them to food, as **salt**, to bring out its taste.

**A solid molecular covalent compound**    Water is a molecular covalent compound. When you cool it below 0 °C it becomes a solid: ice.

In ice, the water molecules are held in a regular lattice like this. But the forces between them are weak.

The lattice grows in all directions, giving a crystal of ice. These grew in an ice-tray in a freezer.

We use ice to keep drinks cool, and food fresh. (The reactions that cause food to decay are slower in the cold.)

So both types of compounds have a regular lattice structure in the solid state, and form crystals. But they differ in two key ways:

- In ionic solids the particles (ions) are charged, and the forces between them are strong.
- In molecular covalent solids the particles (molecules) are not charged, and the forces between them are weak.

These differences lead to very different properties, as you will see next.

**About crystals**
- A regular arrangement of particles in a lattice always leads to crystals.
- The particles can be atoms, ions, or molecules.

## The properties of ionic compounds

1 **Ionic compounds have high melting and boiling points.**
   For example:

   | Compound | Melting point/°C | Boiling point/°C |
   |---|---|---|
   | sodium chloride, NaCl | 801 | 1413 |
   | magnesium oxide, MgO | 2852 | 3600 |

   This is because the ionic bonds are very strong. It takes a lot of heat energy to break up the lattice. So ionic compounds are solid at room temperature.

   Note that magnesium oxide has a far higher melting and boiling point than sodium chloride does. This is because its ions have double the charge ($Mg^{2+}$ and $O^{2-}$ compared with $Na^+$ and $Cl^-$), so its ionic bonds are stronger.

2 **Ionic compounds are usually soluble in water.**
   The water molecules are able to separate the ions from each other. The ions then move apart, surrounded by water molecules.

3 **Ionic compounds conduct electricity, when melted or dissolved in water.**
   A solid ionic compound will not conduct electricity. But when it melts, or dissolves in water, the ions become free to move. Since they are charged, they can then conduct electricity.

▲ Magnesium oxide is used to line furnaces in steel works, because of its high melting point, 2852 °C. (By contrast, iron melts at 1538 °C.)

## The properties of covalent compounds

1 **Molecular covalent compounds have low melting and boiling points.**
   For example:

   | Compound | Melting point/°C | Boiling point/°C |
   |---|---|---|
   | carbon monoxide, CO | −199 | −191 |
   | hexane, $C_6H_{14}$ | −95 | 69 |

   This is because the attraction between the molecules is low. So it does not take much energy to break up the lattice and separate them from each other. That explains why many molecular compounds are liquids or gases at room temperature – and why many of the liquids are **volatile** (evaporate easily).

2 **Covalent compounds tend to be insoluble in water.**
   But they do dissolve in some solvents, for example tetrachloromethane.

3 **Covalent compounds do not conduct electricity.**
   There are no charged particles, so they cannot conduct.

▲ The covalent compound carbon monoxide is formed when petrol burns in the limited supply of air in a car engine. And it is poisonous.

---

1 The particles in solids usually form a *regular lattice*. Explain what that means, in your own words.
2 Which type of particles make up the lattice, in:
   a ionic compounds?   b molecular compounds?
3 Solid sodium chloride will not conduct electricity, but a solution of sodium chloride will conduct. Explain this.
4 A compound melts at 20 °C.
   a What kind of structure do you think it has? Why do you think so?
   b Will it conduct electricity at 25 °C? Give a reason.
5 Describe the arrangement of the molecules in ice. How will the arrangement change as the ice warms up?

ATOMS COMBINING

## 4.8 Giant covalent structures

### Not all covalent solids are molecular
In all the solids in this table, the atoms are held together by covalent bonds. But compare their melting points. What do you notice?

| Substance | Melting point/°C |
|---|---|
| ice | 0 |
| phosphorus | 44 |
| sulfur | 115 |
| silicon dioxide (silica) | 1710 |
| carbon (as diamond) | 3550 |

The first three substances are molecular solids. Their molecules are held in a lattice by weak forces – so the solids melt easily, at low temperatures.

But diamond and silica are different. Their melting points show that *they* are not molecular solids with weak lattices. In fact they exist as **giant covalent structures**, or **macromolecules**.

▲ Diamond: so hard that it is used to edge wheels for cutting stone.

### Diamond – a giant covalent structure
Diamond is made of carbon atoms, held in a strong lattice:

A carbon atom forms covalent bonds to *four* others, as shown above. Each outer atom then bonds to three more, and so on.

Eventually billions of carbon atoms are bonded together, in a giant covalent structure. This shows just a very tiny part of it.

The result is a single crystal of diamond. This one has been cut, shaped, and polished, to make it sparkle.

Diamond has these properties:
1 It is very hard, because each atom is held in place by four strong covalent bonds. In fact it is the hardest substance on Earth.
2 For the same reason it has a very high melting point, 3550 °C.
3 It can't conduct electricity because there are no ions or free electrons to carry the charge.

### Silica is similar to diamond
Silica, $SiO_2$, occurs naturally as **quartz**, the main mineral in **sand**. Like diamond, it forms a giant covalent structure, as shown on the right.

Each silicon atom bonds covalently to four oxygen atoms. And each oxygen atom bonds covalently to two silicon atoms. The result is a very hard substance with a melting point of 1710 °C.

▲ Silicon dioxide is made up of oxygen atoms ● and silicon atoms ○. Billions of them bond together like this, to give a giant structure.

## Graphite – a very different giant structure

Like diamond, graphite is made only of carbon atoms. So diamond and graphite are **allotropes** of carbon – two forms of the same element.

Diamond is the hardest solid on Earth. But graphite is one of the softest! This difference is a result of their very different structures:

In graphite, each carbon atom forms covalent bonds to *three* others. This gives rings of *six* atoms.

The rings form flat sheets that lie on top of each other, held together by weak forces.

Under a microscope, you can see the layered structure of graphite quite clearly.

Graphite has these properties:
1. Unlike diamond, it is soft and slippery. That is because the sheets can slide over each other easily.
2. Unlike diamond, it is a good conductor of electricity. That is because each carbon atom has four outer electrons, but forms only three bonds. So the fourth electron is free to move through the graphite, carrying charge.

## Making use of these giant structures

Different properties lead to different uses, as this table shows.

| Substance | Properties | Uses |
| --- | --- | --- |
| diamond | hardest known substance does not conduct | in tools for drilling and cutting |
|  | sparkles when cut | for jewellery |
| graphite | soft and slippery | as a lubricant for engines and locks |
|  | soft and dark in colour | for pencil 'lead' (mixed with clay) |
|  | conducts electricity | for electrodes, and connecting brushes in generators |
| silica | hard, can scratch things | in sandpaper |
|  | hard, lets light through | for making glass and lenses |
|  | high melting point | in bricks for lining furnaces |

▲ Pencil 'lead' is a mixture of graphite and clay.

## Q

1. The covalent compound ethanol melts at −114 °C. Is it a molecular compound, or a giant structure? Explain.
2. Diamond and graphite are *allotropes* of carbon. What does that mean?
3. Why is diamond so hard?
4. Why do diamond and graphite have such very different properties? Draw diagrams to help you explain.
5. a Explain why silica has a high melting point.
   b See if you can suggest a reason why its melting point is lower than diamond's.

ATOMS COMBINING

## 4.9 The bonding in metals

### Clues from melting points

Compare these melting points:

| Structure | Examples | Melting point /°C |
|---|---|---|
| molecular | carbon dioxide | −56 |
| | water | 0 |
| giant ionic | sodium chloride | 801 |
| | magnesium oxide | 2852 |
| giant covalent | diamond | 3550 |
| | silica | 1610 |
| metal | iron | 1535 |
| | copper | 1083 |

The table shows clearly that:

- **molecular substances have low melting points.** That is because the forces between molecules in the lattice are weak.
- **giant structures such as sodium chloride and diamond have much higher melting points.** That is because the bonds between ions or atoms within giant structures are very strong.

Now look at the metals. They too have high melting points – much higher than for carbon dioxide or water. This gives us a clue that they too might be giant structures. And so they are, as you'll see below.

▲ Equipment for measuring melting points in the school lab. It can heat substances up to 300°C – so no good for sodium chloride!

### The structure of metals

In metals, the atoms are packed tightly together in a regular lattice. The tight packing allows outer electrons to separate from their atoms. The result is a lattice of ions in a 'sea' of electrons that are free to move.

Look at copper:

The copper ions are held together by their attraction to the free electrons between them.
The strong forces of attraction are called **metallic bonds**.

The regular arrangement of ions results in **crystals** of copper. This shows the crystals in a piece of copper, magnified 1000 times. (They are all at different angles.)

The copper crystals are called **grains**. A lump of copper like this one consists of millions of grains joined together. You need a microscope to see them.

**The metallic bond is the attraction between metal ions and free electrons.**

It is the same with all metals. The ions sit in a lattice, held together by their strong attraction to the free electrons. And because the ions are in a regular pattern, metals are crystalline.

**Delocalised electrons**
The electrons that move freely in the metal lattice are not tied to any one ion. So they are called **delocalised**.

# ATOMS COMBINING

## Explaining some key properties of metals

In Unit 3.5 you read about the properties of metals. We can now explain some of those properties. Look at these examples.

**1 Metals usually have high melting points.**
That is because it takes a lot of heat energy to break up the lattice, with its strong metallic bonds. Copper melts at 1083 °C, and nickel at 1455 °C. (But there are exceptions. Sodium melts at only 98 °C, for example. And mercury melts at −39 °C, so it is a liquid at room temperature.)

**2 Metals are malleable and ductile.**
*Malleable* means they can be bent and pressed into shape. *Ductile* means they can be drawn out into wires. This is because the layers can slide over each other. This diagram represents any metal lattice:

▲ Metals: malleable, ductile, and sometimes very glamorous – like this silver bracelet.

The layers can slide without breaking the metallic bond, because the electrons are free to move too.

**3 Metals are good conductors of heat.**
That is because the free electrons take in heat energy, which makes them move faster. They quickly transfer the heat through the metal structure:

**4 Metals are good conductors of electricity.**
That is because the free electrons can move through the lattice carrying charge, when a voltage is applied across the metal.

Silver is the best conductor of all the metals. Copper is next – but it is used much more than silver because it is cheaper.

▲ What uses of metals can you see in this scene?

## Q

1 Describe in your own words the structure of a metal.
2 What is a *metallic bond*?
3 What does *malleable* mean?
4 Explain why metals can be drawn out into wires without breaking.
5 a Explain why metals can conduct electricity.
  b Would you expect molten metals to conduct? Give a reason.
6 Because metals are malleable, we use some of them to make saucepans. Give two other examples of uses of metals that depend on:
  a their malleability    b their ductility
  c their ability to conduct electricity
7 Mercury forms ions with a charge of 2+. It goes solid (freezes) at −39 °C. Try drawing a diagram to show the structure of solid mercury.

ATOMS COMBINING

# Checkup on Chapter 4

## Revision checklist

### Core curriculum
Make sure you can ...
- [ ] explain the difference between:
  - an *element* and a *compound*
  - a *compound* and a *mixture*
- [ ] say what the signs of a chemical change are
- [ ] explain why:
  - atoms of Group 0 elements do not form bonds
  - atoms of other elements do form bonds
- [ ] explain the difference between an *ionic bond* and a *covalent bond*
- [ ] draw a diagram to show how an ionic bond forms between atoms of sodium and chlorine
- [ ] explain what a *molecule* is
- [ ] say that non-metal atoms form covalent bonds with each other (except for the noble gas atoms)
- [ ] draw diagrams to show the covalent bonding in:
  hydrogen    chlorine    water
  methane    hydrogen chloride
- [ ] give three ways in which ionic and molecular compounds differ in their properties, and explain these differences
- [ ] describe the giant covalent structures of graphite and diamond, and sketch them
- [ ] explain how their structures lead to different uses for diamond and graphite, with examples

### Extended curriculum
Make sure you can also ...
- [ ] show how ionic bonds form between atoms of other metals and non-metals
- [ ] describe the lattice structure of ionic compounds
- [ ] work out the formulae of ionic compounds, from the charges on the ions
- [ ] draw diagrams to show the covalent bonding in nitrogen, oxygen, ammonia, methanol, carbon dioxide, and ethene
- [ ] describe metallic bonding, and draw a sketch for it
- [ ] explain how the structure and bonding in metals enables them to be malleable, ductile, and good conductors of heat and electricity
- [ ] describe the structure of silicon dioxide
- [ ] explain why silicon dioxide and diamond have similar properties
- [ ] give examples of uses for silicon dioxide

## Questions

### Core curriculum

1. This question is about the ionic bond formed between the metal lithium (proton number 3) and the non-metal fluorine (proton number 9).
   a. How many electrons does a lithium atom have? Draw a diagram to show its electron structure.
   b. How does a metal atom obtain a stable outer shell of electrons?
   c. Draw the structure of a lithium ion, and write the symbol for it, showing its charge.
   d. How many electrons does a fluorine atom have? Draw a diagram to show its electron structure.
   e. How does a non-metal atom become an ion?
   f. Draw the structure of a fluoride ion, and write a symbol for it, showing its charge.
   g. Draw a diagram to show what happens when a lithium atom reacts with a fluorine atom.
   h. Write a word equation for the reaction between lithium and fluorine.

2. This diagram represents a molecule of a certain gas.
   a. Name the gas, and give its formula.
   b. What do the symbols • and × represent?
   c. Which type of bonding holds the atoms together?
   d. Name another compound with this type of bonding.

3. Hydrogen bromide is a compound of the two elements hydrogen and bromine. It melts at $-87\,°C$ and boils at $-67\,°C$. It has the same type of bonding as hydrogen chloride.
   a. Is hydrogen bromide a solid, a liquid, or a gas at room temperature ($20\,°C$)?
   b. Is hydrogen bromide molecular, or does it have a giant structure? What is your evidence?
   c. i. Which type of bond is formed between the hydrogen and bromine atoms, in hydrogen bromide?
      ii. Draw a diagram of the bonding between the atoms, showing only the outer electrons.
   d. Write a formula for hydrogen bromide.
   e. i. Name two other compounds with bonding similar to that in hydrogen bromide.
      ii. Write formulae for these two compounds.

4 These are some properties of substances A to G.

| Substance | Melting point /°C | Electrical conductivity | | Solubility in water |
|---|---|---|---|---|
| | | solid | liquid | |
| A | −112 | poor | poor | insoluble |
| B | 680 | poor | good | soluble |
| C | −70 | poor | poor | insoluble |
| D | 1495 | good | good | insoluble |
| E | 610 | poor | good | soluble |
| F | 1610 | poor | poor | insoluble |
| G | 660 | good | good | insoluble |

a Which of the seven substances are metals? Give reasons for your choice.
b Which of the substances are ionic compounds? Give reasons for your choice.
c Two of the substances have very low melting points, compared with the rest. Explain why these could *not* be ionic compounds.
d Two of the substances are molecular. Which two are they?
e i Which substance is a giant covalent structure?
  ii What other name is used to describe this type of structure? (Hint: starts with *m*.)
f Name the type of bonding found in:
  i B   ii C   iii E   iv F

### Extended curriculum

5 Aluminium and nitrogen react to form an ionic compound called aluminium nitride. These show the electron arrangement for the two elements:

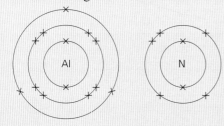

a Answer these questions for an aluminium atom.
  i Does it gain or lose electrons, to form an ion?
  ii How many electrons are transferred?
  iii Is the ion formed positive, or negative?
  iv What charge does the ion have?
b Now repeat a, but for a nitrogen atom.
c i Give the electron distribution for the ions formed by the two atoms. (2 + …)
  ii What do you notice about these distributions? Explain it.
d Name another non-metal that will form an ionic compound with aluminium, in the same way as nitrogen does.

6 Silicon lies directly below carbon in Group IV of the Periodic Table. Here is some data for silicon, carbon (in the form of diamond), and their oxides.

| Substance | Symbol or formula | Melting point/°C | Boiling point/°C |
|---|---|---|---|
| carbon | C | 3730 | 4530 |
| silicon | Si | 1410 | 2400 |
| carbon dioxide | $CO_2$ | (turns to gas at −78°C) | |
| silicon dioxide | $SiO_2$ | 1610 | 2230 |

a In which state are the two *elements* at room temperature (20°C)?
b Which type of structure does carbon (diamond) have: giant covalent, or molecular?
c Which type of structure would you expect to find in silicon? Give reasons.
d In which state are the two oxides, at room temperature?
e Which type of structure has carbon dioxide?
f Does silicon dioxide have the same structure as carbon dioxide? What is your evidence?

7 The compound zinc sulfide has a structure like this:

— sulfide ion ($S^{2-}$)
— zinc ion ($Zn^{2+}$)

a Which does the diagram represent: a giant structure, or a molecular structure?
b Which type of bonding does zinc sulfide have?
c Look carefully at the structure. How many:
  i sulfur ions are joined to each zinc ion?
  ii zinc ions are joined to each sulfur ion?
d i From c, deduce the formula of zinc sulfide.
  ii Is this formula consistent with the charges on the two ions? Explain your answer.
e Name another metal and non-metal that will form a compound with a similar formula.

8 The properties of metals can be explained by the structure and bonding within the metal lattice.
a Describe the bonding in metals.
b Use the bonding to explain why metals:
  i are good conductors of electricity
  ii are malleable and flexible

REACTING MASSES, AND CHEMICAL EQUATIONS

## 5.1 The names and formulae of compounds

### The names of compounds
Many compounds contain just two elements. If you know which elements they are, you can usually name the compound. Just follow these rules:

- When the compound contains a metal and a non-metal:
  - the name of the metal is given first
  - and then the name of the non-metal, but ending with *-ide*.
  *Examples*: sodium chloride, magnesium oxide, iron sulfide.

- When the compound is made of two non-metals:
  - if one is hydrogen, that is named first
  - otherwise the one with the lower group number comes first
  - and then the name of the other non-metal, ending with *-ide*.
  *Examples*: hydrogen chloride, carbon dioxide.

But some compounds have 'everyday' names that give you no clue about the elements in them. Water, methane, and ammonia are examples. You just have to remember their formulae!

▲ That very common compound, water. Your body is full of it. Which elements does it contain?

### Finding formulae from the structure of compounds
Every compound has a formula as well as a name. The formula is made up of the symbols for the elements, and often has numbers too.

The formula of a compound is related to its structure. For example:

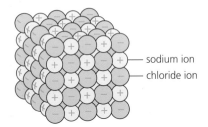

Sodium chloride forms a giant structure with one sodium ion for every chloride ion. So its formula is NaCl.

Water is made up of molecules in which two hydrogen atoms are bonded to an oxygen atom. So its formula is $H_2O$.

Silicon dioxide (silica) forms a giant structure in which there are two oxygen atoms for every silicon atom. So its formula is $SiO_2$.

Note the difference:

- In giant structures like sodium chloride and silicon dioxide, the formula tells you the *ratio* of the ions or atoms in the compound.
- In a molecular compound, the formula tells you *exactly* how many atoms are bonded together in each molecule.

### Valency
But you don't need to draw the structure of a compound to work out its formula. You can work it out quickly if you know the valency of the elements:

**The valency of an element is the number of electrons its atoms lose, gain or share, to form a compound.**

Look at this table. (You can check the groups in the Periodic Table on page 31.)

| Elements | In forming a compound, the atoms ... | So the valency of the element is ... | Examples of compounds formed (those in blue are covalent, with shared electrons) |
|---|---|---|---|
| Group I | lose 1 electron | 1 | sodium chloride, NaCl |
| Group II | lose 2 electrons | 2 | magnesium chloride, $MgCl_2$ |
| Group III | lose 3 electrons | 3 | aluminium chloride, $AlCl_3$ |
| Group IV | share 4 electrons | 4 | methane, $CH_4$ |
| Group V | gain or share 3 electrons | 3 | ammonia, $NH_3$ |
| Group VI | gain or share 2 electrons | 2 | magnesium oxide, MgO; water, $H_2O$ |
| Group VII | gain or share 1 electron | 1 | sodium chloride, NaCl; hydrogen chloride, HCl |
| Group 0 | (do not form compounds) | – | none |
| hydrogen | lose or share 1 electron | 1 | hydrogen bromide, HBr |
| transition elements | can lose different numbers of electrons | variable | iron(II) chloride, $FeCl_2$; iron(III) chloride, $FeCl_3$ copper(I) chloride, CuCl; copper(II) chloride, $CuCl_2$ |

## Writing formulae using valencies

This is how to write the formula of a compound, using valencies:

1 Write down the valencies of the two elements.
2 Write down their symbols, in the same order as the elements in the name.
3 Add numbers after the symbols if you need to, to balance the valencies.

*Example 1* What is the formula of hydrogen sulfide?

1 Valencies: hydrogen, 1; sulfur (Group VI), 2
2 HS (valencies not balanced)
3 The formula is **$H_2S$** ($2 \times 1$ and 2, so the valencies are now balanced)

*Example 2* What is the formula of aluminium oxide?

1 Valencies: aluminium (Group III), 3; oxygen (Group VI), 2
2 AlO (valencies not balanced)
3 The formula is **$Al_2O_3$** ($2 \times 3$ and $3 \times 2$, so the valencies are now balanced)

▲ Hydrogen sulfide is a very poisonous colourless gas. It smells of rotten eggs.

## Writing formulae by balancing charges

In an ionic compound, the total charge is zero. So you can also work out the formula of an ionic compound by balancing the charges on its ions. To find out how to do this, turn to Unit 4.4.

**Q**

The Periodic Table on page 31 will help you with these.
1 Write the chemical name for water (ending in *-ide*).
2 Name the compounds containing only these elements:
   a sodium and fluorine    b fluorine and hydrogen
   c sulfur and hydrogen    d bromine and beryllium
3 Why does silica have the formula $SiO_2$?
4 Decide whether this formula is correct. If it is not correct, write it correctly.
   a $HBr_2$    b ClNa    c $Cl_3Ca$    d $Ba_2O$
5 Write the correct formula for barium iodide.
6 See if you can give a name and formula for a compound that forms when phosphorus reacts with chlorine.

REACTING MASSES, AND CHEMICAL EQUATIONS

## 5.2 Equations for chemical reactions

### Equations for two sample reactions

1  **The reaction between carbon and oxygen**  When carbon is heated in oxygen, they react together to form carbon dioxide. Carbon and oxygen are the **reactants**. Carbon dioxide is the **product**.
You could show the reaction using a diagram, like this:

 +  →

1 atom of carbon     1 molecule of oxygen     1 molecule of carbon dioxide

> **Note**
> Reactants are sometimes called reagents.

or by a word equation, like this:

carbon + oxygen ⟶ carbon dioxide

or by a **symbol equation**, which gives symbols and formulae:

$C + O_2 \rightarrow CO_2$

2  **The reaction between hydrogen and oxygen**  Hydrogen and oxygen react together to give water. The diagram is:

    →

2 molecules of hydrogen     1 molecule of oxygen     2 molecules of water

and you can use it to write the symbol equation:

$2H_2 + O_2 \rightarrow 2H_2O$

### Symbol equations must be balanced

Now look at the number of atoms on each side of this equation:

$2H_2 + O_2 \rightarrow 2H_2O$

| On the left: | On the right: |
|---|---|
| 4 hydrogen atoms | 4 hydrogen atoms |
| 2 oxygen atoms | 2 oxygen atoms |

The number of each type of atoms is the same on both sides of the arrow. This is because atoms do not *disappear* during a reaction – they are just *rearranged*, as shown in the diagram of the molecules, in **2** above.

When the number of each type of atom is the same on both sides, the symbol equation is **balanced**. If it is not balanced, it is not correct.

### Adding state symbols

Reactants and products may be solids, liquids, gases, or in solution. You can show their states by adding **state symbols** to the equations:

(s) for solid        (l) for liquid
(g) for gas         (aq) for aqueous solution (solution in water)

For the two reactions above, the equations with state symbols are:

$C\,(s) + O_2\,(g) \rightarrow CO_2\,(g)$
$2H_2\,(g) + O_2\,(g) \rightarrow 2H_2O\,(l)$

▲ The reaction between hydrogen and oxygen gives out so much energy that it is used to power rockets. The reactants are carried as liquids in the fuel tanks.

# REACTING MASSES, AND CHEMICAL EQUATIONS

## How to write the equation for a reaction

These are the steps to follow, when writing an equation:
1 Write the equation in words.
2 Now write it using symbols. Make sure all the formulae are correct.
3 Check that the equation is balanced, for each type of atom in turn. *Make sure you do not change any formulae.*
4 Add the state symbols.

*Example 1* Calcium burns in chlorine to form calcium chloride, a solid. Write an equation for the reaction, using the steps above.

1 calcium + chlorine ⟶ calcium chloride
2 $Ca + Cl_2 \longrightarrow CaCl_2$
3 Ca: 1 atom on the left and 1 atom on the right.
  Cl: 2 atoms on the left and 2 atoms on the right.
  The equation is balanced.
4 $Ca\,(s) + Cl_2\,(g) \longrightarrow CaCl_2\,(s)$

▲ Calcium chloride absorbs water, so it is used to dry gases. The glass cylinder above is packed with calcium chloride, and the gas is piped up through it.

*Example 2* Hydrogen chloride is formed by burning hydrogen in chlorine. Write an equation for the reaction.

1 hydrogen + chlorine ⟶ hydrogen chloride
2 $H_2 + Cl_2 \longrightarrow HCl$
3 H: 2 atoms on the left and 1 atom on the right.
  Cl: 2 atoms on the left and 1 atom on the right.
  The equation is *not* balanced. It needs another molecule of hydrogen chloride on the right. So a 2 is put *in front of* the HCl.
  $H_2 + Cl_2 \longrightarrow 2HCl$
  The equation is now balanced. Do you agree?
4 $H_2\,(g) + Cl_2\,(g) \longrightarrow 2HCl\,(g)$

*Example 3* Magnesium burns in oxygen to form magnesium oxide, a white solid. Write an equation for the reaction.

1 magnesium + oxygen ⟶ magnesium oxide
2 $Mg + O_2 \longrightarrow MgO$
3 Mg: 1 atom on the left and 1 atom on the right.
  O: 2 atoms on the left and 1 atom on the right.
  The equation is *not* balanced. Try this:
  $Mg + O_2 \longrightarrow 2MgO$ (The 2 goes *in front of* the MgO.)
  Another magnesium atom is now needed on the left:
  $2Mg + O_2 \longrightarrow 2MgO$
  The equation is balanced.
4 $2Mg\,(s) + O_2\,(g) \longrightarrow 2MgO\,(s)$

▲ Magnesium burning in oxygen.

---

**Q**

1 What do + and ⟶ mean, in an equation?
2 Balance the following equations:
  a $Na\,(s) + Cl_2\,(g) \longrightarrow NaCl\,(s)$
  b $H_2\,(g) + I_2\,(g) \longrightarrow HI\,(g)$
  c $Na\,(s) + H_2O\,(l) \longrightarrow NaOH\,(aq) + H_2\,(g)$
  d $NH_3\,(g) \longrightarrow N_2\,(g) + H_2\,(g)$
  e $C\,(s) + CO_2\,(g) \longrightarrow CO\,(g)$
  f $Al\,(s) + O_2\,(g) \longrightarrow Al_2O_3\,(s)$
3 Aluminium burns in chlorine to form aluminium chloride, $AlCl_3$, a solid. Write a balanced equation for the reaction.

REACTING MASSES, AND CHEMICAL EQUATIONS

## 5.3 The masses of atoms, molecules, and ions

### Relative atomic mass

A single atom weighs hardly anything. You can't use scales to weigh it. But scientists do need a way to compare the masses of atoms. So this is what they did.

First, they chose an atom of carbon-12 to be the standard atom. They fixed its mass as exactly 12 atomic mass units. (It has 6 protons and 6 neutrons, as shown on the right. They ignored the electrons.)

Then they compared all the other atoms with this standard atom, in a machine called a mass spectrometer, and found values for their masses. Like this:

$^{12}_{6}C$    6 protons   6 electrons   6 neutrons

▲ An atom of carbon-12. It is the main isotope of carbon. (See page 34.)

This is the standard atom, $^{12}_{6}C$ or carbon-12. Its mass is taken as exactly 12.

This magnesium atom is twice as heavy as the carbon-12 atom. So its mass is 24.

This hydrogen atom has only one-twelfth the mass of the carbon-12 atom. So its mass is 1.

**The mass of an atom compared with the carbon-12 atom is called its relative atomic mass, or $A_r$.**

The small **r** stands for *relative to the mass of a carbon-12 atom*.

So the $A_r$ of hydrogen is 1, and the $A_r$ of magnesium is 24.

### $A_r$ and isotopes

As you saw on page 34, the atoms of an element are not always identical. Some may have extra neutrons. Different atoms of the same element are called **isotopes**. Chlorine has two isotopes:

| Name | Protons | Neutrons | Nucleon number | % of chlorine atoms like this |
|---|---|---|---|---|
| chlorine-35 | 17 | 18 | 35 | 75% |
| chlorine-37 | 17 | 20 | 37 | 25% |

Cl — this one has a mass of 35

Cl — this one has a mass of 37

▲ The two isotopes of chlorine.

We need to take all the natural isotopes of an element into account, to work out the relative atomic mass. This is the formula to use:

**relative atomic mass ($A_r$) of an element =**
       (% × nucleon number for the first isotope)
      + (% × nucleon number for the second isotope)
      … and so on, for all its natural isotopes

The calculation for chlorine is given on the right. It shows that the relative atomic mass of chlorine is **35.5**.

**The relative atomic mass ($A_r$) of an element is the average mass of its isotopes compared to an atom of carbon-12.**

For most elements, $A_r$ is very close to a whole number. It is usually rounded off to a whole number, to make calculations easier.

$A_r$ **for chlorine**
Using the formula on the left, the relative atomic mass of chlorine …

$= 75\% \times 35 + 25\% \times 37$

$= \dfrac{75}{100} \times 35 + \dfrac{25}{100} \times 37$

(changing % to fractions)

$= 26.25 + 9.25$

$= \mathbf{35.5}$

## $A_r$ values for some common elements

| Element | Symbol | $A_r$ | Element | Symbol | $A_r$ |
|---|---|---|---|---|---|
| hydrogen | H | 1 | chlorine | Cl | 35.5 |
| carbon | C | 12 | potassium | K | 39 |
| nitrogen | N | 14 | calcium | Ca | 40 |
| oxygen | O | 16 | iron | Fe | 56 |
| sodium | Na | 23 | copper | Cu | 64 |
| magnesium | Mg | 24 | zinc | Zn | 65 |
| sulfur | S | 32 | iodine | I | 127 |

> **Finding the mass of an ion**
> mass of sodium atom = 23, so mass of sodium ion = 23 since a sodium ion is just a sodium atom minus an electron (which has negligible mass).
>
> **An ion has the same mass as the atom from which it is made.**

## Finding the masses of molecules and ions

Using $A_r$ values, it is easy to work out the mass of any molecule or group of ions. Read the blue panel on the right above, then look at these examples:

Hydrogen gas is made of molecules. Each molecule contains 2 hydrogen atoms, so its mass is 2. ($2 \times 1 = 2$)

The formula for water is $H_2O$. Each water molecule contains 2 hydrogen atoms and 1 oxygen atom, so its mass is 18. ($2 \times 1 + 16 = 18$)

mass of ion = mass of its atom

Sodium chloride (NaCl) forms a giant structure with 1 sodium ion for every chloride ion. So the mass of a 'unit' of sodium chloride is 58.5. ($23 + 35.5 = 58.5$)

If the substance is made of molecules, its mass found in this way is called the **relative molecular mass**, or $M_r$. So the $M_r$ for hydrogen is 2, and for water is 18.

But if the substance is made of *ions*, its mass is called the **relative formula mass**, which is also $M_r$ for short. So the $M_r$ for NaCl is 58.5.

This table gives two more examples of how to calculate $M_r$ values.

| Substance | Formula | Atoms in formula | $A_r$ of atoms | $M_r$ |
|---|---|---|---|---|
| ammonia | $NH_3$ | 1N<br>3H | N = 14<br>H = 1 | $1 \times 14 = 14$<br>$3 \times 1 = \underline{3}$<br>Total = **17** |
| magnesium nitrate | $Mg(NO_3)_2$ | 1Mg<br>2N<br>6O | Mg = 24<br>N = 14<br>O = 16 | $1 \times 24 = 24$<br>$2 \times 14 = 28$<br>$6 \times 16 = \underline{96}$<br>Total = **148** |

1 a What does *relative atomic mass* mean?
  b Why does it have the word *relative*?
2 What is the $A_r$ of the iodide ion, $I^-$?
3 The relative molecular mass and formula mass are both called $M_r$ for short. What is the difference between them?
4 Work out the $M_r$ for each of these, and say whether it is the relative molecular mass or the relative formula mass:
  a oxygen, $O_2$  b iodine, $I_2$  c methane, $CH_4$
  d chlorine, $Cl_2$  e butane, $C_4H_{10}$  f ethanol, $C_2H_5OH$
  g ammonium sulfate $(NH_4)_2SO_4$

REACTING MASSES, AND CHEMICAL EQUATIONS

## 5.4 Some calculations about masses and %

### Two laws of chemistry

If you know the *actual* amounts of two substances that react, you can:
- predict other amounts that will react
- say how much product will form.

You just need to remember these two laws of chemistry:

1 **Elements always react in the same ratio, to form a given compound.**
For example, when carbon burns in oxygen to form carbon dioxide:
**6 g** of carbon combines with **16 g** of oxygen, so
**12 g** of carbon will combine with **32 g** of oxygen, and so on.

2 **The total mass does not change, during a chemical reaction.
So total mass of reactants = total mass of products.**
So **6 g** of carbon and **16 g** of oxygen give **22 g** of carbon dioxide.
**12 g** of carbon and **32 g** of oxygen give **44 g** of carbon dioxide.

▲ A model of the carbon dioxide molecule. The amounts of carbon and oxygen that react to give this compound are always in the same ratio.

### Calculating quantities

Calculating quantities is quite easy, using the laws above.

***Example*** 64 g of copper reacts with 16 g of oxygen to give the black compound copper(II) oxide.

a What mass of copper will react with 32 g of oxygen?
64 g of copper reacts with 16 g of oxygen, so
2 × 64 g or **128 g** of copper will react with 32 g of oxygen.

b What mass of oxygen will react with 32 g of copper?
16 g of oxygen reacts with 64 g of copper, so
$\frac{16}{2}$ or **8 g** of oxygen will react with 32 g of copper.

c What mass of copper(II) oxide will be formed, in **b**?
**40 g** of copper(II) oxide will be formed. (32 + 8 = 40)

d How much copper and oxygen will give 8 g of copper(II) oxide?
64 g of copper and 16 g of oxygen give 80 g of copper(II) oxide, so
$\frac{64}{10}$ of copper and $\frac{16}{10}$ g of oxygen will give 8 g of copper(II) oxide, so
**6.4 g of copper** and **1.6 g of oxygen** are needed.

### Percentages: a reminder

Calculations in chemistry often involve percentages. Remember:
- **The full amount of anything is 100%.**
- **To change a fraction to a %, just multiply it by 100.**

***Example 1*** Change the fractions $\frac{1}{2}$ and $\frac{18}{25}$ to percentages.

$\frac{1}{2} \times 100 = \mathbf{50\%}$ $\qquad$ $\frac{18}{25} \times 100 = \mathbf{72\%}$

***Example 2*** Give 19 % as a fraction.
$19\% = \frac{19}{100}$

▲ The French scientist Antoine Lavoisier (1743–1794) was the first to state that the total mass does not change, during a reaction. He was executed during the French Revolution, when he was 51.

## Calculating the percentage composition of a compound

The **percentage composition** of a compound tells you how much of each element it contains, *as a percentage of the total mass*. This is how to work it out:

1 Write down the formula of the compound.
2 Using $A_r$ values, work out its molecular or formula mass ($M_r$).
3 Write the mass of the element as a fraction of the $M_r$.
4 Multiply the fraction by 100, to give a percentage.

*Example* Calculate the percentage of oxygen in sulfur dioxide.

1 The formula of sulfur dioxide is $SO_2$.
2 The $M_r$ of the compound is 64, as shown on the right.
3 Mass of oxygen as a fraction of the total $= \dfrac{32}{64}$
4 Mass of oxygen as a percentage of the total $= \dfrac{32}{64} \times 100 = 50\%$
So the compound is **50% oxygen**.
This means it is also 50% sulfur (100% − 50% = 50%).

> $M_r$ for sulfur dioxide, $SO_2$
> $A_r$: S = 32, O = 16.
> So the $M_r$ is:
> 1 S = 32
> 2 O = 2 × 16 = 32
> Total = 64

## Calculating % purity

A **pure** substance has nothing else mixed with it.
But substances often contain unwanted substances, or **impurities**.
Purity is usually given as a percentage. This is how to work it out:

$$\% \text{ purity of a substance} = \dfrac{\text{mass of pure substance in it}}{\text{total mass}} \times 100\%$$

*Example* Impure copper is refined (purified), to obtain pure copper for use in computers. 20 tonnes of copper gave 18 tonnes of pure copper, on refining.

a What was the % purity of the copper before refining?
% purity of the copper $= \dfrac{18 \text{ tonnes}}{20 \text{ tonnes}} \times 100\% = \mathbf{90\%}$
So the copper was **90% pure**.

b How much pure copper will 50 tonnes of the impure copper give?
The impure copper is 90% pure.
90% is $\dfrac{90}{100}$.
So 50 tonnes of it will give $\dfrac{90}{100} \times 50$ tonnes or **45 tonnes** of pure copper.

▲ Copper of high purity is needed for circuit boards like this one, in computers.

**Q**

1 Magnesium burns in chlorine to give magnesium chloride, $MgCl_2$. In an experiment, 24 g of magnesium was found to react with 71 g of chlorine.
   a How much magnesium chloride was obtained in the experiment?
   b How much chlorine will react with 12 g of magnesium?
   c How much magnesium chloride will form, in **b**?

2 Methane has the formula $CH_4$. Work out the % of carbon and hydrogen in it. ($A_r$: C = 12, H = 1)

3 In an experiment, a sample of lead(II) bromide was made. It weighed 15 g. But the sample was found to be impure. In fact it contained only 13.5 g of lead(II) bromide.
   a Calculate the % purity of the sample.
   b What mass of impurity was present in the sample?

REACTING MASSES, AND CHEMICAL EQUATIONS

# Checkup on Chapter 5

## Revision checklist

### Core curriculum
Make sure you can …
- [ ] name a simple compound, when you are given the names of the two elements that form it
- [ ] work out the formula of a compound from a drawing of its structure
- [ ] work out the formula of a simple compound when you know the two elements in it, by balancing their valencies
- [ ] work out the formula of an ionic compound by balancing the charges of the ions, so that the total charge is zero
- [ ] write the equation for a reaction:
  - as a word equation
  - as a symbol equation
- [ ] balance a symbol equation
- [ ] say what the state symbols mean: $(s)$, $(l)$, $(g)$, $(aq)$
- [ ] define *aqueous*
- [ ] explain that the carbon-12 atom is taken as the standard, for working out masses of atoms
- [ ] say what these two symbols mean: $A_r$   $M_r$
- [ ] work out $M_r$ values, given the $A_r$ values
- [ ] explain the difference between *relative formula mass* and *relative molecular mass* (both known as $M_r$ for short)
- [ ] explain that the $A_r$ value of an element is the average value for all its isotopes
- [ ] predict other amounts of reactants that will react, when you are given some actual amounts
- [ ] calculate:
  - the mass of a product, when you are given the masses of the reactants that combine to form it
  - the percentage of an element in a compound, using the formula and $A_r$ values
  - the percentage purity of a substance, when you are given the total mass of the impure substance, and the amount of the pure substance in it

## Questions

### Core curriculum
If you are not sure about symbols for the elements, you can check the Periodic Table on page 314.

1. Write the formulae for these compounds:
   - **a** water
   - **b** carbon monoxide
   - **c** carbon dioxide
   - **d** sulfur dioxide
   - **e** sulfur trioxide
   - **f** sodium chloride
   - **g** magnesium chloride
   - **h** hydrogen chloride
   - **i** methane
   - **j** ammonia

2. You can work out the formula of a compound from the ratio of the different atoms in it. Sodium carbonate has the formula $Na_2CO_3$ because it contains 2 atoms of sodium for every 1 atom of carbon and 3 atoms of oxygen.
   Deduce the formula for each compound **a** to **h**:

   | | Compound | Ratio in which the atoms are combined in it |
   |---|---|---|
   | a | lead oxide | 1 of lead, 2 of oxygen |
   | b | lead oxide | 3 of lead, 4 of oxygen |
   | c | potassium nitrate | 1 of potassium, 1 of nitrogen, 3 of oxygen |
   | d | nitrogen oxide | 2 of nitrogen, 1 of oxygen |
   | e | nitrogen oxide | 2 of nitrogen, 4 of oxygen |
   | f | sodium hydrogen carbonate | 1 of sodium, 1 of hydrogen, 1 of carbon, 3 of oxygen |
   | g | sodium sulfate | 2 of sodium, 1 of sulfur, 4 of oxygen |
   | h | sodium thiosulfate | 2 of sodium, 2 of sulfur, 3 of oxygen |

3. For each compound, write down the ratio of atoms present:
   - **a** copper(II) oxide, $CuO$
   - **b** copper(I) oxide, $Cu_2O$
   - **c** aluminium chloride, $AlCl_3$
   - **d** nitric acid, $HNO_3$
   - **e** calcium hydroxide, $Ca(OH)_2$
   - **f** ethanoic acid, $CH_3COOH$
   - **g** ammonium nitrate, $NH_4NO_3$
   - **h** ammonium sulfate, $(NH_4)_2SO_4$
   - **i** sodium phosphate, $Na_3(PO_4)_2$
   - **j** hydrated iron(II) sulfate, $FeSO_4.7H_2O$
   - **k** hydrated cobalt(II) chloride, $CoCl_2.6H_2O$

**4** Write the chemical formulae for the compounds with the structures shown below:

a  H—Br

b  Cl—P—Cl
       |
       Cl

c  H—O—O—H

d  H—C≡C—H

e  H  \\N=N/  H
    H  /    \\ H

f  F   F
    \\ | /
    F—Xe—F
    / | \\
      F

g  O   O
    \\\\ //
     S
    / \\
  H—O   O—H

h      O
       ||
       S
      / \\
   H—O   O—H

**5** This shows the structure of an ionic compound.
   a  Name the compound.
   b  What is the simplest formula for it?

**6** This shows the structure of a molecular compound.
   a  Name the compound.
   b  What is the simplest formula for it?

**7** Write these as word equations:
   a  $Zn + 2HCl \longrightarrow ZnCl_2 + H_2$
   b  $Na_2CO_3 + H_2SO_4 \longrightarrow Na_2SO_4 + CO_2 + H_2O$
   c  $2Mg + CO_2 \longrightarrow 2MgO + C$
   d  $ZnO + C \longrightarrow Zn + CO$
   e  $Cl_2 + 2NaBr \longrightarrow 2NaCl + Br_2$
   f  $CuO + 2HNO_3 \longrightarrow Cu(NO_3)_2 + H_2O$

**8** Balance these equations:
   a  $N_2 + ... O_2 \longrightarrow ... NO_2$
   b  $K_2CO_3 + ... HCl \longrightarrow ... KCl + CO_2 + H_2O$
   c  $C_3H_8 + ... O_2 \longrightarrow ... CO_2 + 4H_2O$
   d  $Fe_2O_3 + ...CO \longrightarrow ...Fe + ...CO_2$
   e  $Ca(OH)_2 + ... HCl \longrightarrow CaCl_2 + ... H_2O$
   f  $2Al + ... HCl \longrightarrow 2AlCl_3 + ... H_2$

**9** Copy and complete these equations:
   a  $MgSO_4 + ........ \longrightarrow MgSO_4.6H_2O$
   b  $... C + ........ \longrightarrow 2CO$
   c  $2CuO + C \longrightarrow 2Cu + ........$
   d  $C_2H_6 \longrightarrow ........ + H_2$
   e  $ZnO + C \longrightarrow Zn + ........$
   f  $NiCO_3 \longrightarrow NiO + ........$
   g  $CO_2 + ........ \longrightarrow CH_4 + O_2$
   h  $NaOH + HNO_3 \longrightarrow NaNO_3 + ........$
   i  $C_2H_6 \longrightarrow C_2H_4 + ........$

**10** Calculate $M_r$ for these compounds.
($A_r$ values are given at the top of page 315.)
   a  water, $H_2O$
   b  ammonia, $NH_3$
   c  ethanol, $CH_3CH_2OH$
   d  sulfur trioxide, $SO_3$
   e  sulfuric acid, $H_2SO_4$
   f  hydrogen chloride, HCl
   g  phosphorus(V) oxide, $P_2O_5$

**11** Calculate the relative formula mass for these ionic compounds. ($A_r$ values are given on page 315.)
   a  magnesium oxide, MgO
   b  lead sulfide, PbS
   c  calcium fluoride, $CaF_2$
   d  sodium chloride, NaCl
   e  silver nitrate, $AgNO_3$
   f  ammonium sulfate, $(NH_4)_2SO_4$
   g  potassium carbonate, $K_2CO_3$
   h  hydrated iron(II) sulfate, $FeSO_4.7H_2O$

**12** Iron reacts with excess sulfuric acid to give iron(II) sulfate. The equation for the reaction is:
$Fe + H_2SO_4 \longrightarrow FeSO_4 + H_2$
5 g of iron gives 13.6 g of iron(II) sulfate.
   a  Using excess acid, how much iron(II) sulfate can be obtained from:
      i  10 g of iron?   ii  1 g of iron?
   b  How much iron will be needed to make 136 g of iron(II) sulfate?
   c  A 10 g sample of impure iron(II) sulfate contains 8 g of iron(II) sulfate. Calculate the percentage purity of the iron(II) sulfate.

**13** Aluminium is extracted from the ore bauxite, which is impure aluminium oxide.
1 tonne (1000 kg) of the ore was found to have this composition:
   aluminium oxide 825 kg
   iron(III) oxide 100 kg
   sand 75 kg
   a  What percentage of this ore is impurities?
   b  1 tonne of the ore gives 437 kg of aluminium.
      i  How much aluminium will be obtained from 5 tonnes of the ore?
      ii  What mass of sand is in this 5 tonnes?
   c  What will the percentage of aluminium oxide in the ore be, if all the iron(III) oxide is removed, leaving only the aluminium oxide and sand?

USING MOLES

# 6.1 The mole

## What is a mole?

As you saw on page 70, the masses of atoms are found by comparing them with the carbon-12 atom:

This is an atom of carbon-12. It is chosen as the standard atom. Its $A_r$ is taken as 12. Then other atoms are compared with it.

This is a magnesium atom. It is twice as heavy as a carbon-12 atom, so its $A_r$ is 24. So it follows that …

… 24 g of magnesium contains the *same number of atoms* as 12 g of carbon-12. 24 g of magnesium is called **a mole** of magnesium atoms.

**A mole of a substance is the amount that contains the same number of units as the number of carbon atoms in 12 grams of carbon-12.**
These units can be atoms, or molecules, or ions, as you will see.

## The Avogadro constant

Thanks to the work of the Italian scientist Avogadro, we know that 12 g of carbon-12 contains **602 000 000 000 000 000 000 000** carbon atoms!

This huge number is called **the Avogadro constant**.
It is written in a short way as $6.02 \times 10^{23}$. (The $10^{23}$ tells you to move the decimal point 23 places to the right, to get the full number.)

So 1 mole of magnesium atoms contains $6.02 \times 10^{23}$ magnesium atoms.

## More examples of moles

| Sodium is made of single sodium atoms. Its symbol is Na. Its $A_r$ is **23**.  | Iodine is made of iodine molecules. Its formula is $I_2$. Its $M_r$ is **254**.  | Water is made of water molecules. Its formula is $H_2O$. Its $M_r$ is **18**.  |
|---|---|---|
|  |  |  |
| This is **23 grams** of sodium. It contains $6.02 \times 10^{23}$ sodium atoms, or **1 mole of sodium atoms**. | This is **254 grams** of iodine. It contains $6.02 \times 10^{23}$ iodine molecules, or **1 mole of iodine molecules**. | The beaker contains **18 grams** of water, or $6.02 \times 10^{23}$ water molecules, or **1 mole of water molecules**. |

So you can see that:
**One mole of a substance is obtained by weighing out the $A_r$ or $M_r$ of the substance, in grams.**

## Finding the mass of a mole

You can find the mass of one mole of any substance by these steps:
1. Write down the symbol or formula of the substance.
2. Find its $A_r$ or $M_r$.
3. Express that mass in grams (g).

This table shows three more examples:

| Substance | Symbol or formula | $A_r$ | $M_r$ | Mass of 1 mole |
|---|---|---|---|---|
| helium | He | He = 4 | exists as single atoms | **4 grams** |
| oxygen | $O_2$ | O = 16 | 2 × 16 = 32 | **32 grams** |
| ethanol | $C_2H_5OH$ | C = 12<br>H = 1<br>O = 16 | 2 × 12 = 24<br>6 × 1 = 6<br>1 × 16 = 16<br>46 | **46 grams** |

## Some calculations on the mole

These equations will help you:

**Mass of a given number of moles = mass of 1 mole × number of moles**

**Number of moles in a given mass = $\dfrac{\text{mass}}{\text{mass of 1 mole}}$**

### Use the calculation triangle

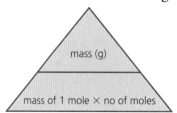

▲ Cover the one you want to find – and you will see how to calculate it.

*Example 1* Calculate the mass of 0.5 moles of bromine atoms.
The $A_r$ of bromine is 80, so 1 mole of bromine atoms has a mass of 80 g.
So 0.5 moles of bromine atoms has a mass of 0.5 × 80 g, or **40 g**.

*Example 2* Calculate the mass of 0.5 moles of bromine molecules.
A bromine *molecule* contains 2 atoms, so its $M_r$ is 160.
So 0.5 moles of bromine molecules has a mass of 0.5 × 160 g, or **80 g**.

$A_r = 80$
$M_r = 160$

*Example 3* How many moles of oxygen molecules are in 64 g of oxygen?
The $M_r$ of oxygen is 32, so 32 g of it is 1 mole.
Therefore 64 g is $\dfrac{64}{32}$ moles, or **2 moles** of oxygen molecules.

$A_r = 16$
$M_r = 32$

---

**Q**

1. How many atoms are in 1 mole of atoms?
2. How many molecules are in 1 mole of molecules?
3. What name is given to the number $6.02 \times 10^{23}$?
4. Find the mass of 1 mole of:
   a hydrogen atoms  b iodine atoms
   c chlorine atoms  d chlorine molecules
5. Find the mass of 2 moles of:
   a oxygen atoms  b oxygen molecules
6. Find the mass of 3 moles of ethanol, $C_2H_5OH$.
7. How many moles of molecules are there in:
   a 18 grams of hydrogen, $H_2$?
   b 54 grams of water?
8. Sodium chloride is made up of $Na^+$ and $Cl^-$ ions.
   a How many sodium ions are there in 58.5 g of sodium chloride? ($A_r$: Na = 23; Cl = 35.5.)
   b What is the mass of 1 mole of chloride ions?

## USING MOLES

# 6.2 Calculations from equations, using the mole

## What an equation tells you
When carbon burns in oxygen, the reaction can be shown as:

 +  →

1 atom of carbon     1 molecule of oxygen     1 molecule of carbon dioxide

or in a short way, using the symbol equation:

$$C\ (s) + O_2\ (g) \longrightarrow CO_2\ (g)$$

This equation tells you that:

| 1 carbon atom | reacts with | 1 molecule of oxygen | to give | 1 molecule of carbon dioxide |

Now suppose there is 1 *mole* of carbon atoms. Then we can say that:

| 1 mole of carbon atoms | reacts with | 1 mole of oxygen molecules | to give | 1 mole of carbon dioxide molecules |

So from the equation, we can tell how many moles react.
But moles can be changed to grams, using $A_r$ and $M_r$.
The $A_r$ values are: C = 12, O = 16.
So the $M_r$ values are: $O_2$ = 32, $CO_2$ = (12 + 32) = 44, and we can write:

| 12 g of carbon | reacts with | 32 g of oxygen | to give | 44 g of carbon dioxide |

Since substances always react in the same ratio, this also means that:

| 6 g of carbon | reacts with | 16 g of oxygen | to give | 22 g of carbon dioxide |

and so on.
So we have gained a great deal of information from the equation.
In fact you can obtain the same information from any equation.

**From the equation for a reaction you can tell:**
- how many moles of each substance take part
- how many grams of each substance take part.

## Reminder: the total mass does not change
Look what happens to the total mass, during the reaction above:

    mass of carbon and oxygen at the start:    12 g + 32 g = **44 g**
    mass of carbon dioxide at the end:                    **44 g**

The total mass has not changed, during the reaction. This is because no atoms have disappeared. They have just been rearranged.
That is one of the two laws of chemistry that you met on page 72:
**The total mass does not change, during a chemical reaction.**

▲ Iron and sulfur reacting: the total mass is the same before and after.

# USING MOLES

## Calculating masses from equations

These are the steps to follow:

1 Write the balanced equation for the reaction. (It gives *moles*.)
2 Write down the $A_r$ or $M_r$ for each substance that takes part.
3 Using $A_r$ or $M_r$, change the moles in the equation to *grams*.
4 Once you know the theoretical masses from the equation, you can then find any *actual* mass.

*Example*  Hydrogen burns in oxygen to form water. What mass of oxygen is needed for 1 g of hydrogen, and what mass of water is obtained?

1 The equation for the reaction is: $2H_2 (g) + O_2 (g) \longrightarrow 2H_2O (l)$
2 $A_r$: H = 1, O = 16.  $M_r$: $H_2$ = 2, $O_2$ = 32, $H_2O$ = 18.
3 So, for the equation, the amounts in grams are:

$2H_2 (g) \quad + \quad O_2 (g) \quad \longrightarrow \quad 2H_2O (l)$
$2 \times 2$ g  $\qquad$  32 g  $\qquad$  $2 \times 18$ g  $\quad$ or
4 g  $\qquad\qquad$  32 g  $\qquad\qquad$  36 g

4 But you start with only 1 g of hydrogen, so the *actual* masses are:

1 g  $\qquad$  32/4 g  $\qquad$  36/4 g  $\quad$ or
1 g  $\qquad$  8 g  $\qquad\quad$  9 g

So **1 g** of hydrogen needs **8 g** of oxygen to burn, and gives **9 g** of water.

▲ These models show how the atoms are rearranged, when hydrogen burns in oxygen. Which colour is oxygen?

## Working out equations, from masses

If you know the actual masses that react, you can work out the equation for the reaction. Just change the masses to moles.

*Example*  Iron reacts with a solution of copper(II) sulfate ($CuSO_4$) to give copper and a solution of iron sulfate. The formula for the iron sulfate could be either $FeSO_4$ or $Fe_2(SO_4)_3$. 1.4 g of iron gave 1.6 g of copper. Write the correct equation for the reaction.

1 $A_r$: Fe = 56, Cu = 64.
2 Change the masses to moles of atoms:

$\dfrac{1.4}{56}$ moles of iron atoms gave $\dfrac{1.6}{64}$ moles of copper atoms, or

0.025 moles of iron atoms gave 0.025 moles of copper atoms, so
1 mole of iron atoms gave 1 mole of copper atoms.

3 So the equation for the reaction must be:

$\quad$ Fe  $+$  $CuSO_4$  $\longrightarrow$  Cu  $+$  $FeSO_4$

4 Add the state symbols to complete it:

$\quad$ Fe $(s)$  $+$  $CuSO_4 (aq)$  $\longrightarrow$  Cu $(s)$  $+$  $FeSO_4 (aq)$

▲ Iron wool reacting with copper(II) sulfate solution. Iron is more reactive than copper so displaces the copper (deep pink) from solution.

---

**Q**

**1** The reaction between magnesium and oxygen is:
$2Mg (s) + O_2 (g) \rightarrow 2MgO (s)$
  **a** Write a word equation for the reaction.
  **b** How many moles of magnesium atoms react with 1 mole of oxygen molecules?
  **c** The $A_r$ values are: Mg = 24, O = 16.
  How many grams of oxygen react with:
   **i** 48 g of magnesium?  **ii** 12 g of magnesium?

**2** Copper(II) carbonate breaks down on heating, like this:
$CuCO_3 (s) \rightarrow CuO (s) + CO_2 (g)$
  **a** Write a word equation for the reaction.
  **b** Find the mass of 1 mole of each substance taking part in the reaction. ($A_r$: Cu = 64, C = 12, O = 16.)
  **c** When 31 g of copper(II) carbonate is used:
   **i** how many grams of carbon dioxide form?
   **ii** what mass of solid remains after heating?

## USING MOLES

# 6.3 Reactions involving gases

### A closer look at some gases

Imagine five very large flasks, each with a volume of 24 dm³. Each is filled with a different gas. Each gas is at room temperature and pressure, or **rtp**.

(We take **room temperature and pressure** as the standard conditions for comparing gases; rtp is 20 °C and 1 atmosphere.)

If you weighed the gas in the five flasks, you would discover something amazing. There is exactly 1 mole of each gas!

So we can conclude that:

**1 mole of every gas occupies the same volume, at the same temperature and pressure. At room temperature and pressure, this volume is 24 dm³.**

This was discovered by Avogadro, in 1811. So it is often called **Avogadro's Law**. It does not matter whether a gas exists as atoms or molecules, or whether its atoms are large or small. The law still holds.

The volume occupied by 1 mole of a gas is called its **molar volume**.
**The molar volume of a gas is 24 dm³ at rtp.**

### Another way to look at it

Look at these two gas jars.
**A** is full of nitrogen dioxide, $NO_2$.
**B** is full of oxygen, $O_2$.

The two gas jars have identical volumes, and the gases are at the same temperature and pressure.

You cannot see the gas molecules – let alone count them. But, from Avogadro's Law, you can say that the two jars contain the same number of molecules.

**Remember**
24 dm³ = 24 litres
= 24 000 cm³

Imagine a ball about 36 cm in diameter. Its volume is about 24 dm³.

# USING MOLES

## Calculating gas volumes from moles and grams

Avogadro's Law makes it easy to work out the volumes of gases.

***Example 1*** What volume does 0.25 moles of a gas occupy at rtp?
1 mole occupies 24 dm³ so
0.25 moles occupies 0.25 × 24 dm³ = 6 dm³
so 0.25 moles of any gas occupies **6 dm³** (or **6000 cm³**) at rtp.

***Example 2*** What volume does 22 g of carbon dioxide occupy at rtp?
$M_r$ of carbon dioxide = 44, so
44 g = 1 mole, so
22 g = 0.5 mole
so the volume occupied = 0.5 × 24 dm³ = **12 dm³**.

## Calculating gas volumes from equations

From the equation for a reaction, you can tell how many *moles* of a gas take part. Then you can use Avogadro's Law to work out its *volume*. In these examples, all volumes are measured at rtp.

***Example 1*** What volume of hydrogen will react with 24 dm³ of oxygen to form water?

1  The equation for the reaction is:   $2H_2 (g) + O_2 (g) \longrightarrow 2H_2O (l)$
2  So 2 volumes of hydrogen react with 1 of oxygen, or
   2 × 24 dm³         react with 24 dm³, so
   **48 dm³** of hydrogen will react.

***Example 2*** When sulfur burns in air it forms sulfur dioxide. What volume of this gas is produced when 1 g of sulfur burns? ($A_r$: S = 32.)

1  The equation for the reaction is:   $S (s) + O_2 (g) \longrightarrow SO_2 (g)$
2  32 g of sulfur atoms = 1 mole of sulfur atoms, so
   $1 g = \frac{1}{32}$ mole or 0.03125 moles of sulfur atoms.
3  1 mole of sulfur atoms gives 1 mole of sulfur dioxide molecules so
   0.03125 moles         give 0.03125 moles.
4  1 mole of sulfur dioxide molecules has a volume of 24 dm³ at rtp so
   0.03125 moles has a volume of 0.03125 × 24 dm³ at rtp, or 0.75 dm³.
   So **0.75 dm³** (or **750 cm³**) of sulfur dioxide are produced.

### Use the calculation triangle

▲ Cover the one you want to find – and you will see how to calculate it.

▲ Sulfur dioxide is one of the gases given out in volcanic eruptions. These scientists are collecting gas samples on the slopes of an active volcano.

---

**Q**

($A_r$: O = 16, N = 14, H = 1, C = 12.)

1  What does *rtp* mean? What values does it have?
2  What does *molar volume* mean, for a gas?
3  What is the molar volume of neon gas at rtp?
4  For any gas, calculate the volume at rtp of:
   **a** 7 moles   **b** 0.5 moles   **c** 0.001 moles
5  Calculate the volume at rtp of:
   **a** 16 g of oxygen ($O_2$)   **b** 1.7 g of ammonia ($NH_3$)
6  You burn 6 grams of carbon in plenty of air:
   $C (s) + O_2 (g) \longrightarrow CO_2 (g)$
   **a** What volume of gas will form (at rtp)?
   **b** What volume of oxygen will be used up?
7  If you burn the carbon in limited air, the reaction is different: $2C (s) + O_2 (g) \longrightarrow 2CO (g)$
   **a** What volume of gas will form this time?
   **b** What volume of oxygen will be used up?

USING MOLES

## 6.4 The concentration of a solution

### What does 'concentration' mean?

A

Solution A contains 2.5 grams of copper(II) sulfate in 1 dm³ of water. So its concentration is **2.5 g/dm³**.

B

Solution B contains 25 grams of copper(II) sulfate in 1 dm³ of water. So its concentration is **25 g/dm³**.

C

Solution C contains 125 grams of copper(II) sulfate in 0.5 dm³ of water. So its concentration is **250 g/dm³**.

**The concentration of a solution is the amount of solute, in grams or moles, that is dissolved in 1 dm³ of solution.**

### Finding the concentration in moles

*Example*  Find the concentrations of A and C above, in moles per dm³.

First, change the mass of the solute to moles.
The formula mass of copper(II) sulfate is 250, as shown on the right.
So 1 mole of the compound has a mass of 250 g.

**Solution A** has 2.5 g of the compound in 1 dm³ of solution.

2.5 g  moles  0.01 moles

so its concentration is **0.01 mol/dm³**.

Note the unit of concentration: **mol/dm³**. This is often shortened to **M**, so the concentration of solution A can be written as **0.01 M**.

**Solution C** has 250 g of the compound in 1 dm³ of solution.
250 g  1 mole
so its concentration is **1 mol/dm³**, or **1 M** for short.

A solution that contains 1 mole of solute per dm³ of solution is often called a **molar solution**. So C is a molar solution.

> $M_r$ for copper(II) sulfate
> Its formula is $CuSO_4 \cdot 5H_2O$.
> This has 1 Cu, 1 S, 9 O, and 10 H.
> So the formula mass is:
> 1 Cu = 1 × 64 = 64
> 1 S  = 1 × 32 = 32
> 9 O  = 9 × 16 = 144
> 10 H = 10 × 1 = 10
> Total = 250

In general, to find the concentration of a solution in moles per dm³:

$$\text{concentration (mol/dm}^3\text{)} = \frac{\text{amount of solute (mol)}}{\text{volume of solution (dm}^3\text{)}}$$

Use the equation above to check that the last column in this table is correct:

| Amount of solute (mol) | Volume of solution (dm³) | Concentration of solution (mol/dm³) |
|---|---|---|
| 1.0 | 1.0 | 1.0 |
| 0.2 | 0.1 | 2.0 |
| 0.5 | 0.2 | 2.5 |
| 1.5 | 0.3 | 5.0 |

**Remember**
- 1 dm³ = 1 litre
  = 1000 cm³
  = 1000 ml
- All these mean the same thing:
  moles per dm³
  mol/dm³
  mol dm⁻³
  moles per litre

## Finding the amount of solute in a solution

If you know the concentration of a solution, and its volume:

- you can work out how much solute it contains, in moles. Just rearrange the equation from the last page:
  **amount of solute (mol) = concentration (mol/dm³) × volume (dm³)**
- you can then convert moles to grams, by multiplying the number of moles by $M_r$.

## Use the calculation triangle

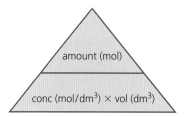

▲ Cover the one you want to find – and you will see how to calculate it. To draw this triangle, remember that **a**lligators **c**hew **v**isitors!

## Sample calculations

The table shows four solutions, with different volumes and concentrations. Check that you understand the calculations that give the masses of solute in the bottom row.

| solution | sodium hydroxide NaOH | sodium thiosulfate $Na_2S_2O_3$ | lead nitrate $Pb(NO_3)_2$ | silver nitrate $AgNO_3$ |
|---|---|---|---|---|
| | 2 dm³ | 250 cm³ | 100 cm³ | 25 cm³ |
| concentration (mol/dm³) | 1 | 2 | 0.1 | 0.05 |
| amount of solute present (moles) | 1 × 2 = 2 | 2 × $\frac{250}{1000}$ = 0.5 | 0.1 × $\frac{100}{1000}$ = 0.01 | 0.05 × $\frac{25}{1000}$ = 0.00125 |
| $M_r$ | 40 | 158 | 331 | 170 |
| mass of solute present (g) | 80 | 79 | 3.31 | 0.2125 |

## Q

1. How many moles of solute are in:
   a. 500 cm³ of solution, of concentration 2 mol/dm³?
   b. 2 litres of solution, of concentration 0.5 mol/dm³?
2. What is the concentration of a solution containing:
   a. 4 moles in 2 dm³ of solution?
   b. 0.3 moles in 200 cm³ of solution?
3. Different solutions of salt X are made up. What volume of:
   a. a 4 mol/dm³ solution contains 2 moles of X?
   b. a 6 mol/dm³ solution contains 0.03 moles of X?
4. The $M_r$ of sodium hydroxide is 40. How many grams of sodium hydroxide are there in:
   a. 500 cm³ of a molar solution?
   b. 25 cm³ of a 0.5 M solution?
5. What is the concentration in moles per litre of:
   a. a sodium carbonate solution containing 53 g of the salt ($Na_2CO_3$) in 1 litre?
   b. a copper(II) sulfate solution containing 62.5 g of the salt ($CuSO_4.5H_2O$) in 1 litre?

USING MOLES

## 6.5 Finding the empirical formula

### What a formula tells you about moles and masses

The formula of carbon dioxide is **$CO_2$**. Some molecules of it are shown on the right. You can see that:

| 1 carbon atom | combines with | 2 oxygen atoms | so |

| 1 mole of carbon atoms | combines with | 2 moles of oxygen atoms |

$A_r$: C = 12, O = 16

Moles can be changed to grams, using $A_r$ and $M_r$. So we can write:

| 12 g of carbon | combines with | 32 g of oxygen |

In the same way:
6 g of carbon combines with 16 g of oxygen
24 kg of carbon combines with 64 kg of oxygen, and so on.
The masses of substances that combine are *always in the same ratio*.

Therefore, from the formula of a compound, you can tell:

- how many moles of the different atoms combine
- how many grams of the different elements combine.

### Finding the empirical formula

From the formula of a compound you can tell what masses of the elements combine. But you can also do things the other way round.
**If you know what masses combine, you can work out the formula.**
These are the steps:

| Find the masses that combine (in **grams**) by experiment. | → | Change grams to **moles of atoms**. | → | This tells you the **ratio** in which atoms combine. | → | So you can write a **formula**. |

A formula found in this way is called the **empirical formula**.
**The empirical formula shows the simplest ratio in which atoms combine.**

*Example 1* 32 grams of sulfur combine with 32 grams of oxygen to form an oxide of sulfur. What is its empirical formula?

Draw up a table like this:

| Elements that combine | sulfur | oxygen |
|---|---|---|
| Masses that combine | 32 g | 32 g |
| Relative atomic masses ($A_r$) | 32 | 16 |
| Moles of atoms that combine | 32/32  1 | 32/16  2 |
| Ratio in which atoms combine | 1:2 | |
| **Empirical formula** | **$SO_2$** | |

So the empirical formula of the oxide that forms is **$SO_2$**.

▲ Sulfur combines with oxygen when it burns.

# USING MOLES

*Example 2* An experiment shows that compound Y is 80% carbon and 20% hydrogen. What is its empirical formula?

Y is 80% carbon and 20% hydrogen. So 100 g of Y contains 80 g of carbon and 20 g of hydrogen. Draw up a table like this:

| Elements that combine | carbon | hydrogen |
|---|---|---|
| Masses that combine | 80 g | 20 g |
| Relative atomic masses ($A_r$) | 12 | 1 |
| Moles of atoms that combine | 80/12  6.67 | 20/1  20 |
| Ratio in which atoms combine | 6.67 : 20 or 1:3 in its simplest form | |
| **Empirical formula** | $CH_3$ | |

So the empirical formula of Y is **$CH_3$**.

But we can tell right away that the *molecular* formula for Y must be different. (A carbon atom does not bond to only 3 hydrogen atoms.) You will learn how to find the molecular formula from the empirical formula in the next unit.

▲ Empirical formulae are found by experiment – and that usually involves weighing.

## An experiment to find the empirical formula

To work out the empirical formula, you need to know the masses of elements that combine. *The only way to do this is by experiment*.

For example, magnesium combines with oxygen to form magnesium oxide. The masses that combine can be found like this:

1  Weigh a crucible and lid, empty. Then add a coil of magnesium ribbon and weigh it again, to find the mass of the magnesium.
2  Heat the crucible. Raise the lid carefully at intervals to let oxygen in. The magnesium burns brightly.
3  When burning is complete, let the crucible cool (still with its lid on). Then weigh it again. The increase in mass is due to oxygen.

The results showed that 2.4 g of magnesium combined with 1.6 g of oxygen. Draw up a table again:

| Elements that combine | magnesium | oxygen |
|---|---|---|
| Masses that combine | 2.4 g | 1.6 g |
| Relative atomic masses ($A_r$) | 24 | 16 |
| Moles of atoms that combine | 2.4/24  0.1 | 1.6/16  0.1 |
| Ratio in which atoms combine | 1:1 | |
| **Empirical formula** | MgO | |

$A_r = 24$   $A_r = 16$

So the empirical formula for the oxide is **MgO**.

**Q**

1  **a**  How many atoms of hydrogen combine with one carbon atom to form methane, $CH_4$?
   **b**  How many grams of hydrogen combine with 12 grams of carbon to form methane?
2  What does the word *empirical* mean? (Check the glossary?)
3  56 g of iron combine with 32 g of sulfur to form iron sulfide. Find the empirical formula for iron sulfide. ($A_r$: Fe = 56, S = 32.)
4  An oxide of sulfur is 40% sulfur and 60% oxygen. What is its empirical formula?

USING MOLES

# 6.6 From empirical to final formula

## The formula of an ionic compound

You saw in the last unit that the empirical formula shows the *simplest ratio* in which atoms combine.

The diagram on the right shows the structure of sodium chloride. The sodium and chlorine atoms are in the ratio 1:1 in this compound. So its empirical formula is NaCl.

**The formula of an ionic compound is the same as its empirical formula.**

In the experiment on page 85, the empirical formula for magnesium oxide was found to be MgO. So the formula for magnesium oxide is also **MgO**.

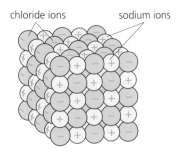

▲ The structure of sodium chloride.

## The formula of a molecular compound

The gas ethane is one of the alkane family of compounds. An ethane molecule is drawn on the right. It contains only hydrogen and carbon atoms, so ethane is a **hydrocarbon**.

From the drawing you can see that the ratio of carbon to hydrogen atoms in ethane is 2:6. The simplest ratio is therefore 1:3.
So the *empirical* formula of ethane is $CH_3$. (It is compound Y on page 85.) But its *molecular* formula is $C_2H_6$.

**The molecular formula shows the *actual* numbers of atoms that combine to form a molecule.**

The molecular formula is more useful than the empirical formula, because it gives you more information.

For some molecular compounds, both formulae are the same. For others they are different. Compare them for the alkanes in the table on the right. What do you notice?

▲ An ethane molecule.

| Alkane | Molecular formula | Empirical formula |
|---|---|---|
| methane | $CH_4$ | $CH_4$ |
| ethane | $C_2H_6$ | $CH_3$ |
| propane | $C_3H_8$ | $C_3H_8$ |
| butane | $C_4H_{10}$ | $C_2H_5$ |
| pentane | $C_5H_{12}$ | $C_5H_{12}$ |
| hexane | $C_6H_{14}$ | $C_3H_7$ |

## How to find the molecular formula

To find the molecular formula for an unknown compound, you need to know these:

- the **relative molecular mass** of the compound ($M_r$). This can be found using a mass spectrometer.
- its **empirical formula**. This is found by experiment, as on page 85.
- its **empirical mass**. This is the mass calculated using the empirical formula and $A_r$ values.

Once you know those, you can work out the molecular formula by following these steps:

> **To find the molecular formula:**
> 
> i Calculate $\dfrac{M_r}{\text{empirical mass}}$ for the compound. This gives a number, $n$.
> 
> ii Multiply the numbers in the empirical formula by $n$.

Let's look at two examples.

▲ A mass spectrometer, for finding relative molecular mass. It compares the mass of a molecule with the mass of a carbon-12 atom, using an electric field.

# USING MOLES

## Calculating the molecular formula

***Example 1*** A molecular compound has the empirical formula HO. Its relative molecular mass is 34. What is its molecular formula? ($A_r$: H = 1, O = 16.)

For the empirical formula HO, the empirical mass = 17. But $M_r$ = 34.

So $\dfrac{M_r}{\text{empirical mass}} = \dfrac{34}{17} = 2$

So the molecular formula is 2 × HO, or **$H_2O_2$**.
So the compound is hydrogen peroxide.
Note how you write the 2 *after* the symbols, when you multiply.

***Example 2*** Octane is a hydrocarbon – it contains only carbon and hydrogen. It is 84.2% carbon and 15.8% hydrogen by mass. Its $M_r$ is 114. What is its molecular formula?

1 **First find the empirical formula for the compound.**
   From the %, we can say that in 100 g of octane, 84.2 g is carbon and 15.8 g is hydrogen.
   So 84.2 g of carbon combines with 15.8 g of hydrogen.
   Changing masses to moles:
   $\dfrac{84.2}{12}$ moles of carbon atoms combine with $\dfrac{15.8}{1}$ moles of hydrogen atoms, or
   7.02 moles of carbon atoms combine with 15.8 moles of hydrogen atoms, so
   **1 mole** of carbon atoms combines with $\dfrac{15.8}{7.02}$ or **2.25 moles** of hydrogen atoms.
   So the atoms combine in the ratio of 1: 2.25 or **4:9**.
   (Give the ratio as *whole* numbers, since only whole atoms combine.)
   The empirical formula of octane is therefore **$C_4H_9$**.

2 **Then use $M_r$ to find the molecular formula.**
   For the empirical formula ($C_4H_9$), the empirical mass = 57.
   But $M_r$ = 114.

   So $\dfrac{M_r}{\text{empirical mass}} = \dfrac{114}{57} = 2$

   So the molecular formula of octane is 2 × $C_4H_9$ or **$C_8H_{18}$**.

▲ Using hydrogen peroxide solution to clean a hospital floor. Hydrogen peroxide acts as a bleach, and kills germs.

▲ Octane is one of the main ingredients in gasoline (petrol). When it burns in the engine, it gives out lots of energy to move that car.

**Q**

1 In the ionic compound magnesium chloride, magnesium and chlorine atoms combine in the ratio 1:2. What is the formula of magnesium chloride?

2 In the ionic compound aluminium fluoride, aluminium and fluorine atoms combine in the ratio 1:3. What is the formula of aluminium fluoride?

3 What is the difference between an empirical formula and a molecular formula? Can they ever be the same?

4 What is the empirical formula of benzene, $C_6H_6$?

5 A compound has the empirical formula $CH_2$. Its $M_r$ is 28. What is its molecular formula?

6 A hydrocarbon is 84% carbon, by mass. Its relative molecular mass is 100. Find:
   **a** its empirical formula   **b** its molecular formula

7 An oxide of phosphorus has an $M_r$ value of 220. It is 56.4% phosphorus. Find its molecular formula.

USING MOLES

## 6.7 Finding % yield and % purity

### Yield and purity

The **yield** is the amount of product you obtain from a reaction. Suppose you own a factory that makes paint or fertilisers. You will want the highest yield possible, for the lowest cost!

Now imagine your factory makes medical drugs, or flavouring for foods. The yield will still be important – but the **purity** of the product may be even more important. Impurities could harm people.

In this unit you'll learn how to calculate the % yield from a reaction, and remind yourself how to calculate the % purity of the product obtained.

### Finding the % yield

You can work out % yield like this:

$$\% \text{ yield} = \frac{\text{actual mass obtained}}{\text{calculated mass}} \times 100\%$$

**Example** The medical drug aspirin is made from salicyclic acid. 1 mole of salicylic acid gives 1 mole of aspirin:

$$C_7H_6O_3 \xrightarrow{\text{chemicals}} C_9H_8O_4$$
salicylic acid → aspirin

In a trial, 100.0 grams of salicylic acid gave 121.2 grams of aspirin. What was the % yield?

1  $A_r$: C = 12, H = 1, O = 16.

   So $M_r$: salicyclic acid = 138, aspirin = 180.

2  138 g of salicylic acid = 1 mole

   so 100 g = $\frac{100}{138}$ mole = 0.725 moles

3  1 mole of salicylic acid gives 1 mole of aspirin
   so 0.725 moles give 0.725 moles of aspirin
   or 0.725 × 180 g = 130.5 g

   So 130.5 g is the **calculated mass** for the reaction.

4  But the **actual mass** obtained in the trial was 121.2 g.

   So % yield = $\frac{121.2}{130.5}$ g × 100 = **92.9%**

This is a high yield – so it is worth continuing with those trials.

### Finding the % purity

When you make something in a chemical reaction, and separate it from the final mixture, it will not be pure. It will have impurities mixed with it – for example small amounts of unreacted substances, or another product. You can work out the % purity of the product you obtained like this:

$$\% \text{ purity of a product} = \frac{\text{mass of the pure product}}{\text{mass of the impure product obtained}} \times 100\%$$

▲ Everything is carefully controlled in a chemical factory, to give a high yield – and as quickly as possible.

▲ For some products, a very high level of purity is essential – for example when you are creating new medical drugs.

Below are examples of how to work out the % purity.

***Example 1*** Aspirin is itself an acid. (Its full name is acetylsalicylic acid.) It is neutralised by sodium hydroxide in this reaction:

$C_9H_8O_4$ (aq) + NaOH (aq) ⟶ $C_9H_7O_4Na$ (aq) + $H_2O$ (l)

Some aspirin was prepared in the lab. Through titration, it was found that 4.00 g of the aspirin were neutralised by 17.5 cm³ of 1M sodium hydroxide solution. How pure was the aspirin sample?

1  $M_r$ of $C_9H_8O_4$ = 180   ($A_r$: C = 12, H = 1, O = 16)
2  17.5 cm³ of 1M sodium hydroxide contain $\frac{17.5}{1000}$ moles or 0.0175 moles of NaOH
3  1 mole of NaOH reacts with 1 mole of $C_9H_8O_4$ so 0.0175 moles react with 0.0175 moles.
4  0.0175 moles of $C_9H_8O_4$ = 0.0175 × 180 g or 3.15 g of aspirin.
5  But the mass of the aspirin sample was 4 g.

So % purity of the aspirin = $\frac{3.15}{4}$ × 100% or **78.75%**.

This is far from acceptable for medical use. The aspirin could be purified by crystallisation. Repeated crystallisation might be needed.

***Example 2*** Chalk is almost pure calcium carbonate. 10 g of chalk was reacted with an excess of dilute hydrochloric acid. 2280 cm³ of carbon dioxide gas was collected at room temperature and pressure (rtp). What was the purity of the sample?

You can work out its purity from the volume of carbon dioxide given off. The equation for the reaction is:

$CaCO_3$ (s) + 2HCl (aq) ⟶ $CaCl_2$ (aq) + $H_2O$ (l) + $CO_2$ (g)

1  $M_r$ of $CaCO_3$ = 100   ($A_r$: Ca = 40, C = 12, O = 16.)
2  1 mole of $CaCO_3$ gives 1 mole of $CO_2$ and
   1 mole of gas has a volume of 24 000 cm³ at rtp.
3  So 24 000 cm³ of gas is produced by 100 g of calcium carbonate and 2280 cm³ is produced by $\frac{2280}{24000}$ × 100 g or **9.5 g**.

So there is 9.5 g of calcium carbonate in the 10 g of chalk.

So the % purity of the chalk = $\frac{9.5}{10}$ g × 100 = **95%**.

> **Purity check!**
> You can check the purity of a sample by measuring its melting and boiling points, and comparing them with the values for the pure product.
> - Impurities lower the melting point and raise the boiling point.
> - The more impurity present, the greater the change.

▲ White chalk cliffs on the Danish island of Mon. Chalk forms in the ocean floor, over many millions of years, from the hard parts of tiny marine organisms.

1  Define the term:  **a** % yield  **b** % purity
2  100 g of aspirin was obtained from 100 g of salicylic acid. What was the % yield?
3  17 kg of aluminium was produced from 51 kg of aluminium oxide ($Al_2O_3$) by electrolysis. What was the percentage yield? ($A_r$: Al = 27, O = 16.)
4  Some seawater is evaporated. The sea salt obtained is found to be 86% sodium chloride. How much sodium chloride could be obtained from 200 g of this salt?
5  A 5.0 g sample of dry ice (solid carbon dioxide) turned into 2400 cm³ of carbon dioxide gas at rtp. What was the percentage purity of the dry ice? ($M_r$ of $CO_2$ = 44.)

USING MOLES

# Checkup on Chapter 6

## Revision checklist
### Extended curriculum
Make sure you can ...
- ☐ explain what a *mole* of atoms or molecules or ions is, and give examples
- ☐ say what the *Avogadro constant* is
- ☐ do these calculations, using $A_r$ and $M_r$:
  - find the mass of 1 mole of a substance
  - change moles to masses
  - change masses to moles
- ☐ use the idea of the mole to:
  - calculate the masses of reactants or products, from the equation for a reaction
  - work out the equation for a reaction, given the masses of the reactants and products
- ☐ define *molar volume* and *rtp*
- ☐ calculate the volume that a gas will occupy at rtp, from its mass, or number of moles
- ☐ calculate the volume of gas produced in a reaction, given the equation and the mass of one substance
- ☐ explain what *concentration of a solution* means and give examples, using grams and moles
- ☐ state the units used for concentration
- ☐ explain what a *molar solution* is
- ☐ work out:
  - the concentration of a solution, when you know the amount of solute dissolved in it
  - the amount of solute dissolved in a solution, when you know its concentration
- ☐ explain what the *empirical formula* of a substance is
- ☐ work out the empirical formula, from the masses that react
- ☐ work out the correct formula, using the empirical formula and $M_r$
- ☐ define % yield
- ☐ calculate the % yield for a reaction, from the equation and the actual mass of product obtained
- ☐ define % purity
- ☐ calculate the % purity of a product, given the mass of the impure product, and the mass of pure product it contains

## Questions
### Extended curriculum

1. Iron is obtained by reducing iron(III) oxide using the gas carbon monoxide. The reaction is:
   $Fe_2O_3\ (s) + 3CO\ (g) \longrightarrow 2Fe\ (s) + 3CO_2\ (g)$
   a. Write a word equation for the reaction.
   b. What is the formula mass of iron(III) oxide? ($A_r$: Fe = 56, O = 16.)
   c. How many moles of $Fe_2O_3$ are there in 320 kg of iron(III) oxide? (1 kg = 1000 g.)
   d. How many moles of Fe are obtained from 1 mole of $Fe_2O_3$?
   e. From **c** and **d**, find how many moles of iron atoms are obtained from 320 kg of iron(III) oxide.
   f. How much iron (in kg) is obtained from 320 kg of iron(III) oxide?

2. With strong heating, calcium carbonate undergoes thermal decomposition:
   $CaCO_3\ (s) \longrightarrow CaO\ (s) + CO_2\ (g)$
   a. Write a word equation for the change.
   b. How many moles of $CaCO_3$ are in 50 g of calcium carbonate? ($A_r$: Ca = 40, C = 12, O = 16.)
   c. i. What mass of calcium oxide is obtained from the thermal decomposition of 50 g of calcium carbonate, assuming a 40% yield?
      ii. What mass of carbon dioxide will be given off at the same time?
      iii. What volume will this gas occupy at rtp?

3. Nitroglycerine is used as an explosive. The equation for the explosion reaction is:
   $4C_3H_5(NO_3)_3\ (l) \longrightarrow$
   $\qquad 12CO_2\ (g) + 10H_2O\ (l) + 6N_2\ (g) + O_2\ (g)$
   a. How many moles does the equation show for:
      i. nitroglycerine?
      ii. gas molecules produced?
   b. How many moles of gas molecules are obtained from 1 mole of nitroglycerine?
   c. What is the total volume of gas (at rtp) obtained from 1 mole of nitroglycerine?
   d. What is the mass of 1 mole of nitroglycerine? ($A_r$: H = 1, C = 12, N = 14, O = 16.)
   e. What will be the total volume of gas (at rtp) from exploding 1 kg of nitroglycerine?
   f. Using your answers above, try to explain *why* nitroglycerine is used as an explosive.

90

4 Nitrogen monoxide reacts with oxygen like this:
$2NO\ (g) + O_2\ (g) \longrightarrow 2NO_2\ (g)$
   a How many moles of oxygen molecules react with 1 mole of nitrogen monoxide molecules?
   b What volume of oxygen will react with 50 cm³ of nitrogen monoxide?
   c Using the volumes in b, what is:
     i the total volume of the two reactants?
     ii the volume of nitrogen dioxide formed?

5 2 g (an excess) of iron is added to 50 cm³ of 0.5 M sulfuric acid. When the reaction is over, the reaction mixture is filtered. The mass of the unreacted iron is found to be 0.6 g. ($A_r$: Fe = 56.)
   a What mass of iron took part in the reaction?
   b How many moles of iron atoms took part?
   c How many moles of sulfuric acid reacted?
   d Write the equation for the reaction, and deduce the charge on the iron ion that formed.
   e What volume of hydrogen (calculated at rtp) bubbled off during the reaction?

6 27 g of aluminium burns in chlorine to form 133.5 g of aluminium chloride. ($A_r$: Al = 27, Cl = 35.5.)
   a What mass of chlorine is present in 133.5 g of aluminium chloride?
   b How many moles of chlorine atoms is this?
   c How many moles of aluminium atoms are present in 27 g of aluminium?
   d Use your answers for parts b and c to find the simplest formula of aluminium chloride.
   e 1 dm³ of an aqueous solution is made using 13.35 g of aluminium chloride. What is its concentration in moles per dm³?

7 You have to prepare some 2 M solutions, with 10 g of solute in each. What volume of solution will you prepare, for each solute below?
($A_r$: H = 1, Li = 7, N = 14, O = 16, Mg = 24, S = 32.)
   a lithium sulfate, $Li_2SO_4$
   b magnesium sulfate, $MgSO_4$
   c ammonium nitrate, $NH_4NO_3$

8 Phosphorus forms two oxides, which have the empirical formulae $P_2O_3$ and $P_2O_5$.
   a Which oxide contains the higher percentage of phosphorus? ($A_r$: P = 31, O = 16.)
   b What mass of phosphorus will combine with 1 mole of oxygen molecules ($O_2$) to form $P_2O_3$?
   c What is the molecular formula of the oxide that has a formula mass of 284?
   d Suggest a molecular formula for the other oxide.

9 Zinc and phosphorus react to give zinc phosphide. 9.75 g of zinc combines with 3.1 g of phosphorus.
   a Find the empirical formula for the compound. ($A_r$: Zn = 65, P = 31.)
   b Calculate the percentage of phosphorus in it.

10 110 g of manganese was extracted from 174 g of manganese oxide. ($A_r$: Mn = 55, O = 16.)
   a What mass of oxygen is there in 174 g of manganese oxide?
   b How many moles of oxygen atoms is this?
   c How many moles of manganese atoms are there in 110 g of manganese?
   d Give the empirical formula of manganese oxide.
   e What mass of manganese can obtained from 1000 g of manganese oxide?

11 Find the molecular formulae for these compounds. ($A_r$: H = 1, C = 12, N = 14, O = 16.)

| Compound | $M_r$ | Empirical formula | Molecular formula |
|---|---|---|---|
| a hydrazine | 32 | $NH_2$ | |
| b cyanogen | 52 | CN | |
| c nitrogen oxide | 92 | $NO_2$ | |
| d glucose | 180 | $CH_2O$ | |

12 Hydrocarbons A and B both contain 85.7% carbon. Their molar masses are 42 and 84 g respectively.
   a Which elements does a hydrocarbon contain?
   b Calculate the empirical formulae of A and B.
   c Calculate the molecular formulae of A and B.

13 Mercury(II) oxide breaks down on heating:
$2HgO\ (s) \longrightarrow 2Hg\ (l) + O_2\ (g)$
   a Calculate the mass of 1 mole of mercury(II) oxide. ($A_r$: O = 16, Hg = 201)
   b How much mercury and oxygen *could* be obtained from 21.7 g of mercury(II) oxide?
   c Only 19.0 g of mercury was collected. Calculate the % yield of mercury for this experiment.

14 A 5-g sample of impure magnesium carbonate is reacted with an excess of hydrochloric acid:
$MgCO_3\ (s) + 2HCl\ (aq) \longrightarrow MgCl_2\ (aq) + H_2O\ (l) + CO_2\ (g)$
1250 cm³ of carbon dioxide is collected at rtp.
   a How many moles of $CO_2$ are produced?
   b What mass of pure magnesium carbonate would give this volume of carbon dioxide? ($A_r$: C = 12, O = 16, Mg = 24.)
   c Calculate the % purity of the 5-g sample.

# REDOX REACTIONS

## 7.1 Oxidation and reduction

### Different groups of reactions
Thousands of different reactions go on around us, in labs, and factories, and homes. We can divide them into different groups. For example two of the groups are **neutralisation reactions** and **precipitation reactions**.

One big group is the **redox reactions**, in which **oxidation** and **reduction** occur. We focus on those in this chapter.

### Oxidation: oxygen is gained
Magnesium burns in air with a dazzling white flame. A white ash is formed. The reaction is:

magnesium + oxygen ⟶ magnesium oxide
$2Mg\,(s) + O_2\,(g) \longrightarrow 2MgO\,(s)$

The magnesium has gained oxygen. We say it has been oxidised.

**A gain of oxygen is called *oxidation*. The substance has been oxidised.**

▲ Magnesium burning in oxygen.

### Reduction: oxygen is lost
Now look what happens when hydrogen is passed over heated copper(II) oxide. The black compound turns pink:

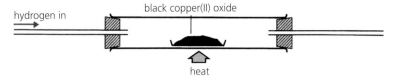

This reaction is taking place:

copper(II) oxide + hydrogen ⟶ copper + water
$CuO\,(s) + H_2\,(g) \longrightarrow Cu\,(s) + H_2O\,(l)$

This time the heated substance is *losing* oxygen. It is being **reduced**.

**A loss of oxygen is called *reduction*. The substance is reduced.**

▲ Iron occurs naturally in the earth as iron(III) oxide, $Fe_2O_3$. This is **reduced** to iron in the blast furnace. Here, molten iron runs out from the bottom of the furnace.

▲ And here, iron is being **oxidised** to iron(III) oxide again! We call this process **rusting**. It is ruining the bikes. The formula for rust is $Fe_2O_3.2H_2O$.

# REDOX REACTIONS

## Oxidation and reduction take place together

Look again at the reaction between copper(II) oxide and hydrogen.

Copper(II) oxide loses oxygen, and hydrogen gains oxygen:

$$CuO\,(s) + H_2\,(g) \longrightarrow Cu\,(s) + H_2O\,(l)$$

(oxidation arrow from CuO to H₂O; reduction arrow from CuO to Cu)

So the copper(II) oxide is reduced, and the hydrogen is oxidised.

**Oxidation and reduction *always* take place together.**
**So the reaction is called a *redox reaction*.**

▲ A redox reaction that cooks our food. The gas reacts with the oxygen in air, giving out heat.

## Two more examples of redox reactions

**The reaction between calcium and oxygen** Calcium burns in air with a red flame, to form the white compound calcium oxide. It is easy to see that calcium has been oxidised. But oxidation and reduction *always* take place together, which means oxygen has been reduced:

$$2Ca\,(s) + O_2\,(g) \longrightarrow 2CaO\,(s)$$

**The reaction between hydrogen and oxygen** Hydrogen reacts explosively with oxygen, to form water. Hydrogen is oxidised, and oxygen is reduced:

$$2H_2\,(g) + O_2\,(g) \longrightarrow 2H_2O\,(l)$$

▲ Roaming around on redox. The burning of petrol is a redox reaction. So is the 'burning' of glucose in our cells. It reacts with oxygen to give us energy, in a process called **respiration**.

### Those burning reactions
- Another name for burning is **combustion**.
- Combustion is a redox reaction.
- For example, when an element burns in oxygen, it is oxidised to its oxide.

---

**1** Copy and complete the statements:
  **a** Oxidation means …
  **b** Reduction means …
  **c** Oxidation and reduction always …

**2** Magnesium reacts with sulfur dioxide like this:

$$2Mg\,(s) + SO_2\,(g) \longrightarrow 2MgO\,(s) + S\,(s)$$

Copy the equation, and use labelled arrows to show which substance is oxidised, and which is reduced.

**3** Explain where the term *redox* comes from.

**4** Many people cook with natural gas, which is mainly methane, $CH_4$. The equation for its combustion is:

$$CH_4\,(g) + 2O_2\,(g) \longrightarrow CO_2\,(g) + 2H_2O\,(l)$$

Show that this is a redox reaction.

**5** Write down the equation for the reaction between magnesium and oxygen. Use labelled arrows to show which element is oxidised, and which is reduced.

# REDOX REACTIONS

## 7.2 Redox and electron transfer

### Another definition for oxidation and reduction

When magnesium burns in oxygen, magnesium oxide is formed:

$$2Mg\,(s) + O_2\,(g) \longrightarrow 2MgO\,(s)$$

The magnesium has clearly been oxidised. Oxidation and reduction *always* take place together, so the oxygen must have been reduced. But how? Let's see what is happening to the electrons:

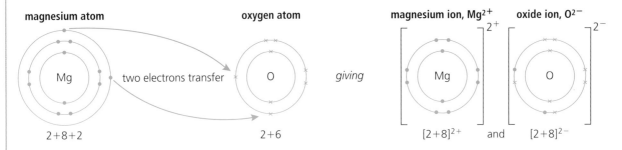

During the reaction, each magnesium atom loses two electrons and each oxygen atom gains two. This leads us to a new definition:

**If a substance loses electrons during a reaction, it has been oxidised.**

**If it gains electrons, it has been reduced.**

**The reaction is a redox reaction.**

> **Remember OILRIG!**
> **O**xidation **I**s **L**oss of electrons.
> **R**eduction **I**s **G**ain of electrons.

### Writing half-equations to show the electron transfer

You can use **half-equations** to show the electron transfer in a reaction. One half-equation shows electron loss, and the other shows electron gain.

This is how to write the half-equations for the reaction above:

1 **Write down each reactant, with the electrons it gains or loses.**
   magnesium: $\quad Mg \longrightarrow Mg^{2+} + 2e^-$
   oxygen: $\quad O + 2e^- \longrightarrow O^{2-}$

2 **Check that each substance is in its correct form (ion, atom or molecule) on each side of the arrow. If it is not, correct it.**
   Oxygen is not in its correct form on the left above. It exists as molecules, so you must change O to $O_2$. That means you must also double the number of electrons and oxide ions:
   oxygen: $\quad O_2 + 4e^- \longrightarrow 2O^{2-}$

3 **The number of electrons must be the same in both equations. If it is not, multiply one (or both) equations by a number, to balance them.**
   So we must multiply the magnesium half-equation by 2.
   magnesium: $\quad 2Mg \longrightarrow 2Mg^{2+} + 4e^-$
   oxygen: $\quad O_2 + 4e^- \longrightarrow 2O^{2-}$
   The equations are now balanced, each with 4 electrons.

> **Two ways to show oxidation**
> You can show oxidation (the loss of electrons) in two ways:
> $Mg \longrightarrow Mg^{2+} + 2e^-$
> or
> $Mg - 2e^- \longrightarrow Mg^{2+}$
> Both are correct!

# REDOX REACTIONS

## Redox without oxygen

Our definition of redox reactions is now much broader:
**Any reaction in which electron transfer takes place is a redox reaction.**
So the reaction does not have to include oxygen! Look at these examples:

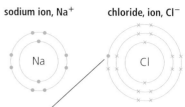

1 **The reaction between sodium and chlorine**
   The equation is:
   $2Na\,(s) + Cl_2\,(g) \longrightarrow 2NaCl\,(s)$

   The sodium atoms give electrons to the chlorine atoms, forming ions as shown on the right. So sodium is oxidised, and chlorine is reduced.

   the sodium atom has lost an electron to the chlorine atom

   So the reaction is a redox reaction. Look at the half-equations:
   sodium: $2Na \longrightarrow 2Na^+ + 2e^-$ (oxidation)
   chorine: $Cl_2 + 2e^- \longrightarrow 2Cl^-$ (reduction)

2 **The reaction between chlorine and potassium bromide**
   When chlorine gas is bubbled through a colourless solution of potassium bromide, the solution goes orange due to this reaction:
   $Cl_2\,(g) + 2KBr\,(aq) \longrightarrow 2KCl\,(aq) + Br_2\,(aq)$
       colourless                 orange

   Bromine has been **displaced**. The half-equations for the reaction are:
   chlorine: $Cl_2 + 2e^- \longrightarrow 2Cl^-$ (reduction)
   bromide ion: $2Br^- \longrightarrow Br_2 + 2e^-$ (oxidation)

## From half-equations to the ionic equation

Adding the balanced half-equations gives the **ionic equation** for the reaction.
**An ionic equation shows the ions that take part in the reaction.**

For example, for the reaction between chlorine and potassium bromide:

$$Cl_2 + 2e^- \longrightarrow 2Cl^-$$
$$2Br^- \longrightarrow Br_2 + 2e^-$$
$$\overline{Cl_2 + \cancel{2e^-} + 2Br^- \longrightarrow 2Cl^- + Br_2 + \cancel{2e^-}}$$

The electrons cancel, giving the ionic equation for the reaction:
$$Cl_2 + 2Br^- \longrightarrow 2Cl^- + Br_2$$

## Redox: a summary

**Oxidation** is gain of oxygen, or loss of electrons.
**Reduction** is loss of oxygen, or gain of electrons.
Oxidation and reduction always take place together, in a **redox reaction**.

▲ Bromine being displaced by chlorine, from a colourless solution of potassium bromide. The solution goes orange.

1 Give a *full* definition for: **a** oxidation    **b** reduction
2 What does a *half-equation* show?
3 Potassium and chlorine react to form potassium chloride.
   **a** It is a redox reaction. Explain why.
   **b** See if you can write the balanced half-equations for it.
4 Bromine displaces iodine from a solution of potassium iodide.
   **a** Write the balanced half-equations for this reaction.
   **b** Add the half-equations, to give the ionic equation for the reaction.

REDOX REACTIONS

## 7.3 Redox and changes in oxidation state

### What does oxidation state mean?
**Oxidation state** tells you how many electrons each atom of an element has gained, lost, or shared, in forming a compound.

As you will see, oxidation states can help you to identify redox reactions.

### The rules for oxidation states
1  Each atom in a formula has an oxidation state.
2  The oxidation state is usually given as a Roman numeral. Note these Roman numerals:

| number | 0 | 1 | 2 | 3 | 4 | 5 | 6 | 7 |
|---|---|---|---|---|---|---|---|---|
| Roman numeral | 0 | I | II | III | IV | V | VI | VII |

3  Where an element is not combined with other elements, its atoms are in oxidation state 0.
4  Many elements have the same oxidation state in most or all their compounds. Look at these:

| Element | Usual oxidation state in compounds |
|---|---|
| hydrogen | +I |
| sodium and the other Group I metals | +I |
| calcium and the other Group II metals | +II |
| aluminium | +III |
| chlorine and the other Group VII non-metals, in compounds without oxygen | −I |
| oxygen (except in peroxides) | −II |

▲ The element sodium: oxidation state 0 (zero).

5  But atoms of transition elements can have variable oxidation states in their compounds. Look at these:

| Element | Common oxidation states in compounds |
|---|---|
| iron | +II and +III |
| copper | +I and +II |
| manganese | +II, +IV, and +VII |
| chromium | +III and +VI |

So for these elements, the oxidation state is included in the compound's name. For example iron(III) chloride, copper(II) oxide.

6  Note that in any formula, the oxidation states must add up to zero. Look at the formula for magnesium chloride, for example:

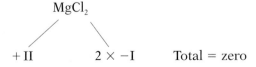

$MgCl_2$

+II    2 × −I    Total = zero

So you could use oxidation states to check that formulae are correct.

▲ Copper in its three oxidation states:
A – copper metal, 0
B – copper(I) chloride, +I
C – copper(II) chloride, +II

# REDOX REACTIONS

## Oxidation states change during redox reactions

Look at the equation for the reaction between sodium and chlorine:

$$2Na\ (s) + Cl_2\ (g) \longrightarrow 2NaCl\ (s)$$
$$\phantom{2N}0 \phantom{a\ (s) +\ } 0 \phantom{Cl_2\ (g) \longrightarrow\ 2N} +I\ -I$$

The oxidation states are also shown, using the rules on page 96. Notice how they have changed during the reaction.

Each sodium atom loses an electron during the reaction, to form an $Na^+$ ion. So sodium is oxidised, and its oxidation state rises from 0 to $+I$.
Each chlorine atom gains an electron, to form a $Cl^-$ ion. So chlorine is reduced, and its oxidation state falls from 0 to $-I$.

**If oxidation states change during a reaction, it is a redox reaction.**

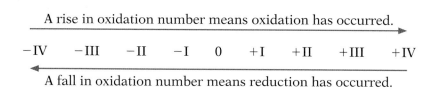

A rise in oxidation number means oxidation has occurred.
$-IV \quad -III \quad -II \quad -I \quad 0 \quad +I \quad +II \quad +III \quad +IV$
A fall in oxidation number means reduction has occurred.

▲ Sodium burning in chlorine, to form sodium chloride.

## Using oxidation states to identify redox reactions

**Example 1** Iron reacts with sulfur to form iron(II) sulfide:

$$Fe\ (s) + S\ (s) \longrightarrow FeS\ (s)$$
$$\phantom{F}0 \phantom{e\ (s) +\ }0 \phantom{S\ (s) \longrightarrow Fe}+II\ -II$$

The oxidation states are shown, using the rules on page 96.
There has been a change in oxidation states. So this is a redox reaction.

**Example 2** When chlorine is bubbled through a solution of iron(II) chloride, iron(III) choride is formed. The equation and oxidation states are:

$$2FeCl_2\ (aq) + Cl_2\ (aq) \longrightarrow 2FeCl_3\ (aq)$$
$$+II\ -I \phantom{aaaaaaaaa} 0 \phantom{aaaaaaaaaa} +III\ -I$$

There has been a change in oxidation states. So this is a redox reaction.

**Example 3** When ammonia and hydrogen chloride gases mix, they react to form ammonium chloride. The equation and oxidation states are:

$$NH_3\ (g) + HCl\ (g) \longrightarrow NH_4Cl\ (s)$$
$$-III\ +I \phantom{aaaa} +I\ -I \phantom{aaaaa} -III\ +I\ -I$$

There has been no change in oxidation states. So this is *not* a redox reaction.

▲ Iron filings reacting with sulfur. You need heat to start the reaction off – but then it gives out heat.

## Q

**1** **a** Write a word equation for this reaction:
$$2H_2\ (g) + O_2\ (g) \longrightarrow 2H_2O\ (l)$$
 **b** Now copy out the chemical equation from **a**. Below each symbol write the oxidation state of the atoms.
 **c** Is the reaction a redox reaction? Give evidence.
 **d** Say which substance is oxidised, and which reduced.

**2** Repeat the steps in question **1** for each of these equations:
 **i** $2KBr\ (s) \longrightarrow 2K\ (s) + Br_2\ (l)$
 **ii** $2KI\ (aq) + Cl_2\ (g) \longrightarrow 2KCl\ (aq) + I_2\ (aq)$

**3** **a** Read point **6** on page 96.
 **b** Using the idea in point **6**, work out the oxidation state of the carbon atoms in carbon dioxide, $CO_2$.
 **c** Carbon burns in oxygen to form carbon dioxide. Write a chemical equation for the reaction.
 **d** Now using oxidation states, show that this is a redox reaction, and say which substance is oxidised, and which is reduced.

**4** *Every reaction between two elements is a redox reaction.* Do you agree with this statement? Explain.

# REDOX REACTIONS

## 7.4 Oxidising and reducing agents

### What are oxidising and reducing agents?

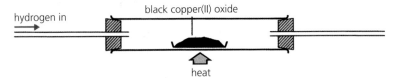

When hydrogen reacts with heated copper(II) oxide, the reaction is:

copper(II) oxide + hydrogen ⟶ copper + water
$CuO\ (s)\ \ \ +\ \ \ H_2\ (g)\ \longrightarrow\ Cu\ (s) + H_2O\ (l)$

The copper(II) oxide is **reduced** to copper by reaction with hydrogen. So hydrogen acts as a **reducing agent**.

The hydrogen is itself **oxidised** to water, in the reaction. So copper(II) oxide acts as an **oxidising agent**.

**An oxidising agent oxidises another substance – and is itself reduced.
A reducing agent reduces another substance – and is itself oxidised.**

> **Oxidants and reductants**
> - Oxidising agents are also called **oxidants**.
> - Reducing agents are called **reductants**.

### Oxidising and reducing agents in the lab

Some substances have a strong drive to gain electrons. So they are strong oxidising agents. They readily oxidise other substances by taking electrons from them. Examples are oxygen and chlorine.

Some substances are strong reducing agents, readily giving up electrons to other substances. Examples are hydrogen, carbon monoxide, and reactive metals like sodium.

Some oxidising and reducing agents show a colour change when they react. This makes them useful in lab tests. Look at these three examples.

> **Remember OILRIG!**
> **O**xidation **I**s **L**oss of electrons.
> **R**eduction **I**s **G**ain of electrons.

### 1 Potassium manganate(VII): an oxidising agent

Manganese is a transition element. Like other transition elements, it can exist in different oxidation states. (Look back at point **5** on page 96.)

Potassium manganate(VII) is a purple compound. Its formula is $KMnO_4$. In it, the manganese is in oxidation state +VII. But it is much more stable in oxidation state +II. So it is strongly driven to reduce its oxidation state to +II, by gaining electrons.

That is why potassium manganate(VII) acts as a powerful oxidising agent. It takes electrons from other substances, in the presence of a little acid. It is itself reduced in the reaction – with a colour change:

$$MnO_4^-\ (aq) \xrightarrow{\text{reduction}} Mn^{2+}\ (aq)$$
manganate(VII) ion  →  manganese(II) ion
(purple)                  (colourless)

▲ Adding potassium manganate(VII) to an unknown liquid. The purple colour is fading, so the liquid must contain a reducing agent.

This colour change means that potassium manganate(VII) can be used to test for the presence of a reducing agent. If a reducing agent is present, the purple colour will fade.

## 2 Potassium dichromate(VI): an oxidising agent

Chromium is also a transition element, and can exist in different oxidation states. In potassium dichromate(VI) it is in oxidation state +VI. But oxidation state +III is the most stable.

So potassium dichromate(VI) is a strong oxidising agent, in the presence of acid. It reacts to gain electrons and reduce the oxidation state to +III. Once again there is a colour change on reduction:

$$\underset{\substack{\text{dichromate(VI) ion} \\ \text{(orange)}}}{Cr_2O_7^{2-}\,(aq)} \xrightarrow{\text{reduction}} \underset{\substack{\text{chromium(III) ion} \\ \text{(green)}}}{2Cr^{3+}\,(aq)}$$

▲ A and B show the colour change from orange to green when potassium dichromate(VI) is reduced.

This colour change means that potassium dichromate(VI) can be used to test for the presence of reducing agents.

Outside the lab, it is used to test for alcohol (ethanol) on a driver's breath, in the **breathalyser test**. It oxidises ethanol to ethanal:

$$\underset{\text{ethanol}}{C_2H_5OH} \xrightarrow{K_2Cr_2O_7} \underset{\text{ethanal}}{CH_3CHO}$$

So a colour change proves that the driver had been drinking.

## 3 Potassium iodide: a reducing agent

When potassium iodide solution is added to hydrogen peroxide, in the presence of sulfuric acid, this redox reaction takes place:

$$\underset{\substack{\text{hydrogen} \\ \text{peroxide}}}{H_2O_2\,(aq)} + \underset{\substack{\text{potassium} \\ \text{iodide}}}{2KI\,(aq)} + H_2SO_4\,(aq) \longrightarrow \underset{\text{iodine}}{I_2\,(aq)} + \underset{\substack{\text{potassium} \\ \text{sulfate}}}{K_2SO_4\,(aq)} + 2H_2O\,(l)$$

▲ A breathalyser test. The device contains potassium dichromate(VI), which is orange. Alcohol on the breath causes a colour change to green.

You can see that the hydrogen peroxide loses oxygen: it is reduced. The potassium iodide acts as a reducing agent. At the same time the potassium iodide is oxidised to iodine. This causes a colour change:

$$\underset{\text{colourless}}{2I^-\,(aq)} \xrightarrow{\text{oxidation}} \underset{\text{red-brown}}{I_2\,(aq)}$$

So potassium iodide is used to test for the presence of an oxidising agent.

▲ This shows the red-brown colour you get when potassium iodide is oxidised by an oxidising agent.

---

**Q**

1. What is:
   a an oxidising agent?  b a reducing agent?
2. Identify the oxidising and reducing agents in these reactions, by looking at the gain and loss of oxygen:
   a $2Mg\,(s) + O_2\,(g) \rightarrow 2MgO\,(s)$
   b $Fe_2O_3\,(s) + 3CO\,(g) \rightarrow 2Fe\,(l) + 3CO_2\,(g)$
3. Now identify the oxidising and reducing agents in these:
   a $2Fe + 3Cl_2 \rightarrow 2FeCl_3$
   b $Fe + CuSO_4 \rightarrow FeSO_4 + Cu$
4. Explain why:
   a potassium manganate(VII) is a powerful oxidising agent
   b potassium iodide is used to test for oxidising agents

# REDOX REACTIONS

# Checkup on Chapter 7

## Revision checklist

### Core curriculum
Make sure you can ...
- [ ] define *oxidation* as a gain of oxygen
- [ ] define *reduction* as a loss of oxygen
- [ ] explain that oxidation and reduction always occur together, and give an example
- [ ] explain what a *redox reaction* is
- [ ] define these terms:
  *oxidising agent*    *reducing agent*
- [ ] identify the oxidising and reducing agents, in reactions involving oxygen

### Extended curriculum
Make sure you can also ...
- [ ] define oxidation and reduction in terms of electron transfer
- [ ] explain these terms:
  *half-equation*    *ionic equation*
- [ ] write balanced half-equations for a redox reaction, to show the electron transfer
- [ ] give the ionic equation for a reaction, by adding the balanced half-equations
- [ ] explain the term *oxidation state*
- [ ] give the usual oxidation state for these elements, in their compounds:
  hydrogen    oxygen    aluminium
  sodium and other Group I metals
  calcium and other Group II metals
  chlorine and other Group VII non-metals
- [ ] tell the oxidation state from a compound's name, for elements with variable oxidation states
- [ ] work out the oxidation state for each element in a compound (they must add up to zero)
- [ ] give the oxidation state for each element present, in the equation for a reaction
- [ ] identify a redox reaction from changes in oxidation states, in the equation
- [ ] explain why some substances are:
  *strong oxidising agents*    *strong reducing agents*
  and give examples
- [ ] explain why potassium manganate(VII) and potassium dichromate(VI) are used in the lab to test for the presence of reducing agents
- [ ] explain why potassium iodide is used in the lab to test for the presence of oxidising agents

## Questions

### Core curriculum

**1** If a substance gains oxygen in a reaction, it has been oxidised. If it loses oxygen, it has been reduced. Oxidation and reduction always take place together, so if one substance is oxidised, another is reduced.
  **a** First, see if you can write a word equation for each redox reaction **A** to **F** below.
  **b** Then, using the ideas above, say which substance is being oxidised, and which is being reduced, in each reaction.
  **A** $Ca\,(s) + O_2\,(g) \longrightarrow 2\,CaO\,(s)$
  **B** $2CO\,(g) + O_2\,(g) \longrightarrow 2CO_2\,(g)$
  **C** $CH_4\,(g) + 2O_2\,(g) \longrightarrow CO_2\,(g) + 2H_2O\,(l)$
  **D** $2CuO\,(s) + C\,(s) \longrightarrow 2Cu\,(s) + CO_2\,(g)$
  **E** $2Fe\,(s) + 3O_2\,(g) \longrightarrow 2Fe_2O_3\,(s)$
  **F** $Fe_2O_3\,(s) + 3CO\,(g) \longrightarrow 2Fe\,(s) + 3CO_2\,(g)$

**2 a** Is this a redox reaction? Give your evidence.
  **A** $2Mg\,(s) + CO_2\,(g) \longrightarrow 2MgO\,(s) + C\,(s)$
  **B** $SiO_2\,(s) + C\,(s) \longrightarrow Si\,(s) + CO_2\,(g)$
  **C** $NaOH\,(aq) + HCl\,(aq) \longrightarrow NaCl\,(aq) + H_2O\,(l)$
  **D** $Fe\,(s) + CuO\,(s) \longrightarrow FeO\,(s) + Cu\,(s)$
  **E** $C\,(s) + PbO\,(s) \longrightarrow CO\,(g) + Pb\,(s)$
  **b** For each redox reaction you identify, name:
    **i** the oxidising agent
    **ii** the reducing agent.

### Extended curriculum

**3** All reactions in which electron transfer take place are redox reactions. This diagram shows the electron transfer during one redox reaction.

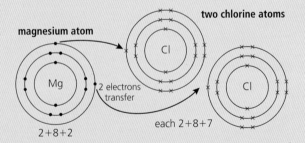

  **a** What is the product of this reaction?
  **b** Write a balanced equation for the full reaction.
  **c i** Which element is being oxidised?
    **ii** Write a half-equation for the oxidation.
  **d i** Which element is being reduced?
    **ii** Write a half-equation for the reduction of this element.

4 Redox reactions involve electron transfer.
  a Fluorine, from Group VII, reacts with lithium, from Group I, to form a poisonous white compound. What is its name?
  b Write a balanced equation for the reaction.
  c Draw a diagram to show the electron transfer that takes place during the reaction.
  d i Which element is oxidised in the reaction?
    ii Write a half-equation for this oxidation.
  e Write a half-equation for the reduction of the other element.

5 Chlorine gas is bubbled into a solution containing sodium bromide. The equation for the reaction is:
  $Cl_2 (g) + 2NaBr (aq) \rightarrow Br_2 (aq) + 2NaCl (aq)$
  a Chlorine takes the place of bromine, in the metal compound. What is this type of reaction called?
  b The compounds of Group I metals are white, and give colourless solutions. What would you see as the above reaction proceeds?
  c i Write a half-equation for the reaction of the chlorine.
    ii Is the chlorine oxidised, or reduced, in this reaction? Explain.
  d Write a half-equation for the reaction of the bromide ion.
  e Reactive elements have a strong tendency to exist as ions. Which is more reactive, chlorine or bromine? Explain why you think so.
  f i Which halide ion could be used to convert bromine back to the bromide ion?
    ii Write the ionic equation for this reaction.

6 Iodine is extracted from seaweed using acidified hydrogen peroxide, in a redox reaction. The ionic equation for the reaction is:
  $2I^- (aq) + H_2O_2 (aq) + 2H^+ (aq) \rightarrow$
  $\phantom{2I^- (aq) + H_2O_2 (aq) + } I_2 (aq) + 2H_2O (l)$
  a In which oxidation state is the iodine in seaweed?
  b There is a colour change in this reaction. Why?
  c i Is the iodide ion oxidised, or reduced?
    ii Write the half-equation for this change.
  d In hydrogen peroxide, the oxidation state of the hydrogen is +I.
    i What is the oxidation state of the oxygen in hydrogen peroxide?
    ii How does the oxidation state of oxygen change during the reaction?
    iii Copy and complete this half-equation for hydrogen peroxide:
    $H_2O_2 (aq) + 2H^+ (aq) + \ldots \rightarrow 2H_2O (l)$

7 The oxidation states in a formula add up to zero.
  a Give the oxidation state of the underlined atom in each formula below:
    i aluminium oxide, $\underline{Al}_2O_3$
    ii ammonia, $\underline{N}H_3$
    iii $H_2\underline{C}O_3$ (aq), carbonic acid
    iv phosphorus trichloride, $\underline{P}Cl_3$
    v copper(I) chloride, $\underline{Cu}Cl$
    vi copper(II) chloride, $\underline{Cu}Cl_2$
  b Now comment on the compounds in v and vi.

8 The oxidising agent potassium manganate(VII) can be used to analyse the % of iron(II) present in iron tablets. Below is an **ionic equation**, showing the ions that take part in the reaction:
  $MnO_4^- (aq) + 8H^+ (aq) + 5Fe^{2+} (aq) \rightarrow$
  $\phantom{MnO_4^- (aq) + } Mn^{2+} (aq) + 5Fe^{3+} (aq) + 4H_2O (l)$
  a What does the $H^+$ in the equation tell you about this reaction? (Hint: check page 150.)
  b Describe the colour change.
  c Which is the reducing reagent in this reaction?
  d How could you tell when all the iron(II) had reacted?
  e Write the half-equation for the iron(II) ions.

9 Potassium chromate(VI) is yellow. In acid it forms orange potassium dichromate(VI). These are the ions that give those colours:

  a What is the oxidation state of chromium in:
    i the yellow compound?
    ii the orange compound?
  b This reaction of chromium ions is not a redox reaction. Explain why.

10 When solutions of silver nitrate and potassium chloride are mixed, a white precipitate forms. The ionic equation for the reaction is:
  $Ag^+ (aq) + Cl^- (aq) \rightarrow AgCl (s)$
  a i What is the name of the white precipitate?
    ii Is it a soluble or insoluble compound?
  b Is the precipitation of silver chloride a redox reaction or not? Explain your answer.
  c When left in light, silver chloride decomposes to form silver and chlorine gas.
    Write an equation for the reaction and show clearly that this is a redox reaction.

ELECTRICITY AND CHEMICAL CHANGE

## 8.1 Conductors and insulators

### Batteries and electric current

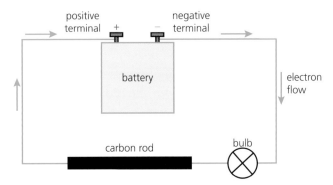

The photograph above shows a battery, a bulb and a rod of graphite joined or **connected** to each other by copper wires. (Graphite is a form of carbon.) This arrangement is called an **electric circuit**.

The bulb is lit: this shows that electricity must be flowing in the circuit. **Electricity is a stream of electrons.**

The diagram shows how the electrons move through the circuit. The battery acts like an electron pump. Electrons leave it through the **negative terminal**. They travel through the wire, bulb, and rod, and enter the battery again through the **positive terminal**.

When the electrons stream through the fine wire in the bulb, they cause it to heat up. It gets white-hot and gives out light.

### Conductors and insulators

In the circuit above, the graphite and copper wire allow electricity to pass through. So they are called conductors.

But if you connect a piece of plastic or ceramic into the circuit, the bulb will not light. Plastic and ceramic do not let electricity pass through them. They are **non-conductors** or **insulators**.

▲ Copper carries the current into the styling iron. Then it flows through wire made of nichrome (a nickel-chromium alloy) which heats up. Meanwhile, the plastic protects you.

### Some uses for conductors and insulators

The cables that carry electricity around the country are made of aluminium and steel. Both are conductors. (Aluminium is a better conductor than steel.)

At pylons, ceramic discs support the bare cables. Since it is an insulator, the ceramic prevents the current from running down the pylon. (Dangerous!)

Copper is used for wiring, at home. It is a very good conductor. But the wires are sheathed in plastic, and plug cases are made of plastic (an insulator), for safety.

# ELECTRICITY AND CHEMICAL CHANGE

## Testing substances to see if they conduct

You can test any substance to see if it conducts, by connecting it into a circuit like the one on page 102. For example:

**Tin.** A strip of tin is connected into the circuit, in place of the graphite rod. The bulb lights, so tin must be a conductor.

**Ethanol.** The liquid is connected into the circuit by placing graphite rods in it. The bulb does not light, so ethanol is a non-conductor.

**Lead bromide.** It does not conduct when solid. But if you melt it, it conducts, and gives off a choking brown vapour.

**The results**   These are the results from a range of tests:

1 **The only solids that conduct are the metals and graphite.**
   These conduct because of their free electrons (pages 61 and 62). The electrons get pumped out of one end of the solid by the battery, while more electrons flow in the other end.
   For the same reason, *molten* metals conduct. (It is hard to test molten graphite, because at room pressure graphite goes from solid to gas.)

2 **Molecular substances are non-conductors.**
   This is because they contain no free electrons, or other charged particles, that can flow through them.
   Ethanol (above) is made of molecules. So is petrol, paraffin, sulfur, sugar, and plastic. These never conduct, whether solid or molten.

3 **Ionic substances do not conduct when solid. But they do conduct when melted or dissolved in water – and *they decompose at the same time*.**
   An ionic substance contains no free electrons. But it does contain **ions**, which have a charge. The ions become free to move when the substance is melted or dissolved, and it is they that conduct the electricity.
   Lead bromide is ionic. It does not conduct when solid, but conducts when it melts. The brown vapour that forms is bromine. Electricity has caused the lead bromide to decompose.

   **Decomposition brought about by electricity is called electrolysis. A liquid that contains ions, and therefore conducts electricity, is called an electrolyte.**
   So molten lead bromide is an electrolyte. Ethanol is a non-electrolyte.

▲ Metals conduct, thanks to their free electrons, which form a current.

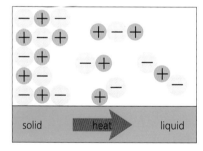

▲ An ionic solid conducts when it melts, because the ions become free to move.

**Q**

1 What is a *conductor* of electricity?
2 Draw a circuit to show how you would test whether mercury conducts.
3 Explain why metals are able to conduct electricity.
4 Naphthalene is a molecular substance. Do you think it will conduct electricity when molten? Explain.
5 What is: **a** an electrolyte?   **b** a non-electrolyte?
   Give *three* examples of each.

ELECTRICITY AND CHEMICAL CHANGE

## 8.2 The principles of electrolysis

### Electrolysis: breaking down by electricity
**Any liquid that contains ions will conduct electricity.**
This is because the ions are free to move. But at the same time, decomposition takes place.

So you can use electricity to break down a substance. The process is called **electrolysis**.

### The electrolysis of molten lead bromide
The diagram on the right shows the apparatus.

- The graphite rods are called **electrodes**.
- The electrode attached to the positive terminal of the battery is also positive. It is called the **anode**.
- The negative electrode is called the **cathode**.

The molten lead bromide contains lead ions ($Pb^{2+}$) and bromide ions ($Br^-$). This shows what happens when the switch is closed:

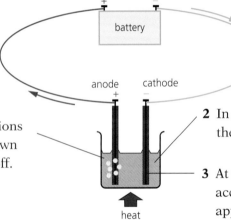

1 Electrons flow from the negative terminal of the battery to the cathode.

5 Electrons flow from the anode to the positive terminal of the battery.

2 In the liquid, the ions move to the electrode of opposite charge.

4 At the anode (+), the $Br^-$ ions give up electrons. Red-brown bromine vapour bubbles off.

3 At the cathode (−), the $Pb^{2+}$ ions accept electrons. Lead begins to appear below the cathode.

The result is that the lead bromide has **decomposed**:

lead bromide ⟶ lead + bromine
$PbBr_2\ (l) \longrightarrow Pb\ (l) + Br_2\ (g)$

Note that:
- Electrons carry the current through the wires and electrodes. But the ions carry it through the liquid.
- The graphite electrodes are inert. They carry the current into the liquid, but remain unchanged. (Electrodes made of platinum are also inert.)

**Which electrode is positive?**
Remember **PA**!

**P**ositive **A**node.

### The electrolysis of other molten compounds
The pattern is the same for all molten ionic compounds of two elements:
**Electrolysis breaks the molten ionic compound down to its elements, giving the metal at the cathode, and the non-metal at the anode.**

So it is a very important process. We depend on it to obtain reactive metals such as lithium, sodium, potassium, magnesium, and aluminium, from compounds dug from the Earth.

**Obtaining aluminium**
Find out how electrolysis is used to extract aluminium, on page 201.

# ELECTRICITY AND CHEMICAL CHANGE

## The electrolysis of aqueous solutions

Electrolysis can also be carried out on solutions of ionic compounds in water, because the ions in solutions are free to move. But the result may be different than for the molten compound. Compare these:

| Electrolyte | At the cathode (−) you get … | At the anode (+) you get … |
| --- | --- | --- |
| molten sodium chloride | sodium | chlorine |
| a concentrated solution of sodium chloride | hydrogen | chlorine |

Why the difference? *Because the water itself produces ions*. Although water is molecular, a tiny % of its molecules is split up into ions:

some water molecules ⟶ hydrogen ions + hydroxide ions
$$H_2O\,(l) \longrightarrow H^+\,(aq) + OH^-\,(aq)$$

These ions also take part in the electrolysis, so the products may change.

**Order of reactivity**
potassium
sodium
calcium
magnesium
aluminium
zinc
iron
lead
hydrogen
copper
silver

↑ increasing reactivity

## The rules for the electrolysis of a solution

### At the cathode (−), either a metal or hydrogen forms.

1  The more reactive an element, the more it 'likes' to exist as ions. So if a metal is more reactive than hydrogen, its ions stay in solution and hydrogen bubbles off. (Look at the list on the right.)
2  But if the metal is less reactive than hydrogen, the metal forms.

### At the anode (+), a non-metal other than hydrogen forms.

1  If it is a concentrated solution of a **halide** (a compound containing $Cl^-$, $Br^-$ or $I^-$ ions), then chlorine, bromine, or iodine form.
2  But if the halide solution is dilute, or there is no halide, oxygen forms.

Look at these examples. Do they follow the rules?

| Electrolyte | At the cathode (−) you get… | At the anode (+) you get … |
| --- | --- | --- |
| a concentrated solution of potassium bromide, KBr | hydrogen | bromine |
| a concentrated solution of silver nitrate, $AgNO_3$ | silver | oxygen |
| concentrated hydrochloric acid, HCl | hydrogen ($H^+$ is the only positive ion present) | chlorine |
| a dilute solution of sodium chloride, NaCl | hydrogen | oxygen |

Notice that, in the last example, the water has been decomposed!

**Q**

1  a  Which type of compounds can be electrolysed? Why?
   b  What form must they be in?
2  What does electrolysis of these molten compounds give?
   a  sodium chloride, NaCl
   b  aluminium oxide, $Al_2O_3$
   c  calcium fluoride, $CaF_2$
   d  lead sulfide, PbS
3  Name the products at each electrode, when these aqueous solutions are electrolysed using inert electrodes:
   a  a concentrated solution of magnesium chloride, $MgCl_2$
   b  concentrated hydrochloric acid, HCl
   c  a dilute solution of copper(II) sulfate, $CuSO_4$

ELECTRICITY AND CHEMICAL CHANGE

## 8.3 The reactions at the electrodes

### What happens to ions in the molten lead bromide?
In molten lead bromide, the ions are free to move. This shows what happens to them, when the switch in the circuit is closed:

| **First, the ions move.** | **At the cathode (−):** | **At the anode (+):** |
|---|---|---|
| Opposite charges attract. So the positive lead ions ($Pb^{2+}$) move to the cathode (−). The negative bromide ions ($Br^-$) move to the anode (+). The moving ions carry the current. | the lead ions each receive two electrons and become lead atoms. The **half-equation** is: $Pb^{2+}(l) + 2e^- \longrightarrow Pb(l)$ Lead collects on the electrode and eventually drops off it. | the bromide ions each give up an electron, and become atoms. These then pair up to form molecules. The **half-equation** is: $2Br^-(l) \longrightarrow Br_2(g) + 2e^-$ The bromine gas bubbles off. |
| The free ions move. | Ions gain electrons: **reduction**. | Ions lose electrons: **oxidation**. |
| Remember **OILRIG**: **O**xidation **I**s **L**oss of electrons, **R**eduction **I**s **G**ain of electrons. | Overall, electrolysis is a redox reaction. Reduction takes place at the cathode and oxidation at the anode. ||

The reactions for other molten compounds follow the same pattern.

### For a concentrated solution of sodium chloride

> **Remember RAC!**
> **R**eduction **A**t **C**athode.

This time, ions from water are also present:

The solution contains $Na^+$ ions and $Cl^-$ ions from the salt, and $H^+$ and $OH^-$ ions from water.
The positive ions go to the cathode and the negative ions to the anode.

At the cathode, the $H^+$ ions accept electrons, since hydrogen is less reactive than sodium:
$2H^+(aq) + 2e^- \longrightarrow H_2(g)$
The hydrogen gas bubbles off.

At the anode, the $Cl^-$ ions give up electrons more readily than the $OH^-$ ions do.
$2Cl^-(aq) \longrightarrow Cl_2(aq) + 2e^-$
The chlorine gas bubbles off.

When the hydrogen and chlorine bubble off, $Na^+$ and $OH^-$ ions are left behind – so a solution of sodium hydroxide is formed.

ELECTRICITY AND CHEMICAL CHANGE

## For a dilute solution of sodium chloride

  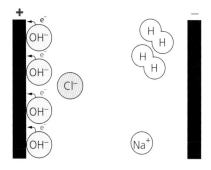

The same ions are present as before. But now the proportion of Na⁺ and Cl⁻ ions is lower, since this is a dilute solution. So the result will be different.

At the cathode, hydrogen 'wins' as before, and bubbles off:

$$4H^+ (aq) + 4e^- \longrightarrow 2H_2 (g)$$

(4 electrons are shown, to balance the half-equation at the anode.)

At the anode, OH⁻ ions give up electrons, since not many Cl⁻ ions are present. Oxygen bubbles off:

$$4OH^- (aq) \longrightarrow$$
$$O_2 (g) + 2H_2O (l) + 4e^-$$

When the hydrogen and oxygen bubble off, the Na⁺ and Cl⁻ ions are left behind. So we *still* have a solution of sodium chloride!
**The overall result is that water has been decomposed.**

## Writing the half-equations for electrode reactions

You may be asked to write half-equations for the reactions at electrodes. This table shows the steps.

| The steps | Example: the electrolysis of molten magnesium chloride |
|---|---|
| **1** First, name the ions present, and the products. | Magnesium ions and chloride ions are present. Magnesium and chlorine form. |
| **2** Write each half-equation correctly.<br>• Give the ion its correct charge.<br>• Remember, positive ions go to the cathode, and negative ions to the anode.<br>• Write the correct symbol for the element that forms. For example, $Cl_2$ for chlorine (not Cl).<br>• The number of electrons in the equation should be the same as the total charge on the ion(s) in it. | Ions: $Mg^{2+}$ and $Cl^-$<br><br>**At the cathode:**<br>$Mg^{2+} + 2e^- \longrightarrow Mg$<br><br>**At the anode:**<br>$2Cl^- \longrightarrow Cl_2 + 2e^-$ (two Cl⁻ ions, so a total charge of 2 −)<br>Note that it is also correct to write the anode reaction as:<br>$2Cl^- - 2e^- \longrightarrow Cl_2$ |
| **3** You could then add the state symbols. | $Mg^{2+} (l) + 2e^- \longrightarrow Mg (l)$<br>$2Cl^- (l) \longrightarrow Cl_2 (g) + 2e^-$ |

**1** At which electrode does reduction always take place?
**2** Give the half-equation for the reaction at the anode, during the electrolysis of these molten compounds:
   **a** potassium chloride   **b** calcium oxide
**3** Give the two half-equations for the electrolysis of:
   **a** a concentrated solution of hydrochloric acid, HCl
   **b** a dilute solution of sodium nitrate, $NaNO_3$
   **c** a dilute solution of copper(II) chloride, $CuCl_2$

ELECTRICITY AND CHEMICAL CHANGE

## 8.4 The electrolysis of brine

### What is brine?
**Brine** is a concentrated solution of sodium chloride, or common salt. It can be obtained by pumping water into salt mines to dissolve the salt, or by evaporating seawater.

Brine might not sound very exciting – but from it, we get chemicals needed for thousands of products we use every day. When it undergoes electrolysis, the overall reaction is:

$$2NaCl\ (aq) + 2H_2O\ (l) \xrightarrow{\text{electrolysis}} 2NaOH\ (aq) + Cl_2\ (g) + H_2\ (g)$$
brine    sodium hydroxide   chlorine   hydrogen

### The electrolysis
The diagram below shows one type of cell used for this electrolysis. The anode is made of titanium, and the cathode of steel. Now look at the diaphragm down the middle of the cell. Its function is to let ions through, but keep the gases apart. (So the cell is called a **diaphragm cell**.)

▲ Inside a salt mine. Many countries have underground salt beds. They were deposited millions of years ago, when the sea drained away from the land.

The ions present are $Na^+$ and $Cl^-$ from the salt, and $H^+$ and $OH^-$ from the water. The reactions at the electrodes are exactly as shown at the bottom of page 106. (Look back at them.)

**At the cathode**   Hydrogen is discharged in preference to sodium:
$$2H^+\ (aq) + 2e^- \longrightarrow H_2\ (g)$$
As usual at the cathode, this is a reduction.

**At the anode**   Chlorine is discharged in preference to oxygen:
$$2Cl^-\ (aq) \longrightarrow Cl_2\ (g) + 2e^-$$
As usual at the anode, this is an oxidation.

The two gases bubble off. $Na^+$ and $OH^-$ ions are left behind, giving a solution of sodium hydroxide. Some of the solution is evaporated to a give a more concentrated solution, and some is evaporated to dryness, giving solid sodium hydroxide.

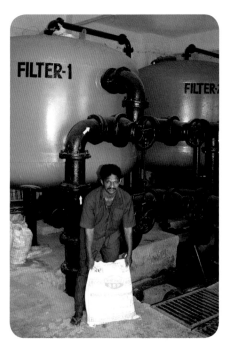

▲ Chlorine has many uses. One is to kill germs in water. Behind the scenes at a swimming pool, this man makes sure there is chlorine in the water.

# ELECTRICITY AND CHEMICAL CHANGE

## What the products are used for

The electrolysis of brine is an important process, because the products are so useful. Look at these:

### Chlorine, a poisonous yellow-green gas
Used for making ...
- the plastic PVC (nearly 1/3 of it used for this)
- solvents for degreasing and drycleaning
- medical drugs (a large % of these involve chlorine)
- weedkillers and pesticides (most of these involve chlorine)
- paints and dyestuffs
- bleaches
- hydrogen chloride and hydrochloric acid

It is also used as a sterilising agent, to kill bacteria in water supplies and swimming pools.

### Sodium hydroxide solution, alkaline and corrosive
Used in making ...
- soaps
- detergents
- viscose (rayon) and other textiles
- paper (like the paper in this book)
- ceramics (tiles, furnace bricks, and so on)
- dyes
- medical drugs

### Hydrogen, a colourless flammable gas
Used ...
- in making nylon
- to make hydrogen peroxide
- to 'harden' vegetable oils to make margarine
- as a fuel in hydrogen fuel cells

Of the three chemicals, chlorine is the most widely used. Around 50 million tonnes of it are produced each year, around the world.

▲ All three products from the electrolysis of brine must be transported with care. Why?

▲ Some hydrogen goes to hydrogen filling stations, for cars with hydrogen fuel cells instead of petrol engines (page 121).

## Q

1. What is brine? Where is it obtained from?
2. Write a word equation for the electrolysis of brine.
3. Draw a rough sketch of the diaphragm cell. Mark in where the oxidation and reduction reactions take place in it, and write the half-equations for them.
4. What is the diaphragm for, in the diaphragm cell?
5. The electrolysis of brine is a very important process.
   a  Explain why.
   b  Give three uses for each of the products.
6. Your job is to keep a brine electrolysis plant running safely and smoothly. Try to think of three or four safety precautions you might need to take.

ELECTRICITY AND CHEMICAL CHANGE

## 8.5 Two more uses of electrolysis

### When electrodes are not inert

A solution of copper(II) sulfate contains blue $Cu^{2+}$ ions, $SO_4^{2-}$ ions, and $H^+$ and $OH^-$ ions from water. Electrolysis of the solution will give different results, *depending on the electrodes*. Compare these:

**A Using inert electrodes (carbon or platinum)**

**At the cathode**  Copper ions are discharged:

$2Cu^{2+} (aq) + 4e^- \longrightarrow 2Cu (s)$

The copper coats the electrode.

**At the anode**  Oxygen bubbles off:

$4OH^- (aq) \longrightarrow 2H_2O (l) + O_2 (g) + 4e^-$

**So copper and oxygen are produced.** This fits the rules on page 105. The blue colour of the solution fades as the copper ions are discharged.

**B Using copper electrodes**

**At the cathode**  Again, copper is formed, and coats the electrode:

$Cu^{2+} (aq) + 2e^- \longrightarrow Cu (s)$

**At the anode**  The anode dissolves, giving copper ions in solution:

$Cu (s) \longrightarrow Cu^{2+} (aq) + 2e^-$

So this time, the electrodes are not inert. The anode dissolves, giving copper ions. These move to the cathode, to form copper. So *copper moves from the anode to the cathode*. The colour of the solution does not fade.

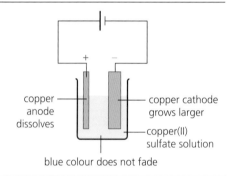

The idea in B leads to two important uses of electrolysis: for **refining** (or purifying) copper, and for **electroplating**.

### Refining copper

The anode is made of impure copper. The cathode is pure copper. The electrolyte is dilute copper(II) sulfate solution.

The copper in the anode dissolves. But the impurities do not dissolve. They just drop to the floor of the cell as a sludge.

A layer of pure copper builds up on the cathode. When the anode is almost gone, the anode and cathode are replaced.

The copper deposited on the cathode is over 99.9% pure.
The sludge may contain valuable metals such as platinum, gold, silver, and selenium. These are recovered and sold.

# ELECTRICITY AND CHEMICAL CHANGE

▲ The purer it is, the better copper is at conducting electricity. Highly refined copper is used for the electrics in cars. A car like this will contain more than 1 km of copper wiring.

▲ A steel tap plated with chromium, to make it look bright and shiny. Chromium does not stick well to steel. So the steel is first electroplated with copper or nickel, and then chromium.

## Electroplating

**Electroplating** means using electricity to coat one metal with another, to make it look better, or to prevent corrosion. For example, steel car bumpers are coated with chromium. Steel cans are coated with tin to make tins for food. And cheap metal jewellery is often coated with silver.

The drawing on the right shows how to electroplate a steel jug with silver. The jug is used as the cathode. The anode is made of silver. The electrolyte is a solution of a soluble silver compound, such as silver nitrate.

**At the anode**   The silver dissolves, forming silver ions in solution:

$$Ag\,(s) \longrightarrow Ag^+\,(aq) + e^-$$

**At the cathode**   The silver ions are attracted to the cathode. There they receive electrons, forming a coat of silver on the jug:

$$Ag^+\,(aq) + e^- \longrightarrow Ag\,(s)$$

When the layer of silver is thick enough, the jug is removed.

▲ Silverplating: electroplating with silver. When the electrodes are connected to a power source, electroplating begins.

> **To electroplate**
> In general, to electroplate an object with metal X, the set-up is:
> **cathode**   –   object to be electroplated
> **anode**   –   metal X
> **electrolyte**   –   a solution of a soluble compound of X.

### Q

1 Copper(II) ions are blue. When copper(II) sulfate solution is electrolysed, the blue solution:
  a loses its colour when carbon electrodes are used
  b keeps its colour when copper electrodes are used.
  Explain each of these observations.
2 If you want to purify a metal by electrolysis, will you make it the anode or the cathode? Why?
3 Describe the process of refining copper.
4 What does *electroplating* mean?
5 Steel cutlery is often electroplated with nickel. Why?
6 You plan to electroplate steel cutlery with nickel.
  a What will you use as the anode?
  b What will you use as the cathode?
  c Suggest a suitable electrolyte.

# ELECTRICITY AND CHEMICAL CHANGE

## Checkup on Chapter 8

### Revision checklist

**Core curriculum**

Make sure you can ...
- define the terms *conductor* and *insulator*
- give examples of how we make use of conductors and insulators
- explain what these terms mean:
  *electrolysis*   *electrolyte*   *electrode*
  *inert electrode*   *anode*   *cathode*
- explain why an ionic compound must be melted, or dissolved in water, for electrolysis
- predict what will be obtained at each electrode, in the electrolysis of a molten ionic compound
- say what *halides* are
- say why the products of electrolysis may be different, when a compound is dissolved in water, rather than melted
- give the general rules for the products at the anode and cathode, in the electrolysis of a solution
- name the product at each electrode, for the electrolysis of:
  - concentrated hydrochloric acid
  - a concentrated solution of sodium chloride
- explain what *electroplating* is, and why it is used
- describe how electroplating is carried out

**Extended curriculum**

Make sure you can also ...
- predict the products, for the electrolysis of halides in dilute and concentrated solutions
- describe the reactions at the electrodes, during the electrolysis of:
  - a molten halide such as lead bromide
  - a dilute solution of a halide such as sodium chloride
  - a concentrated solution of a halide
  and write half-equations for them
- describe the electrolysis of brine, and name the three products, and give some uses for them (you will not be asked for a diagram of the cell)
- describe the differences, when the electrolysis of copper(II) sulfate is carried out:
  - using inert electrodes (carbon or platinum)
  - using copper electrodes
- describe how electrolysis is used to refine impure copper, and say why this is important

### Questions

**Core curriculum**

1 Electrolysis of molten lead bromide is carried out:

a The bulb will not light until the lead bromide has melted. Why not?
b What will be *seen* at the anode?
c Name the substance in **b**.
d What will be formed at the cathode?

2 Six substances A to F were dissolved in water, and connected in turn into the circuit below.
**A** represents an ammeter, which is used to measure current. The table shows the results.

| Substance | Current (amperes) | At cathode (−) | At anode (+) |
|---|---|---|---|
| A | 0.8 | copper | chlorine |
| B | 1.0 | hydrogen | chlorine |
| C | 0.0 | — | — |
| D | 0.8 | copper | oxygen |
| E | 1.2 | hydrogen | oxygen |
| F | 0.7 | silver | oxygen |

a Which solution conducts best?
b Which solution is a non-electrolyte?
c Which solution could be:
  i silver nitrate?   ii copper(II) sulfate?
  iii copper(II) chloride?   iv sodium hydroxide?
  v sugar?   vi concentrated hydrochloric acid?
d Explain how the current is carried:
  i *within* the electrolytes
  ii in the rest of the circuit

112

## Extended curriculum

**3** The electrolysis below produces gases A and B.

**a** Why does the solution conduct electricity?
**b** Identify each gas, and describe a test you could carry out to confirm its identity.
**c** Name one product manufactured from:
  **i** gas A  **ii** gas B
**d i** Write half-equations to show how the two gases are produced.
  **ii** The overall reaction is a *redox* reaction. Explain why.
**e** The solution remaining after the electrolysis will turn litmus paper blue.
  **i** What is the name of this solution?
  **ii** State one chemical property for it.

**4 a** List the ions that are present in concentrated solutions of:
  **i** sodium chloride  **ii** copper(II) chloride
**b** Explain why and how the ions move, when each solution is electrolysed using platinum electrodes.
**c** Write the half-equation for the reaction at:
  **i** the anode  **ii** the cathode
  during the electrolysis of each solution.
**d** Explain why the anode reactions for both solutions are the same.
**e i** The anode reactions will be different if the solutions are made very dilute. Explain why.
  **ii** Write the half-equations for the new anode reactions.
**f** Explain why copper is obtained at the cathode, but sodium is not.
**g** Name another solution that will give the same products as the concentrated solution of sodium chloride does, on electrolysis.
**h** Which solution in **a** could be the electrolyte in an electroplating experiment?

**5** Molten lithium chloride contains lithium ions ($Li^+$) and chloride ions ($Cl^-$).
  **a** Copy the following diagram and use arrows to show which way:
    **i** the ions move when the switch is closed
    **ii** the electrons flow in the wires

  **b i** Write equations for the reaction at each electrode, and the overall reaction.
    **ii** Describe each of the reactions using the terms *reduction*, *oxidation* and *redox*.

**6** This question is about the electrolysis of a dilute aqueous solution of lithium chloride.
  **a** Give the names and symbols of the ions present.
  **b** Say what will be formed, and write a half-equation for the reaction:
    **i** at the anode  **ii** at the cathode
  **c** Name another compound that will give the same products at the electrodes.
  **d** How will the products change, if a concentrated solution of lithium chloride is used?

**7** An experiment is needed, to see if an iron object can be electroplated with chromium.
  **a** Suggest a solution to use as the electrolyte.
  **b i** Draw a labelled diagram of the apparatus that could be used for the electroplating.
    **ii** Show how the *electrons* will travel from one electrode to the other.
  **c** Write half-equations for the reactions at each electrode.
  **d** At which electrode does oxidation take place?
  **e** The concentration of the solution does not change. Why not?

**8** Nickel(II) sulfate ($NiSO_4$) is green. A solution of this salt is electrolysed using nickel electrodes.
  **a** Write a half-equation for the reaction at each electrode.
  **b** At which electrode does reduction take place? Explain your answer
  **c** What happens to the size of the anode?
  **d** The colour of the solution does not change, during the electrolysis. Explain why.
  **e** Suggest one industrial use for this electrolysis.

ENERGY CHANGES, AND REVERSIBLE REACTIONS

## 9.1 Energy changes in reactions

### Energy changes in reactions
**During a chemical reaction, there is always an energy change.**

Energy is given out or taken in. The energy is usually in the form of heat. (But some may be in the form of light and sound.)
So reactions can be divided into two groups: exothermic and endothermic.

### Exothermic reactions
**Exothermic reactions give out energy. So there is a temperature rise.**
Here are three examples:

To start off the reaction between iron and sulfur, you must heat the mixture. But soon it glows red hot – *without* the Bunsen burner!

Mixing silver nitrate and sodium chloride solutions gives a white precipitate of silver chloride – and a temperature rise.

When you add water to lime (calcium oxide) heat is given out, so the temperature rises. Here the rise is being measured.

These reactions can be described as:

**reactants ⟶ products + energy**

The total energy is the same on each side of the arrow, in a reaction. So in exothermic reactions, the products have lower energy than the reactants. This is shown on the **energy level diagram** on the right.

### The energy change
Energy is measured in kilojoule (kJ). For reaction **A** above:

Fe $(s)$ + S $(s)$ ⟶ FeS $(s)$     the energy change = $-100$ kJ

So 100 kJ of energy is given out when the amounts of reactants in the equation (56 g of iron and 32 g of sulfur, or 1 mole of each) react together.
**The minus sign shows that energy is given out.**

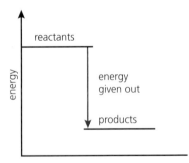

▲ An energy level diagram for an exothermic reaction. The products have lower energy than the reactants.

### Other examples of exothermic reactions
All these are exothermic:
- the neutralisation of acids by alkalis.
- the combustion of fuels. We burn fuels to obtain heat for cooking, heating homes, and so on. The more energy they give out, the better!
- respiration in your body cells. It provides the energy to keep your heart and lungs working, and for warmth and movement.

# Endothermic reactions

**Endothermic reactions take in energy from their surroundings.**
Here are three examples:

When barium hydroxide reacts with ammonium chloride, the temperature falls so sharply that water *under* the beaker will freeze!

Sherbet is citric acid plus the base sodium hydrogen carbonate. The neutralisation that occurs takes in heat – so your tongue cools.

The crucible contains calcium carbonate. If you keep on heating, it will all decompose to calcium oxide and carbon dioxide.

These reactions can be described as:

**reactants + energy ⟶ products**

The energy is transferred from the surroundings: in **D** from the air and wet wood, in **E** from your tongue, and in **F** from the Bunsen burner. Since energy is taken in, the products must have higher energy than the reactants. This is shown on the energy level diagram on the right.

## The energy change

For reaction **F** above:

$CaCO_3 (s) \longrightarrow CaO (s) + CO_2 (g)$   the energy change = + 178 kJ

So 178 kJ of energy is needed to make 100 g (or 1 mole) of $CaCO_3$ decompose. **The plus sign shows that energy is taken in.**

▲ An energy level diagram for an endothermic reaction. The products have higher energy than the reactants.

## Other examples of endothermic reactions

Reactions **D** and **E** above are **spontaneous**. They start off on their own. But many endothermic reactions are like **F**, where energy must be put in to start the reaction *and* keep it going. For example:

- reactions that take place in cooking.
- photosynthesis. This is the process in which plants convert carbon dioxide and water to glucose. It depends on the energy from sunlight.

**Remember!**
*Exo* means *out* (think of Exit)
*Endo* means *in*

---

**Q**

1. Is it exothermic or endothermic?
   a the burning of a candle
   b the reaction between sodium and water
   c the change from raw egg to fried egg
2. Which unit is used to measure energy changes?
3. $2Na (s) + Cl_2 (g) \longrightarrow 2NaCl (s)$
   The energy change for this reaction is − 822.4 kJ.
   What can you conclude about the reaction?
4. Draw an energy level diagram for:
   a an endothermic reaction   b an exothermic reaction

# ENERGY CHANGES, AND REVERSIBLE REACTIONS

## 9.2 Explaining energy changes

### Making and breaking bonds

In a chemical reaction, bonds must first be broken. Then new bonds form. **Breaking bonds takes in energy. Making bonds releases energy.**

**Example 1: an exothermic reaction**
Hydrogen reacts with chlorine in sunshine, to form hydrogen chloride:

▲ Hydrogen burning in chlorine in the lab. Bonds break and new bonds form, giving hydrogen chloride.

**1** First, the bonds in the hydrogen and chlorine molecules must be broken. Energy must be taken in, for this. (Energy from sunshine will do!)

**2** Now new bonds form between hydrogen and chlorine atoms, giving molecules of hydrogen chloride. This step releases energy.

But the energy taken in for step 1 is less than the energy given out in step 2. So this reaction gives out energy, overall. It is exothermic.

**If the energy taken in to break bonds is *less than* the energy released in making bonds, the reaction is exothermic.**

**Example 2: an endothermic reaction**
If you heat ammonia strongly, it breaks down to nitrogen and hydrogen. Here we use lines to show the bonds. (Note the triple bond in nitrogen.)

**1** First, the bonds in ammonia must be broken. Energy must be taken in, for this. (You supply it by heating.)

**2** Now the hydrogen atoms bond together. So do the nitrogen atoms. This releases energy.

This time, the energy taken in for step **1** is greater than the energy given out in step **2**. So the reaction takes in energy, overall. It is **endothermic**.

**If the energy taken in to break bonds is *greater than* the energy released in making bonds, the reaction is endothermic.**

### Bond energies

The energy needed to make or break bonds is called the **bond energy**. Look at the list on the right. 242 kJ must be supplied to break the bonds in a mole of chlorine molecules, to give chlorine atoms. If these atoms join again to form molecules, 242 kJ of energy are given out.

**The bond energy is the energy needed to break bonds, or released when these form. It is given in kJ/mole.**

| Bond energy (kJ/mole) | |
|---|---|
| H–H | 436 |
| Cl–Cl | 242 |
| H–Cl | 431 |
| C–C | 346 |
| C=C | 612 |
| C–O | 358 |
| C–H | 413 |
| O=O | 498 |
| O–H | 464 |
| N≡N | 946 |
| N–H | 391 |

# ENERGY CHANGES, AND REVERSIBLE REACTIONS

## Calculating the energy changes in reactions

So let's calculate the energy change for those reactions on page 116.

> **Calculating energy changes**
> The calculation is always:
> **energy in – energy out**

**1** The exothermic reaction between hydrogen and chlorine

$$H-H + Cl-Cl \longrightarrow 2\,H-Cl$$

**Energy in** to break each mole of bonds:
| | |
|---|---|
| 1 × H—H | 436 kJ |
| 1 × Cl—Cl | 242 kJ |
| Total energy in | 678 kJ |

**Energy out** from the two moles of bonds forming:
| | |
|---|---|
| 2 × H—Cl | 2 × 431 = 862 kJ |

**Energy in – energy out = 678 kJ – 862 kJ = –184 kJ**
So the reaction gives out **184 kJ** of energy, overall.
Its energy level diagram is shown on the right.

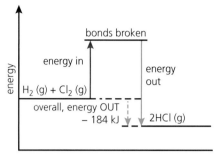

▲ For the hydrogen/chlorine reaction.

**2** The endothermic decomposition of ammonia

$$2\ \begin{array}{c}H\\|\\N-H\\|\\H\end{array} \longrightarrow N\equiv N + 3\,H-H$$

**Energy in** to break the two moles of bonds:
| | |
|---|---|
| 6 × N—H | 6 × 391 = 2346 kJ |

**Energy out** from the four moles of bonds forming:
| | | |
|---|---|---|
| 1 × N≡N | | 946 kJ |
| 3 × H—H | 3 × 436 = | 1308 kJ |
| Total energy out | | 2254 kJ |

**Energy in – energy out = 2346 kJ – 2254 kJ = +92 kJ**
So the reaction takes in **92 kJ** of energy, overall.
Look at its energy level diagram.

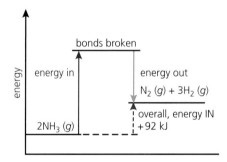

▲ For the decomposition of ammonia.

## Starting a reaction off

To start a reaction, bonds must be broken. As you saw, this needs energy.

- For some reactions, not much energy is needed. Just mix the reactants at room temperature. (For example, reactions **B** and **C** on page 114.)
- Some exothermic reactions need heat from a Bunsen burner just to start bonds breaking. Then the energy given out by the reaction breaks further bonds. (For example, reaction **A** on page 114.)
- But for endothermic reactions like the decomposition of calcium carbonate (reaction **F** on page 115), you must continue heating until the reaction is complete.

▲ One way to get those bonds breaking!

**Q**

**1** Two steps must take place, to go from reactants to products. What are they?

**2** Some reactions are endothermic. Explain why, using the ideas of bond breaking and bond making.

**3** Hydrogen reacts with oxygen. Draw the equation for the reaction as above, with lines to show the bonds.

**4** Now see if you can calculate the energy change for the reaction in **3**, using the bond energy table on page 116.

ENERGY CHANGES, AND REVERSIBLE REACTIONS

## 9.3 Energy from fuels

### What is a fuel?
**A fuel is any substance we use to provide energy.**
We convert the chemical energy in the fuel into another form of energy. We burn most fuels, to obtain their energy in the form of heat.

### The fossil fuels
The **fossil fuels** – **coal, petroleum (oil),** and **natural gas** (methane) – are the main fuels used around the world. We burn them to release heat.

We burn fossil fuels in power stations, to heat water to make steam. A jet of steam drives the turbines that generate electricity.

We burn them in factories to heat furnaces, and in homes for cooking and heating. (Kerosene, from petroleum, is also used in lamps.)

Petrol and diesel (from petroleum) are burned in engines, to give the hot gas that moves the pistons. These then make the wheels turn.

The world uses up enormous quantities of the fossil fuels. For example, nearly 12 million tonnes of petroleum *every day*!

### So what makes a good fuel?
These are the main questions to ask about a fuel:

- **How much heat does it give out?** We want as much heat as possible, per tonne of fuel.
- **Does it cause pollution?** If it causes a lot of pollution, we may be better off without it!
- **Is it easily available?** We need a steady and reliable supply.
- **Is it easy and safe to store and transport?** Most fuels catch fire quite easily, so safety is always an issue.
- **How much does it cost?** The cheaper the better.

The fossil fuels give out a lot of heat. But they cause pollution, with coal the worst culprit. The pollutants include carbon dioxide, which is linked to global warming, and other gases that cause acid rain. (See page 214.)

What about availability? We are using up the fossil fuels fast. Some experts say we could run out of petroleum and gas within 50 years. But there is probably enough coal to last several hundred years.

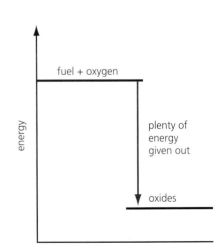

▲ The burning of fuel is an exothermic reaction. The more heat given out the better – as long as the fuel is safe to use.

# ENERGY CHANGES, AND REVERSIBLE REACTIONS

## Two fuels growing in importance

Because of fears about global warming, and dwindling supplies of petroleum and gas, there is a push to use new fuels. Like these two:

**Ethanol** This is an alcohol, with the formula $C_2H_5OH$. It can be made from any plant material. For example, it is made from sugar cane in Brazil, and from corn (maize) in the USA. It is used in car engines, on its own or mixed with petrol. See pages 256 – 257 for more.

**Hydrogen** This gas burns explosively in oxygen, giving out a lot of energy – so it is used to fuel space rockets. It is also used in **fuel cells** (without burning) to give energy in the form of electricity. See page 121 for more.

▲ Filling up with a mixture of 85% ethanol, 15% gasoline.

## Different amounts of heat

Some fuels give out a lot more heat than others. Compare these:

| Fuel | Equation for burning in oxygen | Heat given out per gram of fuel/kJ |
|---|---|---|
| natural gas (methane) | $CH_4 (g) + 2O_2 (g) \rightarrow CO_2 (g) + 2H_2O (l)$ | – 55 |
| ethanol | $C_2H_5OH (l) + 3O_2 (g) \rightarrow 2CO_2 (g) + 3H_2O (l)$ | – 86 |
| hydrogen | $2H_2 (g) + O_2 (g) \rightarrow 2H_2O (g)$ | – 143 |

## Nuclear fuels

**Nuclear fuels** are *not* burned. They contain unstable atoms called **radioisotopes** (page 34). Over time, these break down naturally into new atoms, giving out radiation and a lot of energy.

But you can also *force* radioisotopes to break down, by shooting neutrons at them. That is what happens in a **nuclear power station**. The energy given out is used to heat water, to make jets of steam to drives the turbines for generating electricity.

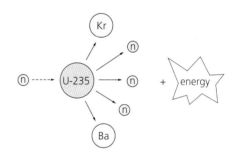

▲ When hit by a neutron, a U-235 atom breaks down to other atoms, giving out a huge amount of energy.

The radioisotope uranium-235 is commonly used in nuclear fuels. When it decays, the new atoms that form are also unstable, and break down further.

Nuclear fuel has two big advantages:

- It gives out huge amounts of energy. A pellet of nuclear fuel the size of a pea can give as much energy as a tonne of coal.
- No carbon dioxide or other polluting gases are formed.

But it is not all good news. An explosion in a nuclear power station could spread radioactive material over a huge area, carried in the wind. The waste material produced in a nuclear power station is also radioactive, and may remain very dangerous for hundreds of years. Finding a place to store it safely is a major problem.

▲ The radiation hazard warning sign.

**1 a** Sketch an energy level diagram that you think shows:
  **i** a good fuel   **ii** a very poor fuel
**b** What else do you need to think about, to decide whether a substance would make a good fuel?

**2** Look at the table above. From *all* the information given, which of the three fuels do you think is best? Explain.

**3** The fuel butane ($C_4H_{10}$) burns to give the same products as methane. Write a balanced equation for its combustion.

ENERGY CHANGES, AND REVERSIBLE REACTIONS

# 9.4 Giving out energy as electricity

## Electricity: a form of energy

**Electricity** is a current of electrons. Like heat, it is a form of energy. When you burn a fuel, chemical energy is converted to heat. But a reaction can also give out energy as electricity.

## Electricity from a redox reaction

Connect a strip of magnesium, a strip of copper, and a light bulb, like this. (Note: no battery!) Nothing happens.

Now stand the strips in a dilute solution of sodium chloride. Something amazing happens: the bulb lights! A current is flowing.

At the same time bubbles of hydrogen start to form on the copper strip, and the magnesium strip begins dissolving.

So what is going on?

1. Magnesium is more reactive than copper. (See the list on the right.) That means it has a stronger drive to form ions. So the magnesium atoms give up electrons, and go into solution as ions:

   $Mg(s) \longrightarrow Mg^{2+}(aq) + 2e^-$     (magnesium is oxidised)

2. The electrons flow along the wire to the copper strip, as a current.

3. The solution contains $Na^+$ and $Cl^-$ ions from sodium chloride, and some $H^+$ and $OH^-$ ions from water. Hydrogen is less reactive than sodium, so the $H^+$ ions accept electrons from the copper strip:

   $2H^+(aq) + 2e^- \longrightarrow H_2(g)$     (hydrogen ions are reduced)

**So a redox reaction is giving out energy in the form of a current.**

## A simple cell

The metal strips, wire, and beaker of solution above form a **simple cell**. Electrons flow from the magnesium strip, so it is called the **negative pole**. The copper strip is the **positive pole**. The solution is the **electrolyte**.

**A simple cell consists of two metals and an electrolyte. The more reactive metal is the negative pole of the cell. Electrons flow from it.**

Other metals can also be used, as long as they differ in reactivity. And any solution can be used, as long as it contains ions.

You could connect a voltmeter into the circuit, to measure the voltage. The bigger the difference in reactivity of the metals, the larger the voltage, and the more brightly the bulb will light. Find out more on page 190.

> **Order of reactivity**
> This shows the order of reactivity of some metals compared to hydrogen:
>
> potassium
> sodium
> calcium
> magnesium
> aluminium
> zinc
> iron
> hydrogen
> copper
> silver
>
> increasing reactivity

> **Remember!**
> - In electrolysis, a current brings about a reaction.
> - In simple cells, reactions produce a current.

ENERGY CHANGES, AND REVERSIBLE REACTIONS

## The hydrogen fuel cell

In the hydrogen fuel cell, hydrogen and oxygen combine without burning. It is a redox reaction. The energy is given out as an electric current.

Like the simple cell, the fuel cell has a negative pole that gives out electrons, a positive pole that accepts them, and an electrolyte.

Both poles are made of carbon.

The negative pole is surrounded by hydrogen, and the positive pole by oxygen (in air). The electrolyte contains OH⁻ ions.

| At the negative pole | At the positive pole |
|---|---|
| Hydrogen loses electrons to the OH⁻ ions. It is oxidised: $$2H_2 (g) + 4OH^- (aq) \rightarrow 4H_2O (l) + 4e^-$$ A current of electrons flows through the wire to the positive pole. You can make use of it on the way. For example, pass it through light bulbs to light your home. | The electrons are accepted by oxygen molecules. Oxygen is reduced to OH⁻ ions: $$O_2 (g) + 2H_2O (l) + 4e^- \rightarrow 4OH^- (aq)$$ But the concentration of OH⁻ ions in the electrolyte does not increase. Why not? |

Adding the two half-equations gives the full equation for the redox reaction:
$$2H_2 (g) + O_2 (g) \rightarrow 2H_2O (l)$$

**So the overall reaction is that hydrogen and oxygen combine to form water.**

## Advantages of the hydrogen fuel cell

- Only water is formed. No pollutants!
- The reaction gives out plenty of energy. 1 kg of hydrogen gives about 2.5 times as much energy as 1 kg of natural gas (methane).
- We will not run out of hydrogen. It can be made by the electrolysis of water with a little acid added. Solar power could provide cheap electricity for this. Scientists also hope to make it from waste plant material, using bacteria.

But there is a drawback. Hydrogen is very flammable. A spark or lit match will cause a mixture of hydrogen and air to explode. So it must be stored safely.

▲ This car has a hydrogen fuel cell instead of a petrol engine.

## Q

1. Can you get electricity from a non-redox reaction? Explain.
2. In a simple cell, which metal gives up electrons to produce the current: the more reactive or less reactive one?
3. A wire connects strips of magnesium and copper, standing in an electrolyte. Bubbles appear at the copper strip. Why?
4. You connect two strips of iron using wire, and stand them in an electrolyte. Will a current flow? Explain your answer.
5. a  In the hydrogen fuel cell, what is the *fuel*?
   b  How are the electrons transferred in this cell?
   c  What type of electrolyte is used?

ENERGY CHANGES, AND REVERSIBLE REACTIONS

# The batteries in your life

## Batteries and you

We depend a lot on batteries. Cars and buses will not start without them. Torches need them. So do mobile phones, laptops, cameras, iPods …

The diagram on the right shows a simple model of a battery (or cell). All batteries contain two solid substances of different reactivity, and an electrolyte. The more reactive substance gives up electrons more readily. These flow out of the battery as an electric current.

Since the more reactive substance provides the electrons, it is called the **negative pole**, or **negative electrode**, or **negative terminal**.

The simple cell, shown on page 120, is the simplest battery of all. But it is not very practical. You could not use it in a torch, for example, and it does not have enough voltage to start a car. You need other types of battery.

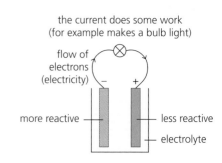

## A torch battery

Torch batteries are 'dry', and easy to carry around:

The metal case is the negative pole. It is usually zinc.

The electrolyte is often sodium or potassium hydroxide, made into a paste that will not leak.

(So these batteries are called alkaline batteries.)

The positive pole is down the middle.

It is often manganese(IV) oxide, packed around a carbon rod. The $Mn^{4+}$ ions accept electrons to become $Mn^{3+}$ ions.

The battery 'dies' when the reactions stop.

▲ Gotcha! Thanks to redox reactions in the torch battery.

## A car battery

A car battery consists of plates of lead and lead(IV) oxide, standing in a solution of sulfuric acid, as shown on the right. This is what happens:

1  The lead plate reacts with the sulfuric acid, giving lead(II) sulfate:

$$Pb\,(s) + H_2SO_4\,(aq) \longrightarrow PbSO_4\,(s) + 2H^+ + 2e^-$$

The lead(II) sulfate coats the plate.

2  The electrons go off through the wire as an electric current. It gets the car's starter motor working.

3  The electrons flow back through the wire to the lead(IV) oxide plate. This also reacts with the acid to form lead(II) sulfate, which coats the plate:

$$PbO_2\,(s) + H_2SO_4\,(aq) + 2H^+ + 2e^- \longrightarrow PbSO_4\,(s) + 2H_2O\,(l)$$

In fact the car battery usually has six sets of plates linked together, giving a total voltage of 12 volts.

▶ A car battery: six sets of linked plates in a plastic container.

## Recharging the car battery

While the car battery is running, the plates are being coated with lead(II) sulfate, and the sulfuric acid is being used up. So if it runs for long enough, the battery will stop working, or 'go flat'.

But it needs to run for only a short time, to start the car. And then something clever happens: electricity generated by the motor causes the reactions to reverse. The lead(II) sulfate on the plates is converted back to lead and lead(IV) oxide, ready for next time.

## A button battery

You probably have a button battery in your watch. Button batteries often use lithium as the negative terminal. Here is a cross-section through one:

▲ Meanwhile, the battery is recharging.

▲ Button batteries come in different sizes, for different uses.

Lithium is a good choice because it is highly reactive: it gives up electrons easily. These flow out through the top of the steel case, to the connection in your watch. They flow back through the lower part of the case, and $Mn^{4+}$ ions accept them, to become $Mn^{3+}$ ions.

## A lithium-ion battery

Lithium-ion batteries are **rechargeable**. So they are used in laptops, mobile phones, and iPods.

The battery consists of thin sheets of lithium cobalt oxide ($LiCoO_2$), and graphite (carbon). The electrolyte is a solution of a lithium salt in an organic solvent. This is how the battery works:

| When it is charging | When you use it |
|---|---|
| When your phone is charging, the graphite becomes negative, and attracts lithium ions from the lithium cobalt oxide. | When you use it, the lithium ions flow back to lithium cobalt oxide, and electrons flow from the graphite to power your phone. |

So your calls and texts depends on those lithium ions moving. Remember that, next time you use your mobile!

▲ Lithium-ion batteries.

▲ Keeping in touch, via lithium ions.

ENERGY CHANGES, AND REVERSIBLE REACTIONS

## 9.5 Reversible reactions

### When you heat copper(II) sulfate crystals ...

The blue crystals above are *hydrated* copper(II) sulphate. On heating, they turn to a white powder. This is *anhydrous* copper(II) sulfate:
$$CuSO_4.5H_2O\ (s) \rightarrow CuSO_4\ (s) + 5H_2O\ (l)$$

The reaction is easy to reverse: add water! The anhydrous copper(II) sulfate gets hot and turns blue. The reaction is:
$$CuSO_4\ (s) + 5H_2O\ (l) \rightarrow CuSO_4.5H_2O\ (s)$$

> **Water of crystallisation**
> - The water in blue copper(II) sulfate crystals is called **water of crystallisation.**
> - *Hydrated* means it has water molecules built into its structure.
> - *Anhydrous* means no water is present.

> **Two tests for water**
> Water will turn:
> - white anhydrous copper(II) sulfate blue
> - blue cobalt chloride paper pink.
>
> Both compounds add on water of crystallisation, giving the colour change. To reverse, just heat!

So the reaction can go in either direction: it is **reversible**.
The reaction we start with (**1** above) is called the **forward** reaction. Reaction **2** is the **back** reaction

We use the symbol $\rightleftharpoons$ instead of a single arrow, to show that a reaction is reversible. So the equation for the reaction above is:

$$CuSO_4.5H_2O\ (s) \rightleftharpoons CuSO_4\ (s) + 5H_2O\ (l)$$

### What about the energy change?
Reaction **1** above requires heat − it is **endothermic**. In **2**, the white powder gets hot and spits when you drip water on it − so that reaction is **exothermic**. It gives out *the same amount of heat* as reaction **1** took in.

**A reversible reaction is endothermic in one direction, and exothermic in the other. The same amount of energy is transferred each time.**

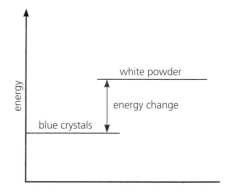

▲ In a reversible reaction, the energy change is the same in both directions.

### Some important reversible reactions
Many important reactions are reversible. Here are some examples:

| Reaction | Comments |
|---|---|
| $N_2\ (g)$ + $3H_2\ (g) \rightleftharpoons 2NH_3\ (g)$<br>nitrogen    hydrogen         ammonia | This is a very important reaction, because ammonia is used to make nitric acid and fertilisers. |
| $2SO_2\ (g)$ + $O_2\ (g) \rightleftharpoons 2SO_3\ (g)$<br>sulfur dioxide   oxygen      sulfur trioxide | This is a key step in the manufacture of sulfuric acid. |
| $CaCO_3\ (s) \rightleftharpoons CaO\ (s) + CO_2\ (g)$<br>calcium carbonate   calcium oxide   carbon dioxide | This is a **thermal decomposition**: it needs heat. Calcium oxide (called lime, or quicklime) has many uses (page 240). |

## Reversible reactions and equilibrium

As you saw in the last table, the reaction between nitrogen and hydrogen to make ammonia is **reversible**:

$$N_2 (g) + 3H_2 (g) \rightleftharpoons 2NH_3 (g)$$

So let's see what happens during the reaction:

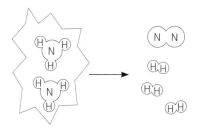

Three molecules of hydrogen react with one of nitrogen to form two of ammonia. So if you put the correct mixture of nitrogen and hydrogen into a closed container …

… will it all turn into ammonia? No! Once a certain amount of ammonia is formed, the system reaches a state of **dynamic equilibrium**. From then on …

… every time two ammonia molecules form, another two break down into nitrogen and hydrogen. So the level of ammonia remains unchanged.

**Equilibrium** means there is *no overall change*. The amounts of nitrogen, hydrogen and ammonia remain steady. But **dynamic** means there is continual change: ammonia molecules continually break down, while new ones form.

**In a closed system, a reversible reaction reaches a state of dynamic equilibrium, where the forward and back reactions take place at the same rate. So there is no overall change.**

The term **dynamic equilibrium** is usually shortened to **equilibrium**.

## A challenge for industry

Imagine you run a factory that makes ammonia. You want the yield of ammonia to be as high as possible.

But the reaction between nitrogen and hydrogen is *never complete*. Once equilibrium is reached, a molecule of ammonia breaks down every time a new one forms.

This is a problem. What can you do to increase the yield of ammonia? You will find out in the next unit.

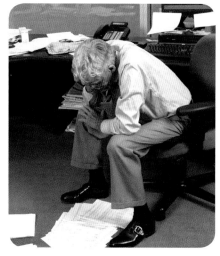

▲ Worried about the yield?

---

**Q**

1. What is a *reversible* reaction?
2. Write a word equation for the reaction between solid copper(II) sulfate and water.
3. How would you turn hydrated copper(II) sulfate into anhydrous copper(II) sulfate?
4. What will you observe if you place pink cobalt chloride paper in warm oven?
5. Explain the term *dynamic equilibrium*.
6. Nitrogen and hydrogen are mixed, to make ammonia.
   a. Soon, two reactions are going on in the mixture. Give the equations for them.
   b. For a time, the rate of the forward reaction is greater than the rate of the back reaction. Has equilibrium been reached? Explain.

ENERGY CHANGES, AND REVERSIBLE REACTIONS

## 9.6 Shifting the equilibrium

### The challenge
Reversible reactions present a challenge to industry, because they *never complete*. Let's look at that reaction between nitrogen and hydrogen again:

$$N_2\,(g) + 3H_2\,(g) \rightleftharpoons 2NH_3\,(g)$$

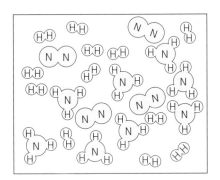

This represents the reaction mixture at equilibrium. The amount of ammonia in it will not increase …

… because every time a new molecule of ammonia forms, another breaks down.

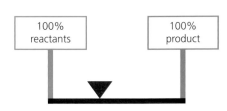

Here the red triangle represents the equilibrium mixture. It is only part way along the scale. Why?

### What can be done?
You want as much ammonia as possible. So how can you increase the yield? This idea, called Le Chatelier's principle, will help you:

**When a reversible reaction is in equilibrium and you make a change, the system acts to oppose the change, and restore equilibrium. A new equilibrium mixture forms.**

A reversible reaction *always* reaches equilibrium, in a closed system. But by changing conditions, you can *shift equilibrium, so* that the mixture contains more product. Let's look at four changes you could make.

▲ A change in reaction conditions has led to a new equilibrium mixture, with more ammonia. *Equilibrium has shifted to the right, to favour the product.*

### 1 Change the temperature
Will raising the temperature help you obtain more ammonia? Let's see.

$$N_2 + 3H_2 \xrightarrow{\text{heat out}} 2NH_3$$
$$2NH_3 \xrightarrow{\text{heat in}} N_2 + 3H_2$$

The forward reaction is exothermic – it gives out heat. The back reaction is endothermic – it takes it in. Heating speeds up *both* reactions …

… but if you heat the equilibrium mixture, it acts to oppose the change. More ammonia breaks down in order to use up the heat you add.

So the reaction reaches equilibrium faster – but the new equilibrium mixture has *less* ammonia. So you are worse off than before.

What if you lower the temperature? The system acts to oppose the change: more ammonia forms, giving out heat. Great! But if the temperature is *too* low, the reaction takes too long to reach equilibrium. Time is money, in a factory. So it is best to choose a moderate temperature.

## 2 Change the pressure

4 molecules → 2 molecules

Pressure is caused by the gas molecules colliding with the walls of the container. *The more molecules present, the higher the pressure.*

When you increase the pressure, the equilibrium mixture acts to oppose this. More ammonia forms, which means fewer molecules.

So the amount of ammonia in the mixture has *increased*. Equilibrium has shifted to the right. Well done. You are on the right track.

### 3 Remove the ammonia
The equilibrium mixture is a balance between nitrogen, hydrogen, and ammonia. Suppose you cool the mixture. Ammonia condenses first, so you can run it off as a liquid. Then warm the remaining nitrogen and hydrogen again. More ammonia will form, to restore the balance.

### 4 Add a catalyst
Iron is a catalyst for this reaction.
**A catalyst speeds up the forward and back reactions equally.**
So the reaction reaches equilibrium faster, which saves you time. But *the amount of ammonia does not change*.

## Choosing the optimum conditions
So to get the best yield of ammonia, it is best to:
- use high pressure, and remove ammonia, to improve the yield
- use a moderate temperature, and a catalyst, to get a decent rate.

Page 227 shows how these ideas are applied in an ammonia factory.

## A note about rate
By now, you should realize that:
- a change in temperature *always* shifts equilibrium.
- a change in pressure will shift equilibrium *only if* the number of molecules is different on each side of the equation.

But how do these changes affect the *rate*? Raising the temperature or pressure increases the rate of *both the forward and back reactions*, so equilibrium is reached faster. (A temperature rise gives the molecules more energy. An increase in pressure forces them closer. So in both cases, the number of successful collisions increases.)

> **A summary, for reversible reactions of gases**
> - Forward reaction *exothermic*: temperature ↑ means yield ↓.
> - Forward reaction *endothermic*: temperature ↑ means yield ↑.
> - *Fewer molecules* on the right-hand side of the equation: pressure ↑ means yield ↑.

> **What about solutions?**
> Many reversible reactions take place in solution:
> **reactants (aq) ⇌ products (aq)**
> You can shift the equilibrium:
> - by adding more of a reactant (increasing its concentration). So more product will form to oppose this change.
> - by changing the temperature. A rise in temperature will favour the endothermic reaction.

---

**Q**

1. The reaction between nitrogen and hydrogen is *reversible*. This causes a problem for the ammonia factory. Why?
2. What is Le Chatelier's principle? Write it down.
3. In manufacturing ammonia, explain why:
   a high pressure is used   b ammonia is removed
4. Sulfur dioxide ($SO_2$) and oxygen react exothermically to form sulfur trioxide ($SO_3$). The reaction is reversible.
   a Write the symbol equation for this reaction.
   b What happens to the yield of sulfur trioxide if you:
      i increase the pressure?   ii raise the temperature?

# ENERGY CHANGES, AND REVERSIBLE REACTIONS

## Checkup on Chapter 9

### Revision checklist

**Core curriculum**

Make sure you can ...
- ☐ explain what these terms mean:
  *exothermic reaction    endothermic reaction*
- ☐ give examples of exothermic and endothermic reactions and draw energy level diagrams for them
- ☐ state the unit used for measuring energy
- ☐ say what the + and − signs mean, in energy values
- ☐ explain what the purpose of a fuel is
- ☐ name the *fossil fuels*, and say how we use them
- ☐ explain what *nuclear fuels* are, and where we use them, and name one
- ☐ give advantages and disadvantages of nuclear fuels
- ☐ say how hydrogen and ethanol are used as fuels
- ☐ explain what a *reversible reaction* is, with examples
- ☐ write the symbol for a reversible reaction
- ☐ describe how to change hydrated copper(II) sulfate to the anhydrous compound, and back again
- ☐ explain why anhydrous copper(II) sulfate can be used to test for the presence of water

**Extended curriculum**

Make sure you can also ...
- ☐ use the idea of bond making and bond breaking to explain why a reaction is exo- or endothermic
- ☐ define *bond energy*
- ☐ calculate the energy change in a reaction, given the equation, and bond energy values
- ☐ describe a *simple cell*, and explain that the current comes from a redox reaction
- ☐ predict which metal will be the negative pole in a simple cell
- ☐ give half-equations for reactions that take place in a simple cell (like the one on page 120)
- ☐ describe the hydrogen fuel cell, and give the overall reaction that takes place in it
- ☐ explain that a reversible reaction never completes, in a closed container – it reaches equilibrium
- ☐ give ways to obtain more product in a gaseous reversible reaction
- ☐ predict the effect of a change in temperature and pressure, for a given reversible reaction
- ☐ say how a catalyst will affect a reversible reaction
- ☐ predict the effect of a change in conditions for a reversible reaction in solution

### Questions

**Core curriculum**

1 Look at this reaction:
   $NaOH\ (aq) + HCl\ (aq) \longrightarrow NaCl\ (aq) + H_2O\ (l)$
   a Which type of reaction is it?
   b It is *exothermic*. What does that mean?
   c What will happen to the temperature of the solution, as the chemicals react?
   d Draw an energy diagram for the reaction.

2 Water at 25 °C was used to dissolve two compounds. The temperature of each solution was measured immediately afterwards.

| Compound | Temperature of solution/°C |
|---|---|
| ammonium nitrate | 21 |
| calcium chloride | 45 |

   a List the apparatus needed for this experiment.
   b Calculate the temperature change on dissolving each compound.
   c i Which compound dissolved exothermically?
     ii How did you decide this?
     iii What can you say about the energy level of its ions in the solution, compared with in the solid compound?
   d For each solution, estimate the temperature of the solution if:
     i the amount of water is halved, but the same mass of compound is used
     ii the mass of the compound is halved, but the volume of water is unchanged
     iii both the mass of the compound, and the volume of water, are halved.

3 Hydrated copper(II) sulfate crystals were heated:

   a What is the ice for?
   b What colour change will occur in the test-tube?
   c The reaction is *reversible*. What does that mean?
   d How would you show that it is reversible?
   e Write the equation for the reversible reaction.

# ENERGY CHANGES, AND REVERSIBLE REACTIONS

**Extended curriculum**

4 The fuel *natural* gas is mostly methane.
Its combustion in oxygen is exothermic:
$$CH_4(g) + 2O_2(g) \longrightarrow CO_2(g) + 2H_2O(l)$$
  a Explain *why* this reaction is exothermic, in terms of bond breaking and bond making.
  b i Copy and complete this energy diagram for the reaction, indicating:
    A the overall energy change
    B the energy needed to break bonds
    C the energy given out by new bonds forming.

   ii Methane will not burn in air until a spark or flame is applied. Why not?
  c When 1 mole of methane burns in oxygen, the energy change is $-890$ kJ.
    i What does the $-$ sign tell you?
    ii Which word describes a reaction with this type of energy change?
  d How much energy is given out when 1 gram of methane burns? ($A_r$: C = 12, H = 1.)

5 Strips of copper foil and magnesium ribbon were cleaned with sandpaper and then connected as shown below. The bulb lit up.

  a Why were the metals cleaned?
  b Name the electrolyte used.
  c Explain why the bulb lit up.
  d Which metal releases electrons into the circuit?
  e In this arrangement, energy is being changed from one form to another. Explain.
  f What is this type of arrangement called?
  g Give reasons why the set-up shown above would not be used as a torch battery.

6 The gas hydrazine, $N_2H_4$, burns in oxygen like this:

$$H_2N-NH_2(g) + O=O(g) \longrightarrow N\equiv N(g) + 2O\begin{smallmatrix}H\\H\end{smallmatrix}(g)$$

  a Count and list the bonds broken in this reaction.
  b Count and list the new bonds formed.
  c Calculate the total energy:
    i required to break the bonds
    ii released when the new bonds form.
    (The bond energies in kJ/mole are: N−H 391; N−N 158; N≡N 945; O−H 464; O=O 498.)
  d Calculate the energy change in the reaction.
  e Is the reaction exothermic, or endothermic?
  f Where is energy transferred from, and to?
  g Comment on the suitability of hydrazine as a fuel.

7 Hydrogen and bromine react reversibly:
$$H_2(g) + Br_2(g) \rightleftharpoons 2HBr(g)$$
  a Which of these will favour the formation of more hydrogen bromide?
    i add more hydrogen    ii remove bromine
    iii remove the hydrogen bromide as it forms
  b Explain why increasing the pressure will have no effect on the amount of product formed.
  c However, the pressure *is* likely to be increased, when the reaction is carried out in industry. Suggest a reason for this.

8 Ammonia is made from nitrogen and hydrogen. The energy change in the reaction is $-92$ kJ/mole. The reaction is reversible, and reaches equilibrium.
  a Write the equation for the reaction.
  b Is the forward reaction endothermic, or exothermic? Give your evidence.
  c Explain why the *yield* of ammonia:
    i rises if you increase the pressure
    ii falls if you increase the temperature
  d What effect does increasing:
    i the pressure   ii the temperature
    have on the *rate* at which ammonia is made?
  e Why is the reaction carried out at 450 °C rather than at a lower temperature?

9 The dichromate ion $Cr_2O_7^{2-}$ and chromate ion $CrO_4^{2-}$ exist in equilibrium, like this:
$$Cr_2O_7^{2-}(aq) + H_2O(l) \rightleftharpoons 2CrO_4^{2-}(aq) + 2H^+(aq)$$
orange                              yellow
  a What would you see if you added dilute acid to a solution containing chromate ions?
  b How would you reverse the change?

THE SPEED OF A REACTION

## 10.1 Rates of reaction

### Fast and slow

Some reactions are **fast** and some are **slow**. Look at these examples:

The precipitation of silver chloride, when you mix solutions of silver nitrate and sodium chloride. This is a very fast reaction.

Concrete setting. This reaction is quite slow. It will take a couple of days for the concrete to fully harden.

Rust forming on an old car. This is usually a very slow reaction. It will take years for the car to rust completely away.

But it is not always enough to know just that a reaction is fast or slow. In factories where they make products from chemicals, they need to know *exactly* how fast a reaction is going, and how long it will take to complete. In other words, they need to know the **rate** of the reaction.

### What is rate?

Rate is a measure of how fast or slow something is. Here are some examples.

This plane has just flown 800 kilometers in 1 hour. It flew at a **rate** of 800 km per hour.

This petrol pump can pump out petrol at a **rate** of 50 litres per minute.

This machine can print newspapers at a **rate** of 10 copies per second.

From these examples you can see that:

**Rate is a measure of the change that happens in a single unit of time.**

Any suitable unit of time can be used – a second, a minute, an hour, even a day.

THE SPEED OF A REACTION

# Rate of a chemical reaction

When zinc is added to dilute sulfuric acid, they react together. The zinc disappears slowly, and a gas bubbles off.

As time goes by, the gas bubbles off more and more slowly. This is a sign that the reaction is slowing down.

Finally, no more bubbles appear. The reaction is over, because all the acid has been used up. Some zinc remains behind.

The gas that bubbles off is hydrogen. The equation for the reaction is:

zinc + sulfuric acid ⟶ zinc sulfate + hydrogen
$Zn\ (s) + H_2SO_4\ (aq) \longrightarrow ZnSO_4\ (aq) + H_2\ (g)$

Both zinc and sulfuric acid get used up in the reaction. At the same time, zinc sulfate and hydrogen form.

You could measure the rate of the reaction, by measuring:

- the amount of zinc used up per minute *or*
- the amount of sulfuric acid used up per minute *or*
- the amount of zinc sulfate produced per minute *or*
- the amount of hydrogen produced per minute.

For this reaction, it is easiest to measure the amount of hydrogen produced per minute, since it is the only gas that forms. It can be collected as it bubbles off, and its volume can be measured.

**In general, to find the rate of a reaction, you should measure:**
**the amount of a reactant used up per unit of time *or***
**the amount of a product produced per unit of time.**

1 Here are some reactions that take place in the home. Put them in order of decreasing rate (the fastest one first).
  a raw egg changing to hard-boiled egg
  b fruit going rotten
  c cooking gas burning
  d bread baking
  e a metal tin rusting
2 Which of these rates of travel is slowest?
  5 kilometres per second
  20 kilometres per minute
  60 kilometres per hour
3 Suppose you had to measure the rate at which zinc is used up in the reaction above. Which of these units would be suitable? Explain your choice.
  a litres per minute
  b grams per minute
  c centimetres per minute
4 Iron reacts with sulfuric acid like this:
  $Fe\ (s) + H_2SO_4\ (aq) \longrightarrow FeSO_4\ (aq) + H_2\ (g)$
  a Write a word equation for this reaction.
  b Write down four different ways in which the rate of the reaction could be measured.

THE SPEED OF A REACTION

# 10.2 Measuring the rate of a reaction

## A reaction that produces a gas

The rate of a reaction is found by measuring the amount of a **reactant** used up per unit of time, or the amount of a **product** produced per unit of time. Look at this reaction:

magnesium + hydrochloric acid ⟶ magnesium chloride + hydrogen
Mg (s) + 2HCl (aq) ⟶ MgCl$_2$ (aq) + H$_2$ (g)

Here hydrogen is the easiest substance to measure, because it is the only gas in the reaction. It bubbles off and can be collected in a **gas syringe**, where its volume is measured.

### The experiment

▲ Testing an explosive substance. The rate of a fast reaction like this, giving a mix of gases, is not easy to measure.

Clean the magnesium with sandpaper. Put dilute hydrochloric acid in the flask. Drop the magnesium into the flask, and insert the stopper and syringe immediately. Start the clock at the same time.

Hydrogen begins to bubble off. It rises up the flask and into the gas syringe, pushing the plunger out:

At the start, no gas has yet been produced or collected. So the plunger is all the way in.

Now the plunger has been pushed out to the 20 cm³ mark. 20 cm³ of gas have been collected.

The volume of gas in the syringe is noted at intervals – for example every half a minute. How will you know when the reaction is complete?

### Typical results

| Time / minutes | 0 | $\frac{1}{2}$ | 1 | $1\frac{1}{2}$ | 2 | $2\frac{1}{2}$ | 3 | $3\frac{1}{2}$ | 4 | $4\frac{1}{2}$ | 5 | $5\frac{1}{2}$ | 6 | $6\frac{1}{2}$ |
|---|---|---|---|---|---|---|---|---|---|---|---|---|---|---|
| Volume of hydrogen / cm³ | 0 | 8 | 14 | 20 | 25 | 29 | 33 | 36 | 38 | 39 | 40 | 40 | 40 | 40 |

This table shows some typical results for the experiment.

You can tell quite a lot from this table. For example, you can see that the reaction lasted about five minutes. But a graph of the results is even more helpful. The graph is shown on the next page.

# A graph of the results

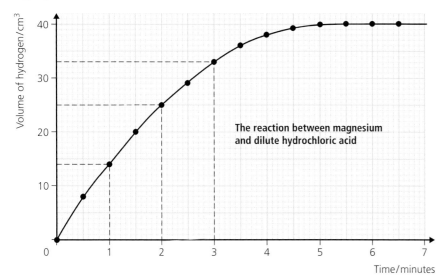

The reaction between magnesium and dilute hydrochloric acid

Notice these things about the results:

1 In the first minute, 14 cm³ of hydrogen are produced.
   So the rate for the first minute is 14 cm³ of hydrogen per minute.
   In the second minute, only 11 cm³ are produced. (25 − 14 = 11)
   So the rate for the second minute is 11 cm³ of hydrogen per minute.
   The rate for the third minute is 8 cm³ of hydrogen per minute.
   So the rate decreases as time goes on.
   **The rate changes all through the reaction. It is greatest at the start, but decreases as the reaction proceeds.**

2 The reaction is fastest in the first minute, and the curve is steepest then. It gets less steep as the reaction gets slower.
   **The faster the reaction, the steeper the curve.**

3 After 5 minutes, no more hydrogen is produced, so the volume no longer changes. The reaction is over, and the curve goes flat.
   **When the reaction is over, the curve goes flat.**

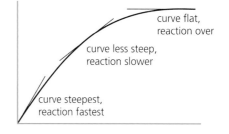

4 Altogether, 40 cm³ of hydrogen are produced in 5 minutes.

   The *average* rate for the reaction $= \dfrac{\text{total volume of hydrogen}}{\text{total time for the reaction}}$

   $= \dfrac{40 \text{ cm}^3}{5 \text{ minutes}}$

   $= \mathbf{8 \text{ cm}^3 \text{ of hydrogen per minute.}}$

Note that this method can be used for *any* reaction where one product is a gas.

---

**Q**

1 For the experiment in this unit, explain why:
   a the magnesium ribbon is cleaned first
   b the clock is started the moment the reactants are mixed
   c the stopper is replaced immediately

2 From the graph at the top of this page, how can you tell when the reaction is over?

3 Look again at the graph at the top of the page.
   a How much hydrogen is produced in the first:
      i  2.5 minutes?    ii  4.5 minutes?
   b How long did it take to get 20 cm³ of hydrogen?
   c What is the rate of the reaction during:
      i  the fourth minute?    ii  the sixth minute?

THE SPEED OF A REACTION

# 10.3 Changing the rate of a reaction (part I)

## Ways to change the rate of a reaction

There are several ways to speed up or slow down a reaction. For example you could change the concentration of a reactant, or the temperature. The *rate* will change - but the *amount of product* you obtain will not.

### 1 By changing concentration

Here you will see how rate changes with the **concentration** of a reactant.

**The method** Repeat the experiment from page 131 twice (A and B below). Keep everything the same each time *except* the concentration of the acid. In B it is *twice as concentrated* as in A.

**The results** Here are both sets of results, shown on the same graph.

The results for experiments A and B

Notice these things about the results:

1 Curve B is steeper than curve A. So the reaction was faster for B.
2 In B, the reaction lasts for 60 seconds. In A it lasts for 120 seconds.
3 Both reactions produced 60 cm³ of hydrogen. Do you agree?
4 So in B the average rate was 1 cm³ of hydrogen per second. (60 ÷ 60)
  In A it was 0.5 cm³ of hydrogen per second. (60 ÷ 120)
  The average rate in B was twice the average rate in A.
  So in this example, doubling the concentration doubled the rate.

These results show that:
**A reaction goes faster when the concentration of a reactant is increased.**
This means you can also slow down a reaction, by reducing concentration.

▲ Bleach reacts with coloured substances. The more concentrated the solution of bleach, the faster this stain will disappear.

THE SPEED OF A REACTION

## 2 By changing temperature

Here you will see how rate changes with the **temperature** of the reactants.

**The method**  Dilute hydrochloric acid and sodium thiosulfate solution react to give a fine yellow precipitate of sulfur. You can follow the rate of the reaction like this:

1. Mark a cross on a piece of paper.
2. Place a beaker containing sodium thiosulfate solution on top of the paper, so that you can see the cross through it, from above.
3. Quickly add hydrochloric acid, start a clock at the same time, and measure the temperature of the mixture.
4. The cross grows fainter as the precipitate forms. Stop the clock the moment you can no longer see the cross. Note the time.
5. Now repeat steps 1–4 several times, changing *only* the temperature. You do this by heating the sodium thiosulfate solution to different temperatures, before adding the acid.

▲ The low temperature in the fridge slows down reactions that make food rot.

View from above the beaker:

The cross grows fainter with time

**The results**  This table shows some typical results:

| Temperature/°C | 20 | 30 | 40 | 50 | 60 |
|---|---|---|---|---|---|
| Time for cross to disappear/seconds | 200 | 125 | 50 | 33 | 24 |

The higher the temperature, the faster the cross disappears

The cross disappears when enough sulfur has formed to hide it. This took 200 seconds at 20 °C, but only 50 seconds at 40 °C. So the reaction is *four times faster* at 40 °C than at 20 °C.

**A reaction goes faster when the temperature is raised. When the temperature increases by 10 °C, the rate generally doubles.**

That it why food cooks much faster in pressure cookers than in ordinary saucepans. (The temperature in a pressure cooker can reach 125 °C.) And if you want to slow a reaction down, of course, you can lower the temperature.

▲ Oh dear. Oven too hot? Reactions faster than expected?

**Q**

1. Look at the graph on the opposite page.
   a. How much hydrogen was obtained after 2 minutes in:
      i. experiment A?   ii. experiment B?
   b. How can you tell which reaction was faster, from the shape of the curves?
2. Explain why experiments A and B both gave the same amount of hydrogen.
3. Copy and complete: A reaction goes …… when the concentration of a …… is increased. It also goes …… when the …… is raised.
4. Raising the temperature speeds up a reaction. Try to give two (new) examples of how this is used in everyday life.
5. What happens to the rate of a reaction when the temperature is *lowered*? How do we make use of this?

# 10.4 Changing the rate of a reaction (part II)

## 3 By changing surface area

In many reactions, one reactant is a solid. The reaction between hydrochloric acid and calcium carbonate (marble chips) is an example. Carbon dioxide gas is produced:

$$CaCO_3\,(s) + 2HCl\,(aq) \longrightarrow CaCl_2\,(aq) + H_2O\,(l) + CO_2\,(g)$$

The rate can be measured using the apparatus on the right.

**The method**   Place the marble in the flask and add the acid. Quickly plug the flask with cotton wool to stop any liquid splashing out. Then weigh it, starting the clock at the same time. Note the mass at regular intervals until the reaction is complete.

Carbon dioxide is a heavy gas. It escapes through the cotton wool, which means that the flask gets lighter as the reaction proceeds. So by weighing the flask at regular intervals, you can follow the rate of reaction.

The experiment is repeated twice. Everything is kept exactly the same each time, except the *surface area* of the marble chips.

For experiment 1, large chips are used. Their surface area is the total area of exposed surface.

For experiment 2, the same *mass* of marble is used – but the chips are small so the surface area is greater.

**The results**   The results of the two experiments are plotted here:

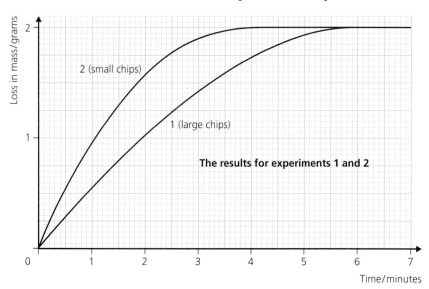

The results for experiments 1 and 2

So what can you conclude about surface area? Did it affect the rate of the reaction?

> **How to draw the graph**
> First you have to find the *loss in mass* at different times:
>
> loss in mass at a given time = mass at start − mass at that time
>
> Then you plot the values for loss in mass against time.

Notice these things about the results:
1. Curve 2 is steeper than curve 1. This shows that the reaction is faster for the small chips.
2. In both experiments, the final loss in mass is 2.0 grams. In other words, 2.0 grams of carbon dioxide are produced each time.
3. For the small chips, the reaction is complete in 4 minutes. For the large chips, it takes 6 minutes.

These results show that:

**The rate of a reaction increases when the surface area of a solid reactant is increased.**

## Explosion!

As you have seen, you can increase the rate of a reaction by increasing:
- the concentration of a reactant
- the temperature
- the surface area of a solid reactant

In some situations, an increase in any of these can lead to a dangerously fast reaction. You get an **explosion**. Here are examples.

**In flour mills** Flour particles are tiny, so flour has a very large surface area. It can also catch fire. In a flour mill, if there is a lot of flour dust in the air, a spark from a machine could be enough to cause an explosion.

For the same reason, explosions are a risk in wood mills, from wood dust, and in silos where wheat and other grains are stored. And in factories that make custard powder, and dried milk. The dust from all these will burn.

**In coal mines** In coal mines, methane ($CH_4$) and other flammable gases collect in the air. At certain concentrations they form an explosive mix with the air. A spark is enough to set off an explosion.

◀ In the old days, miners used candles to see their way underground – which caused many explosions. Now they use sealed lamps powered by batteries.

◀ A fire at a grain silo in Ghent, Belgium, after wheat dust exploded. Several people were injured.

**Q**

1. This question is about the graph on the opposite page. For each experiment find:
   a. the mass of carbon dioxide produced in the first minute
   b. the average rate of production of the gas, for the complete reaction.

2. a. Which has the largest surface area: 1 g of large marble chips, or 1 g of small marble chips?
   b. Which 1 g sample will disappear first when reacted with excess hydrochloric acid? Why?

3. Explain why fine flour dust in the air is a hazard, in flour mills.

# THE SPEED OF A REACTION

## 10.5 Explaining rates

### The collision theory
Magnesium and dilute hydrochloric acid react together like this:

magnesium + hydrochloric acid ⟶ magnesium chloride + hydrogen
$Mg(s)$ + $2HCl(aq)$ ⟶ $MgCl_2(aq)$ + $H_2(g)$

In order for the magnesium and acid particles to react together:
- **the particles must collide with each other, and**
- **the collision must have enough energy to be successful. In other words, enough energy to break bonds to allow reaction to occur.**

This is called the **collision theory**. It is shown by the drawings below.

The particles in the liquid move non-stop. To react, an acid particle must collide with a magnesium atom, and bonds must break.

This collision has enough energy to break bonds. So it is successful. The particles react and new bonds form, giving magnesium chloride and hydrogen.

But this collision did not have enough energy. It was not successful. No bonds were broken. The acid particle just bounced away again.

If there are lots of successful collisions in a given minute, then a lot of hydrogen is produced in that minute. In other words, the rate of reaction is high. If there are not many, the rate of reaction is low.

**The rate of a reaction depends on how many successful collisions there are in a given unit of time.**

### Changing the rate of a reaction
**Why rate increases with concentration**  If the concentration of the acid is increased, the reaction goes faster. It is easy to see why:

In dilute acid, there are not so many acid particles. So there is less chance of an acid particle hitting a magnesium atom.

Here the acid is more concentrated – there are more acid particles. So there is now more chance of a successful collision.

**The more successful collisions there are, the faster the reaction.**

> **Reactions between gases**
> - When you increase the pressure on two reacting gases, it means you squeeze more gas molecules into a given space.
> - So there is a greater chance of successful collisions.
> - So if pressure ↑ then rate ↑ for a gaseous reaction.

That idea also explains why the reaction between magnesium and hydrochloric acid slows down over time:

At the start, there are plenty of magnesium atoms and acid particles. But they get used up in successful collisions.

After a time, there are fewer magnesium atoms, and the acid is less concentrated. So there is less chance of successful collisions.

As a result, the slope of the reaction curve decreases with time, as shown above. It goes flat when the reaction is over.

**Why rate increases with temperature** On heating, *all* the particles take in heat energy.

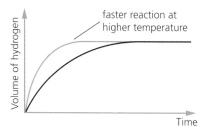

This makes the acid particles move faster – so they collide more often with magnesium particles.

The extra energy also means that more collisions are successful. So the reaction rate increases.

In fact, as you saw earlier, the rate generally doubles for an increase in temperature of 10 °C.

**Why rate increases with surface area** The reaction between the magnesium and acid is much faster when the metal is powdered:

The acid particles can collide only with the magnesium atoms in the outer layer of the metal ribbon.

In the powdered metal, many more atoms are exposed. So the chance of a collision increases.

**Q**

1 Copy and complete: Two particles can react together only if they …… and the …… has enough …… to be …….
2 What is meant by:
   a a successful collision?
   b an unsuccessful collision?
3 Reaction between magnesium and acid speeds up when:
   a the concentration of the acid is doubled. Why?
   b the temperature is raised. Why?
   c the acid is stirred. Why?
   d the metal is ground to a powder. Why?

THE SPEED OF A REACTION

# 10.6 Catalysts

## What is a catalyst?
You saw that a reaction can be speeded up by increasing the temperature, or the concentration of a reactant, or the surface area of a solid reactant.

There is another way to increase the rate of some reactions: use a **catalyst**.

**A catalyst is a substance that speeds up a chemical reaction, but remains chemically unchanged itself.**

## Example: the decomposition of hydrogen peroxide
Hydrogen peroxide is a colourless liquid that breaks down very slowly to water and oxygen:

hydrogen peroxide ⟶ water + oxygen
$2H_2O_2 \, (l) \longrightarrow 2H_2O \, (l) + O_2 \, (g)$

You can show how a catalyst affects the reaction, like this:
1. Pour some hydrogen peroxide into three measuring cylinders. The first one is the control.
2. Add manganese(IV) oxide to the second, and raw liver to the third.
3. Now use a glowing wooden splint to test the cylinders for oxygen. The splint will burst into flame if there is enough oxygen present.

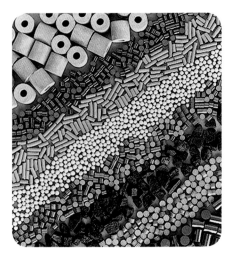

▲ Many different substances can act as catalysts. They are usually made into shapes that offer a very large surface area.

### The results

Since hydrogen peroxide breaks down very slowly, there is not enough oxygen to relight the splint.

Manganese(IV) oxide makes the reaction go thousands of times faster. The splint bursts into flame.

Raw liver also speeds it up. The liquid froths as the oxygen bubbles off – and the splint relights.

So manganese(IV) oxide acts as a catalyst for the reaction. If you add more manganese(IV) oxide, the reaction will go even faster.

Something in the raw liver acts as a catalyst too. That 'something' is an **enzyme** called catalase.

## What are enzymes?
**Enzymes are proteins made by cells, to act as biological catalysts.**
Enzymes are found in every living thing. You have thousands of different enzymes inside you. For example catalase speeds up the decomposition of hydrogen peroxide in your cells, before it can harm you. Amylase in your saliva speeds up the breakdown of the starch in your food.

Without enzymes, most of the reactions that take place in your body would be far too slow at body temperature. You would die.

▲ Enzyme molecules are large and complex, as this model shows.

## How do catalysts work?

For a reaction to take place, the reacting particles must collide with enough energy for bonds to break and reaction to occur.

When a catalyst is present, the reactants are able to react in a way that requires less energy.

This means that more collisions now have enough energy to be successful. So the reaction speeds up. But the catalyst itself is unchanged.

Note that a catalyst must be chosen to suit the particular reaction. It may not work for other reactions.

## Catalysts in the chemical industry

In industry, many reactions need heat. Fuel can be a very big expense.

With a catalyst, a reaction goes faster at a given temperature. So you get the product faster, saving time. Even better, it may go fast enough *at a lower temperature* – which means a lower fuel bill.

So catalysts are very important in the chemical industry. They are often **transition elements** or their **oxides**. Two examples are:

- **iron** used in the manufacture of ammonia
- **vanadium(IV) oxide** used in the manufacture of sulfuric acid.

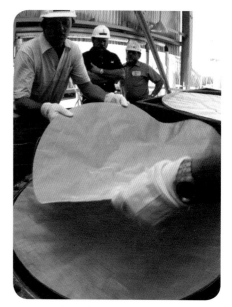

▲ A catalyst of platinum and rhodium, in the form of a gauze, is being fitted into a tank. It will catalyse the production of nitric acid from ammonia and oxygen.

## Making use of enzymes

There are thousands of different enzymes, made by living things. We are finding many uses for them.

For example some bacteria make enzymes that catalyse the breakdown of fat, starch, and proteins. The bacteria can be grown in tanks, in factories. The enzymes are removed, and used in **biological detergents**. In the wash, they help to break down grease, food stains, and blood stains on clothing.

Enzymes work best in conditions like those in the living cells that made them.

- If the temperature gets too high, the enzyme is destroyed or **denatured**. It loses its shape.
- An enzyme also works best in a specific pH range. You can denature it by adding acid or alkali.

▲ Now add a biological detergent? But do not use them for wool or silk, as they cause the proteins in these to break down.

> **More means faster**
> The more catalyst you add, the faster the reaction goes.

1. What is a *catalyst*?
2. Which of these does a catalyst *not* change?
   a  the speed of a reaction
   b  the products that form
   c  the total amount of each product formed
3. Explain what an *enzyme* is, and give an example.
4. Why do our bodies need enzymes?
5. Catalysts are very important in industry. Explain why.
6. Give two examples of catalysts used in the chemical industry.
7. A box of biological detergent had this instruction on the back: *Do not use in a wash above 60 °C.* Suggest a reason.

THE SPEED OF A REACTION

# More about enzymes

## Mainly from microbes
Enzymes are proteins made by living things, to act as catalysts for their own reactions. So we can obtain enzymes from plants, and animals, and microbes such as bacteria and fungi. In fact we get most from microbes.

## Traditional uses for enzymes
We humans have used enzymes for thousands of years. For example …

- **In making bread**  Bread dough contains yeast (a fungus), and sugar. When the dough is left in a warm place, the yeast cells feed on the sugar to obtain energy. Enzymes in the yeast catalyse the reaction, which is called **fermentation**:

  $$C_6H_{12}O_6 \, (aq) \xrightarrow{\text{catalysed by enzymes}} 2C_2H_5OH \, (aq) + 2CO_2 \, (g) + \text{energy}$$
  glucose  ethanol  carbon dioxide

  The carbon dioxide gas makes the dough rise. Later, in the hot oven, the gas expands even more, while the bread sets. So you end up with spongy bread. The heat kills the yeast off.

- **In making yoghurt**  To make yoghurt, special bacteria are added to milk. They feed on the lactose (sugar) in it, to obtain energy. Their enzymes catalyse its conversion to lactic acid and other substances, which turn the milk into yoghurt.

## Making enzymes the modern way
In making bread and yoghurt, the microbes that make the enzymes are present. But in most modern uses for enzymes, they are not. Instead:

- bacteria and other microbes are grown in tanks, in a rich broth of nutrients; so they multiply fast
- then they are killed off, and their enzymes are separated and purified
- the enzymes are sold to factories.

▲ Yeast cells. Cells are living things, but the enzymes they make are not. Enzymes are just chemicals (proteins).

▲ The holes in bread are where carbon dioxide gas expanded.

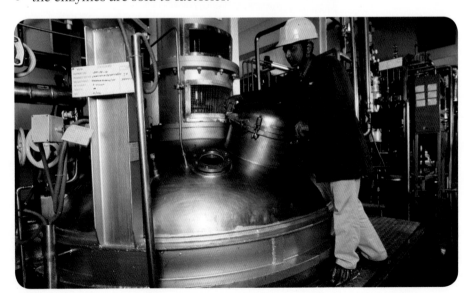

▲ Anyone home? The tank contains bacteria, busy making enzymes. For example it could be the enzyme amylase, that catalyses the conversion of starch to sugar.

▲ The amylase is sold to a company that uses it to make a sweet syrup from corn (maize) flour. The syrup is used in biscuits, cakes, soft drinks, and sauces.

THE SPEED OF A REACTION

## Modern uses of enzymes

Enzymes have many different uses. Here are some common ones:

- **In making soft-centred chocolates**   How do they get the runny centres into chocolates? By using the enzyme **invertase**.

  First they make a paste containing sugars, water, flavouring, colouring, and invertase. Then they dip blobs of it into melted chocolate, which hardens. Inside, the invertase catalyses the breakdown of the sugars to more soluble ones, so the paste goes runny.

  Other enzymes are used in a similar way to 'soften' food, to make tinned food for infants.

- **In making stone-washed denim**   Once, denim was given a worn look by scrubbing it with pumice stone. Now an enzyme does the job.

- **In making biological detergents**   As you saw on page 141, these contain enzymes to catalyse the breakdown of grease and stains.

- **In DNA testing**   Suppose a criminal leaves tiny traces of skin or blood at a crime scene. The enzyme **polymerase** is used to 'grow' the DNA in them, to give enough to identify the criminal.

▲ Thanks to invertase …

▲ Thanks to polymerase …

## How do they work?

This shows how an enzyme molecule catalyses the breakdown of a reactant molecule:

First, the two molecules must fit together like jigsaw pieces. (So the reactant molecule must be the right shape, for the enzyme.)

The 'complex' that forms makes it easy for the reactant molecule to break down. You do not need to provide energy by heating.

When decomposition is complete the molecules of the product move away. Another molecule of the reactant takes their place.

Enzymes are a much more complex shape than the drawing suggests. Even so, this model gives you a good idea of how they work.

## The search for extremophiles

Most of the enzymes we use work around 40 °C, and at a pH not far from 7. In other words, in conditions like those in the cells that made them.

But around the world, scientists are searching high and low for microbes that live in very harsh conditions. For example deep under the ice in Antarctica, or at hot vents in the ocean floor, or in acidic lakes around volcanoes. They call these microbes **extremophiles**.

Why do scientists want them? Because the enzmyes made by these microbes make will work in the same harsh conditions. So they may find a great many uses in industry.

▲ Many bacteria live around hot vents in the ocean floor – in water at up to 400 °C.

THE SPEED OF A REACTION

# 10.7 Photochemical reactions

## Some reactions need light
Some chemical reactions obtain the energy they need from light. They are called **photochemical reactions**. Examples are photosynthesis, and the reactions that occur in film photography.

## Photosynthesis
- **Photosynthesis** is the reaction between carbon dioxide and water, in the presence of chlorophyll and sunlight, to produce glucose:

$$6CO_2\ (g) + 6H_2O\ (l) \xrightarrow[\text{chlorophyll}]{\text{light}} C_6H_{12}O_6\ (aq) + 6O_2\ (g)$$
carbon dioxide   water                    glucose       oxygen

- It takes place in plant leaves. Carbon dioxide enters the leaves through tiny holes called **stomata**.
- **Chlorophyll**, the green pigment in leaves, is a **catalyst** for the reaction.
- The water is taken in from the soil, through the plant's roots.
- Sunlight provides the energy for this endothermic reaction.
- The plant then uses the glucose for energy, and to build the cellulose and other substances it needs for growth.

▲ The stomata of a leaf, magnified by about 700. Carbon dioxide passes in through them, and oxygen passes out.

## Changing the rate of the photosynthesis reaction
Could you change the rate by changing the strength of the light? Let's see.

**The method**   Pondweed is a suitable plant to use for the experiment.

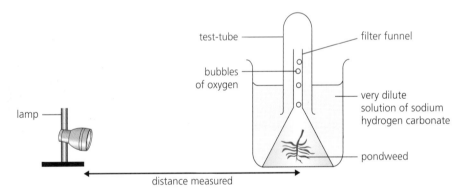

1  Put some pondweed in a beaker containing a very dilute solution of sodium hydrogen carbonate, $NaHCO_3$. (This compound decomposes, giving off carbon dioxide.) Place a funnel over it.
2  Place a test-tube full of the solution over the funnel, as shown.
3  Place the lamp 50 cm from the beaker. (Look at the arrow above.)
4  Let the pondweed adjust to the conditions for 1 minute. Then count the bubbles of oxygen it gives off, over 1 minute. Repeat twice more to get an average value per minute. Record your results.
5  Repeat step 4, with the lamp placed at 40, 30, 20, and 10 cm from the beaker.

You can then plot a graph for your results.

▲ The plant on the right is unhealthy because it did not get enough light – so it made glucose too slowly.

**The results** This graph shows that the number of bubbles per minute *increases* as the lamp is brought closer to the plant.

The closer it is, the greater the strength or **intensity** of the light that reaches the plant. So we can say that the rate of photosynthesis increases as the intensity of the light increases.

That makes sense. Light provides the energy for the reaction. The stronger it is, the more energy it provides. So more molecules of carbon dioxide and water gain enough energy to react.

**A photochemical reaction can be speeded up by increasing the intensity of the light.**

This is true of all photochemical reactions.

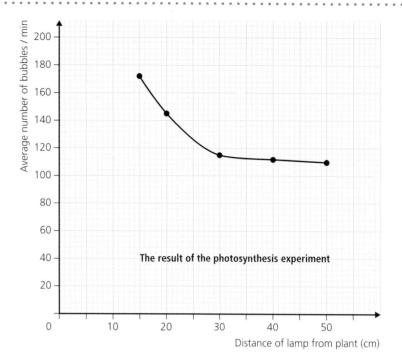

The result of the photosynthesis experiment

## The reactions in film photography

Black-and-white film photography relies on a photochemical reaction. The film is covered with a coating of gel that contains tiny grains of silver bromide. Light causes this to break down:

$$2AgBr\,(s) \longrightarrow 2Ag\,(s) + Br_2\,(l)$$

It is both a photochemical reaction *and* a redox reaction.

The silver ions are **reduced**: $2Ag^+ + 2e^- \longrightarrow 2Ag$ (electron gain)

The bromide ions are **oxidised**: $2Br^- \longrightarrow Br_2 + 2e^-$ (electron loss)

### So how is a photo produced?

1 When you click to take the photo, the camera shutter opens briefly. Light enters and strikes the film. The silver bromide decomposes, giving tiny dark particles of silver. Where brighter light strikes (from brighter parts of the scene), decomposition is faster, giving more silver.

2 Next the film is **developed**: unreacted silver bromide is washed away, leaving clear areas on the film. The silver remains, giving darker areas.

3 Then the film is printed. In this step, light is shone through the film onto photographic paper, which is also coated with silver bromide. The light passes through the clear areas of the film easily, causing the silver bromide to decompose. But the darker areas block light.

4 The unreacted silver bromide is washed from the paper. This leaves a black-and-white image of the original scene, made of silver particles.

▲ A 'negative' portrait in silver particles. What will the printed photo show?

**Q**

1 What is a *photochemical reaction*? Give two examples.
2 a Write down the equation for photosynthesis.
   b What is the purpose of the chlorophyll?
   c Stronger light speeds up photosynthesis. Why?
3 a Why is silver bromide used in photographic film?
   b Its decomposition is a redox reaction. Explain why.
4 *The more intense the light, the faster the photochemical reaction.* Explain how this idea is used in photography with film.

THE SPEED OF A REACTION

# Checkup on Chapter 10

## Revision checklist
### Core curriculum
Make sure you can …
☐ explain what *the rate of a reaction* means
☐ describe a way to measure the rate of a reaction that produces a gas, using a gas syringe
☐ describe a way to measure the rate of a reaction that produces carbon dioxide (a heavy gas), using a balance
☐ give the correct units for the rate of a given reaction (for example $cm^3$ per minute, or grams per minute)
☐ work out, from the graph for a reaction:
  – how long the reaction lasted
  – how much product was obtained
  – the average rate of the reaction
  – the rate in any given minute
☐ give three ways to increase the rate of a reaction
☐ say which of two reactions was faster, by comparing the slope of their curves on a graph
☐ explain why there is a risk of explosions in flour mills and coal mines
☐ explain these terms: *catalyst   enzyme*
☐ explain why enzymes are important in our bodies
☐ explain why catalysts are important in industry
☐ give examples of the use of catalysts, including enzymes

### Extended curriculum
Make sure you can also …
☐ describe the collision theory
☐ use the collision theory to explain *why* the rate of a reaction increases with concentration, temperature, and surface area
☐ explain how catalysts work
☐ say how enzymes can be destroyed
☐ define *photochemical reaction* and give examples
☐ say what *photosynthesis* is, and give the word and chemical equations for it
☐ name the catalyst for photosynthesis
☐ explain why a photochemical reaction can be speeded up by increasing the intensity of the light
☐ give the equation for the photochemical reaction that takes place on black-and-white film and photographic paper
☐ show that this reaction is also a redox reaction

## Questions
### Core curriculum
**1** The rate of the reaction between magnesium and dilute hydrochloric acid can be measured using this apparatus:

  **a** What is the purpose of:
  **i** the test-tube?     **ii** the gas syringe?
  **b** How would you get the reaction to start?

**2** Some magnesium and an *excess* of dilute hydrochloric acid were reacted together. The volume of hydrogen produced was recorded every minute, as shown in the table:

| Time / min | 0 | 1 | 2 | 3 | 4 | 5 | 6 | 7 |
|---|---|---|---|---|---|---|---|---|
| Volume of hydrogen / $cm^3$ | 0 | 14 | 23 | 31 | 38 | 40 | 40 | 40 |

  **a** What does an *excess* of acid mean?
  **b** Plot a graph of the results.
  **c** What is the *rate of reaction* (in $cm^3$ of hydrogen per minute) during:
   **i** the first minute?
   **ii** the second minute?
   **iii** the third minute?
  **d** Why does the rate change during the reaction?
  **e** How much hydrogen was produced in total?
  **f** How long does the reaction last?
  **g** What is the *average rate* of the reaction?
  **h** How could you slow down the reaction, while keeping the amounts of reactants unchanged?

**3** Suggest a reason for each observation below.
  **a** Hydrogen peroxide decomposes much faster in the presence of the enzyme catalase.
  **b** The reaction between manganese carbonate and dilute hydrochloric acid speeds up when some concentrated hydrochloric acid is added.
  **c** Powdered magnesium is used in fireworks, rather than magnesium ribbon.

## THE SPEED OF A REACTION

**4** In two separate experiments, two metals A and B were reacted with an excess of dilute hydrochloric acid. The volume of hydrogen was measured every 10 seconds. These graphs show the results:

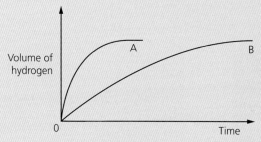

  **a i** Which piece of apparatus can be used to measure the volume of hydrogen produced?
  **ii** What other measuring equipment is needed?
  **b** Which metal, A or B, reacts faster with hydrochloric acid? Give your evidence.
  **c** Sketch and label the curves that will be obtained for metal B if:
  **i** more concentrated acid is used (curve X)
  **ii** the reaction is carried out at a lower temperature (curve Y)

**5** Copper(II) oxide catalyses the decomposition of hydrogen peroxide. 0.5 g of the oxide was added to a flask containing 100 cm³ of hydrogen peroxide solution. A gas was released. It was collected, and its volume noted every 10 seconds.
This table shows the results:

| Time / s | 0 | 10 | 20 | 30 | 40 | 50 | 60 | 70 | 80 | 90 |
|---|---|---|---|---|---|---|---|---|---|---|
| Volume / cm³ | 0 | 18 | 30 | 40 | 48 | 53 | 57 | 58 | 58 | 58 |

  **a** What is a catalyst?
  **b** Draw a diagram of suitable apparatus for this experiment.
  **c** Name the gas that is formed.
  **d** Write a balanced equation for the decomposition of hydrogen peroxide.
  **e** Plot a graph of the volume of gas (vertical axis) against time (horizontal axis).
  **f** Describe how rate changes during the reaction.
  **g** What happens to the concentration of hydrogen peroxide as the reaction proceeds?
  **h** What chemicals are present in the flask after 90 seconds?
  **i** What mass of copper(II) oxide would be left in the flask at the end of the reaction?
  **j** Sketch on your graph the curve that might be obtained for 1.0 g of copper(II) oxide.
  **k** Name one other substance that catalyses this decomposition.

### Extended curriculum

**6** Marble chips (lumps of calcium carbonate) react with hydrochloric acid as follows:
$$CaCO_3\ (s) + 2HCl\ (aq) \longrightarrow CaCl_2\ (aq) + CO_2\ (g) + H_2O\ (l)$$
  **a** What gas is released during this reaction?
  **b** Describe a laboratory method that could be used to investigate the rate of the reaction.
  **c** How will this affect the rate of the reaction?
  **i** increasing the temperature
  **ii** adding water to the acid
  **d** Explain each of the effects in **c** in terms of collisions between reacting particles.
  **e** If the lumps of marble are crushed first, will the reaction rate change? Explain your answer.

**7** Zinc and iodine solution react like this:
$$Zn\ (s) + I_2\ (aq) \longrightarrow ZnI_2\ (aq)$$
The rate of reaction can be followed by measuring the mass of zinc metal at regular intervals, until all the iodine has been used up.
  **a** What will happen to the mass of the zinc, as the reaction proceeds?
  **b** Which reactant is in excess? Explain your choice.
  **c** The reaction rate slows down with time. Why?
  **d** Sketch a graph showing the mass of zinc on the $y$ axis, and time on the $x$ axis.
  **e** How will the graph change if the temperature of the iodine solution is increased by 10 °C?
  **f** Explain your answer to **e** using the idea of collisions between particles.

**8** Some pondweed is placed as shown:

  **a i** Name the gas that collects in the test tube
  **ii** What other product is produced?
  **b** This experiment must be carried out in the light. Why?
  **c** Using the apparatus above, suggest a method by which the rate of reaction could be found.
  **d** What would be the effect of bringing a lamp close to the beaker? Explain your answer.

ACIDS AND BASES

# 11.1 Acids and alkalis

## Acids

One important group of chemicals is called **acids**:

You have probably seen these acids in the lab. They are all solutions of pure compounds in water. They can be dilute, like these, or concentrated.

They must be handled carefully, especially the concentrated solutions, because they are **corrosive**. They can eat away metals, skin, and cloth.

But some acids are not so corrosive, even when concentrated. These are called **weak** acids. Ethanoic acid is one example. It is found in vinegar.

You can tell if something is an acid, by its effect on **litmus**. Litmus is a purple dye. It can be used as a solution, or on paper. **Acids turn litmus red**.

### Some common acids

The main acids you will meet in chemistry are:

| | |
|---|---|
| hydrochloric acid | $HCl\ (aq)$ |
| sulfuric acid | $H_2SO_4\ (aq)$ |
| nitric acid | $HNO_3\ (aq)$ |
| ethanoic acid | $CH_3COOH\ (aq)$ |

But there are many others. For example, lemon and lime juice contain **citric acid**, ant stings contain **methanoic acid**, and fizzy drinks contain **carbonic acid**, formed when carbon dioxide dissolves in water.

## Alkalis

There is another group of chemicals that also affect litmus, but in a different way. They are the **alkalis**.
**Alkalis turn litmus blue**.
Like acids, they must be handled carefully. They too can burn skin.

### Some common alkalis

The pure alkalis are solids – except for ammonia, which is a gas. They are used in the lab as aqueous solutions. The main ones you will meet are:

| | |
|---|---|
| sodium hydroxide | $NaOH\ (aq)$ |
| potassium hydroxide | $KOH\ (aq)$ |
| calcium hydroxide | $Ca(OH)_2\ (aq)$ |
| ammonia | $NH_3\ (aq)$ |

**Remember:**
aci**d** turns litmus re**d**

blue litmus paper goes red in an acid solution

pink litmus paper goes blue in an alkaline solution

▲ Testing with litmus paper.

▲ Common laboratory alkalis. The solution of calcium hydroxide is called limewater.

# ACIDS AND BASES

## Indicators

Litmus is called an **indicator**, because it indicates whether something is an acid or an alkali. This table shows two others. All show a colour change from acid to alkali. That's why they are used!

| Indicator | Colour in acid | Colour in alkali |
|---|---|---|
| litmus | red | blue |
| phenolphthalein | colourless | pink |
| methyl orange | red | yellow |

## Neutral substances

Many substances are not acids or alkalis. They are **neutral**. Examples are pure water, and aqueous solutions of sodium chloride and sugar.

## The pH scale

You can say how acidic or alkaline a solution is using a scale of numbers called the **pH scale**. The numbers go from 0 to 14:

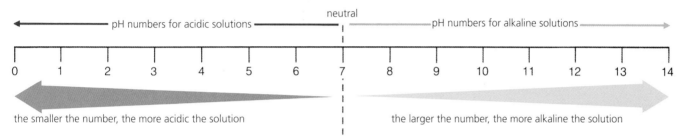

On this scale:

**An acidic solution has a pH number less than 7.**
**An alkaline solution has a pH number greater than 7.**
**A neutral solution has a pH number of exactly 7.**

## Universal indicator paper

You can find the pH of any solution by using **universal indicator**. This is a mixture of dyes. Like litmus, it can be used as a solution, or a paper strip. Its colour changes with pH, as shown here:

1 What does *corrosive* mean?
2 How would you test a substance, to see if it is an acid?
3 Write down the formula for:
   sulfuric acid          nitric acid
   calcium hydroxide      ammonia solution
4 Phenolpthalein is an *indicator*. What does that mean?
5 What does this pH value tell you about the solution?
   a 9   b 4   c 7   d 1   e 10   f 3
6 What colour is universal indicator, in an aqueous solution of sugar? Why?

# ACIDS AND BASES

## 11.2 A closer look at acids and alkalis

### Acids produce hydrogen ions

Hydrogen chloride is a gas, made of molecules. It dissolves in water to give hydrochloric acid. But this is not molecular. In water, the molecules break up or **dissociate** into ions:

$$HCl\ (aq) \longrightarrow H^+\ (aq) + Cl^-\ (aq)$$

So hydrochloric acid contains hydrogen ions. All other solutions of acids do too. The hydrogen ions give them their 'acidity'.
**Solutions of acids contain hydrogen ions.**

> **Remember!**
> Acidic solutions contain hydrogen ions:
>
>
>
> It is what makes them 'acidic'.

### Comparing acids

Since solutions of acids contain ions, they conduct electricity. We can measure how well they conduct using a conductivity meter. We can also check their pH using a pH meter.

Samples of acids of the same concentration were tested. This table gives the results. (The unit of conductivity is the siemens, or S.)

| Acid | For a 0.1 M solution ... | | |
|---|---|---|---|
| | conductivity (µS/cm) | pH | |
| hydrochloric acid | 25 | 1.0 | |
| sulfuric acid | 40 | 0.7 | strong acids |
| nitric acid | 25 | 1.0 | |
| methanoic acid | 2 | 2.4 | |
| ethanoic acid | 0.5 | 2.9 | weak acids |
| citric acid | 4 | 2.1 | |

▲ The conductivity meter measures the current passing through the liquid, carried by ions. (Lemon juice contains hydrogen ions and citrate ions.)

So the acids fall into two groups. The first group shows high conductivity, and low pH. These are **strong acids**. The second group does not conduct nearly so well, and has a higher pH. These are **weak acids**.

### The difference between strong and weak acids

In a solution of hydrochloric acid, all the molecules of hydrogen chloride have become ions:

$$HCl\ (aq) \xrightarrow{100\%} H^+\ (aq) + Cl^-\ (aq)$$

But in weak acids, only some of the molecules have become ions. For example, for ethanoic acid:

$$CH_3COOH\ (aq) \xrightarrow{\text{much less than 100\%}} H^+\ (aq) + CH_3COO^-\ (aq)$$

**In solutions of strong acids, all the molecules become ions.
In solutions of weak acids, only some do.**

So strong acids conduct better because there are more *ions* present. They have a lower pH because there are more *hydrogen ions* present.
**The higher the concentration of hydrogen ions, the lower the pH.**

▲ Strong and weak: the car battery contains sulfuric acid, and the oranges contain citric acid.

# ACIDS AND BASES

## Alkalis produce hydroxide ions

Now let's turn to alkalis, with sodium hydroxide as our example. It is an ionic solid. When it dissolves, all the ions separate:

$$NaOH\ (aq) \longrightarrow Na^+\ (aq) + OH^-\ (aq)$$

So sodium hydroxide solution contains hydroxide ions. The same is true of all alkaline solutions.
**Solutions of alkalis contain hydroxide ions.**

**Remember!**
Alkaline solutions contain hydroxide ions:

$OH^-$

It is what makes them alkaline.

## Comparing alkalis

We can compare the conductivity and pH of alkalis too. Look at these results:

| Alkali | For a 0.1M solution ... | | |
|---|---|---|---|
| | conductivity (µS/cm) | pH | |
| sodium hydroxide | 20 | 13.0 | strong alkali |
| potassium hydroxide | 15 | 13.0 | |
| ammonia solution | 0.5 | 11.1 | weak alkali |

The first two alkalis show high conductivity, and high pH. They are **strong alkalis**. But the ammonia solution shows much lower conductivity, and a lower pH. It is a **weak alkali**.

## Why ammonia solution is different

In sodium hydroxide solution, all the sodium hydroxide exists as ions:

$$NaOH\ (aq) \xrightarrow{100\%} Na^+\ (aq) + OH^-\ (aq)$$

The same is true for potassium hydroxide. But ammonia gas is molecular. When it dissolves in water, this is what happens:

$$NH_3\ (aq) + H_2O\ (l) \xrightarrow{\text{much less than 100\%}} NH_4^+\ (aq) + OH^-\ (aq)$$

Only some of the ammonia molecules form ions. So there are fewer hydroxide ions present than in a sodium hydroxide solution of the same concentration.

The sodium hydroxide solution is a better conductor than the ammonia solution because it contains more *ions*. And it has a higher pH because it contains more *hydroxide ions*.

**The higher the concentration of hydroxide ions, the higher the pH.**

▲ Alkalis react with grease. So the strong alkali sodium hydroxide is used to clear blocked sinks and pipes in homes. What does the drawing tell you?

## Q

1. Write an equation to show what happens when hydrogen chloride dissolves in water.
2. All acids have something in common. What is it?
3. For the table on page 150, explain why ethanoic acid has:
   a lower conductivity
   b a higher pH
   than hydrochloric acid.
4. What do all alkaline solutions have in common?
5. Write an equation to show what happens when ammonia gas dissolves in water.
6. For the table above, explain why the ammonia solution has:
   a lower conductivity
   b a lower pH
   than the potassium hydroxide solution.

ACIDS AND BASES

## 11.3 The reactions of acids and bases

### When acids react
When acids react with metals, bases and carbonates, a **salt** is produced. Salts are ionic compounds. Sodium chloride, NaCl, is an example.

The name of the salt depends on the acid you start with:

| | | |
|---|---|---|
| hydrochloric acid | gives | chlorides |
| sulfuric acid | gives | sulfates |
| nitric acid | gives | nitrates |

### Typical acid reactions
1 **With metals:  acid + metal ⟶ salt + hydrogen**

   For example:

   magnesium + sulfuric acid ⟶ magnesium sulfate + hydrogen

   $Mg(s) + H_2SO_4(aq) \longrightarrow MgSO_4(aq) + H_2(g)$

   So the metal drives the hydrogen out of the acid, and takes its place: it **displaces** hydrogen. A solution of the salt magnesium sulfate is formed.

▲ Magnesium reacting with dilute sulfuric acid. Hydrogen bubbles off.

2 **With bases:  acid + base ⟶ salt + water**

   **Bases** are compounds that react with acid to give *only* a salt and water. **Metal oxides and hydroxides** are bases. **Alkalis** are soluble bases.

   Example for an acid and alkali:

   hydrochloric acid + sodium hydroxide ⟶ sodium chloride + water

   $HCl(aq) + NaOH(aq) \longrightarrow NaCl(aq) + H_2O(l)$

   Example for an acid and insoluble base:

   sulfuric acid + copper(II) oxide ⟶ copper(II) sulfate + water

   $H_2SO_4(aq) + CuO(s) \longrightarrow CuSO_4(aq) + H_2O(l)$

▲ In A, black copper(II) oxide is reacting with dilute sulfuric acid. The solution turns blue as copper(II) sulfate forms. B shows how the final solution will look.

3 **With carbonates:**

   **acid + carbonate ⟶ salt + water + carbon dioxide**

   For example:

   calcium carbonate + hydrochloric acid ⟶ calcium chloride + water + carbon dioxide

   $CaCO_3(s) + 2HCl(aq) \longrightarrow CaCl_2(aq) + H_2O(l) + CO_2(g)$

### Reactions of bases
1 Bases react with acids, as you saw above, giving only a salt and water. That is what identifies a base.

2 Bases such as sodium, potassium and calcium hydroxides react with ammonium salts, driving out ammonia gas. For example:

   calcium hydroxide + ammonium chloride ⟶ calcium chloride + water + ammonia

   $Ca(OH)_2(s) + 2NH_4Cl(s) \longrightarrow CaCl_2(s) + 2H_2O(l) + 2NH_3(g)$

   This reaction is used for making ammonia in the laboratory.

▲ Calcium carbonate reacting with dilute hydrochloric acid. What is that gas?

## Neutralisation

**Neutralisation is a reaction with acid that gives *water* as well as a salt.**
So the reactions of bases and carbonates with acids are neutralisations.
We say the acid is **neutralised**.
But the reactions of acids with metals are not neutralisations. Why not?

### Making use of neutralisation

We often make use of neutralisation outside the lab.
For example, to reduce acidity in soil.

Soil forms when rock is broken up over many years by the action of rain and the weather. It may be acidic because of the type of rock it came from. But rotting vegetation, and heavy use of fertilisers, can also make it acidic.

Most crops grow best when the pH of the soil is close to 7. If the soil is too acidic, crops grow badly or not at all. That could be a disaster for farmers.

So to reduce its acidity, the soil is treated with crushed **limestone**, which is calcium carbonate, or **lime** (calcium oxide) or **slaked lime** (calcium hydroxide). A neutralisation reaction takes place.

▲ The soil is too acidic, so the farmer is spreading lime. It is more soluble than limestone. Is that an advantage – or not?

### Acids and redox reactions

Look again at the three groups of acid reactions.

The reactions of acids with metals are **redox reactions**, because electrons are transferred. For example when magnesium reacts with hydrochloric acid, magnesium ions form. The magnesium is oxidised:

$$Mg\ (s) \longrightarrow Mg^{2+}\ (aq) + 2e^-  \quad \text{(oxidation is loss of electrons)}$$

But in neutralisation reactions, no electrons are transferred. You can check this by looking at the oxidation states in the equation. For example, for the reaction between hydrochloric acid and sodium hydroxide:

$$HCl\ (aq) + NaOH\ (aq) \longrightarrow NaCl\ (aq) + H_2O\ (l)$$
$$+I\ -I \quad\quad +I\ -II\ +I \quad\quad +I\ -I \quad\quad +I\ -II$$

No element changes its oxidation state. So this is not a redox reaction.
In the next unit, you can find out what does go on during neutralisation.

▲ Bee stings are acidic. To neutralise the sting, rub on some baking soda (sodium hydrogen carbonate) or calamine lotion (which contains zinc carbonate).

---

**1** Write a word equation for the reaction of dilute sulfuric acid with: **a** zinc **b** sodium carbonate

**2** Which reaction in question **1** is *not* a neutralisation?

**3** Salts are ionic compounds. Name the salt that forms when calcium oxide reacts with hydrochloric acid, and say which ions it contains.

**4** Zinc oxide is a base. Suggest a way to make zinc nitrate from it. Write a word equation for the reaction.

**5** In what ways are the reactions of hydrochloric acid with calcium oxide and calcium carbonate:
  **a** similar?
  **b** different?

**6 a** Lime can help to control acidity in soil. Why?
  **b** Name one product that will form when it is used.

**7** Zinc reacts with hydrochloric acid to form zinc chloride, $ZnCl_2$. Show that this is a redox reaction.

# 11.4 A closer look at neutralisation

## The neutralisation of an acid by an alkali (a soluble base)

This is a solution of hydrochloric acid. It contains $H^+$ and $Cl^-$ ions. It will turn litmus red.

This is a solution of sodium hydroxide. It contains $Na^+$ and $OH^-$ ions. It will turn litmus blue.

When you mix the two solutions, the $OH^-$ ions and $H^+$ ions join to form **water molecules**. You end up with a neutral solution of sodium chloride, with no effect on litmus.

The overall equation for this neutralisation reaction is:

$$HCl\,(aq) + NaOH\,(aq) \longrightarrow NaCl\,(aq) + H_2O\,(l)$$

### The ionic equation for the reaction

The best way to show what is going on in a neutralisation reaction is to write an **ionic equation** for it.

**The ionic equation shows just the ions that take part in the reaction.**

This is how to write the ionic equation for the reaction above:

1 **First, write down all the ions present in the equation.**

  The drawings above will help you to do that:

  $$H^+\,(aq) + Cl^-\,(aq) + Na^+\,(aq) + OH^-\,(aq) \longrightarrow$$
  $$Cl^-\,(aq) + Na^+\,(aq) + H_2O\,(l)$$

2 **Now cross out any ions that appear, unchanged, on both sides of the equation.**

  $$H^+\,(aq) + \cancel{Cl^-\,(aq)} + \cancel{Na^+\,(aq)} + OH^+\,(aq) \longrightarrow$$
  $$\cancel{Cl^-\,(aq)} + \cancel{Na^+\,(aq)} + H_2O\,(l)$$

  The crossed-out ions are present in the solution, but do not take part in the reaction. So they are called **spectator ions**.

3 **What's left is the ionic equation for the reaction.**

  $$H^+\,(aq) + OH^-\,(aq) \longrightarrow H_2O\,(l)$$

So an $H^+$ ion combines with an $OH^-$ ion to produce a water molecule. This is all that happens during neutralisation.

**During neutralisation, $H^+$ ions combine with $OH^-$ ions to form water molecules.**

But an $H^+$ ion is just a **proton**, as the drawing on the right shows. So, in effect, the acid **donates** (gives) **protons** to the hydroxide ions. The hydroxide ions accept these protons, to form water molecules.

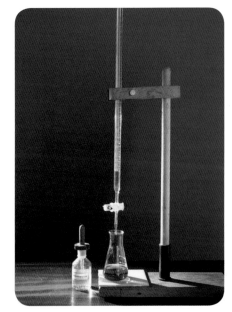

▲ A titration: sodium hydroxide solution was added to hydrochloric acid, from the burette. Neutralisation is complete: the phenolphthalein has turned pink.

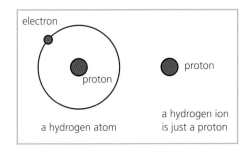

## The neutralisation of an acid by an insoluble base

Magnesium oxide is insoluble. It does not produce hydroxide ions. So how does it neutralise an acid? Like this:

Magnesium oxide is a lattice of magnesium and oxygen ions. It is insoluble in water. But when you add dilute hydrochloric acid …

… the acid donates protons to the oxide ions. The oxide ions accept them, forming water molecules. So the lattice breaks down.

The magnesium ions join the chloride ions in solution. If you evaporate the water you will obtain the salt magnesium chloride.

The equation for this neutralisation reaction is:

$2HCl\ (aq) + MgO\ (s) \longrightarrow MgCl_2\ (aq) + H_2O\ (l)$

The ionic equation for it is:

$2H^+\ (aq) + O^{2-}\ (s) \longrightarrow H_2O\ (l)$

## Proton donors and acceptors

Now compare the ionic equations for the two neutralisations in this unit:

$H^+\ (aq) + OH^-\ (aq) \longrightarrow H_2O\ (l)$

$2H^+\ (aq) + O^{2-}\ (s) \longrightarrow H_2O\ (l)$

In both:
- the protons are donated by the acids
- ions in the bases accept them, forming water molecules.

So this gives us a new definition for acids and bases:

**Acids are proton donors, and bases are proton acceptors.**

▲ Help is at hand. Indigestion is due to excess hydrochloric acid in the stomach. Milk of Magnesia contains magnesium hydroxide, which will neutralise it.

1 a What is an *ionic equation*?
  b Hydrochloric acid is neutralised by a solution of potassium hydroxide.
    What do you expect the ionic equation for this neutralisation reaction to be? Write it down.
2 What are *spectator ions*? Explain in your own words.
3 An H⁺ ion is just a proton. Explain why. (Do a drawing?)
4 a Acids act as *proton donors*. What does that mean?
  b Bases act as *proton acceptors*. Explain what that means.
5 Neutralisation is *not* a redox reaction. Explain why, using the word *proton* in your answer.

6 How to write an ionic equation:
  i Write down all the ions present in the full equation.
  ii Cross out any that are the same on both sides of the equation.
  iii What is left is the ionic equation. Rewrite it neatly.
  a Follow steps i – iii for the reaction between magnesium oxide and hydrochloric acid above.
  b Does your ionic equation match the one shown above? If so, well done!
7 Hydrochloric acid is neutralised by a solution of sodium carbonate. Write the ionic equation for this reaction.

ACIDS AND BASES

## 11.5 Oxides

### What are oxides?
**Oxides** are compounds containing oxygen and another element.
You have seen already that metal oxides act as bases. Here we look more closely at different types of oxides, and their behaviour.

### Basic oxides
Look how these metals react with oxygen:

Magnesium ribbon is lit over a Bunsen flame, and plunged into a jar of oxygen. It burns with a brilliant white flame, leaving a white ash, **magnesium oxide**:

$2Mg\ (s) + O_2\ (g) \rightarrow 2MgO\ (s)$

Hot iron wool is plunged into a gas jar of oxygen. It glows bright orange, and throws out a shower of sparks. A black solid is left in the gas jar. It is **iron(III) oxide**:

$4Fe\ (s) + 3O_2\ (g) \rightarrow 2Fe_2O_3\ (s)$

Copper is too unreactive to catch fire in oxygen. But when it is heated in a stream of the gas, its surface turns black. The black substance is **copper(II) oxide**:

$2Cu\ (s) + O_2\ (g) \rightarrow 2CuO\ (s)$

The more reactive the metal, the more vigorously it reacts.

The copper(II) oxide produced in the last reaction above is insoluble in water. But it does dissolve in dilute acid:

This is dilute hydrochloric acid. It turns blue litmus paper red, like all acids do.

Copper(II) oxide dissolves in it, when it is warmed. But after a time, no more will dissolve.

The resulting liquid has no effect on blue litmus. So the oxide has **neutralised** the acid.

Copper(II) oxide is called a **basic oxide** since it can neutralise an acid:

base + acid → salt + water
$CuO\ (s) + 2HCl\ (aq) \rightarrow CuCl_2\ (aq) + H_2O\ (l)$

Iron(III) oxide and magnesium oxide behave in the same way – they too can neutralise acid, so they are basic oxides.

**In general, metals react with oxygen to form basic oxides.**
**Basic oxides belong to the larger group of compounds called bases.**

## Acidic oxides

Now look how these non-metals react with oxygen:

Powdered carbon is heated over a Bunsen burner until red-hot, then plunged into a jar of oxygen. It glows bright red, and the gas **carbon dioxide** is formed:

$C(s) + O_2(g) \longrightarrow CO_2(g)$

Sulfur catches fire over a Bunsen burner, and burns with a blue flame. In pure oxygen it burns even brighter. The gas **sulfur dioxide** is formed:

$S(s) + O_2(g) \longrightarrow SO_2(g)$

Phosphorus bursts into flame in air or oxygen, without heating. (So it is stored under water!) A white solid, **phosphorus pentoxide**, is formed:

$P_4(s) + 5O_2(g) \longrightarrow P_4O_{10}(s)$

Carbon dioxide is slightly soluble in water. The solution will turn litmus red: it is acidic. The weak acid carbonic acid has formed:

$CO_2(g) + H_2O(l) \longrightarrow H_2CO_3(aq)$

Sulfur dioxide and phosphorus pentoxide also dissolve in water to form acids. So they are all called **acidic oxides**.

**In general, non-metals react with oxygen to form acidic oxides.**

## Amphoteric oxides

Aluminium is a metal, so you would expect aluminium oxide to be a base. In fact it is both acidic *and* basic. It acts as a base with hydrochloric acid:

$Al_2O_3(s) + 6HCl(aq) \longrightarrow 2AlCl_3(aq) + 3H_2O(l)$

But it acts as an acidic oxide with sodium hydroxide, giving a compound called sodium aluminate:

$Al_2O_3(s) + 6NaOH(aq) \longrightarrow 2Na_3AlO_3(aq) + 3H_2O(l)$

So aluminium oxide is called an **amphoteric oxide**.
**An amphoteric oxide will react with both acids and alkalis.**
Zinc oxide is also amphoteric.

▲ Zinc oxide: an amphoteric oxide. It will react with both acid and alkali.

## Neutral oxides

Some oxides of non-metals are neither acidic nor basic: they are **neutral**.
**Neutral oxides do not react with acids or bases.**
The gases carbon monoxide, CO, and dinitrogen oxide, $N_2O$ are neutral. (Other nitrogen oxides are acidic.)

▲ No pain. The neutral oxide dinitrogen oxide ($N_2O$) is used as an anaesthetic by dentists. It is also called **laughing gas**.

**Q**

1. How would you show that magnesium oxide is a base?
2. Copy and complete: Metals usually form .......... oxides while non-metals form .......... oxides.
3. See if you can arrange carbon, phosphorus and sulfur in order of reactivity, using their reaction with oxygen.
4. What colour change would you see, on adding litmus solution to a solution of phosphorus pentoxide?
5. What is an *amphoteric* oxide? Give two examples.
6. Dinitrogen oxide is a *neutral* oxide. It is quite soluble in water. How could you prove it is neutral?

ACIDS AND BASES

## 11.6 Making salts

You can make salts by reacting acids with metals, or insoluble bases, or soluble bases (alkalis), or carbonates.

### Starting with a metal
Zinc sulfate can be made by reacting dilute sulfuric acid with zinc:

$$Zn\ (s) + H_2SO_4\ (aq) \longrightarrow ZnSO_4\ (aq) + H_2\ (g)$$

These are the steps:

**1** Add the zinc to the acid in a beaker. It starts to dissolve, and hydrogen bubbles off. Bubbling stops when all the acid is used up.

**2** Some zinc is still left. (The zinc was *in excess*.) Remove it by filtering. This leaves an aqueous solution of zinc sulfate.

**3** Heat the solution to evaporate some water, to obtain a saturated solution. Leave this to cool. Crystals of zinc sulfate appear.

This method is fine for making salts of magnesium, aluminium, zinc, and iron. But you could not use it with sodium, potassium, or calcium, because these metals react violently with acids.

At the other extreme, the reaction of lead with acids is too slow, and copper, silver and gold do not react at all. (There is more about the reactivity of metals with acids in Unit 13.2.)

### Starting with an insoluble base
Copper will not react with dilute sulfuric acid. So to make copper(II) sulfate, you must start with a base such as copper(II) oxide, which is insoluble. The reaction that takes place is:

$$CuO\ (s) + H_2SO_4\ (aq) \longrightarrow CuSO_4\ (aq) + H_2O\ (l)$$

▲ Crystals of copper(II) sulfate. They are *hydrated*: they contain water molecules in the crystal structure. Their full formula is $CuSO_4.5H_2O$.

The method is quite like the one above:

**1** Add some copper(II) oxide to dilute sulfuric acid. It dissolves on warming, and the solution turns blue. Add more until no more will dissolve ...

**2** ... which means all the acid has now been used up. Remove the excess solid by filtering. This leaves a blue solution of copper(II) sulfate in water.

**3** Heat the solution to obtain a saturated solution. Then leave it to cool. Crystals of copper(II) sulfate form. They look like the crystals in the photo above.

You could also use copper(II) carbonate as the starting compound here.

ACIDS AND BASES

## Starting with an alkali (soluble base)

It is dangerous to add sodium to acid. So to make sodium salts, start with sodium hydroxide. You can make sodium chloride like this:

NaOH (*aq*) + HCl (*aq*) ⟶ NaCl (*aq*) + H$_2$O (*l*)

Both reactants are soluble, and no gas bubbles off. So how can you tell when the reaction is complete? By carrying out a **titration**.

In a titration, one reactant is slowly added to the other in the presence of an indicator. The indicator changes colour when the reaction is complete. So you know how much reactant is needed for a complete reaction. Now you can mix the correct amounts, *without* the indicator.

▲ The phenolphthalein says 'alkaline'.

## The steps in making sodium chloride

You could use **phenolphthalein** as the indicator. It is pink in alkaline solution, but colourless in neutral and acid solutions. These are the steps:

**1** Put 25 cm$^3$ of sodium hydroxide solution into a flask, using a pipette (for accuracy). Add two drops of phenolphthalein.

**2** Add the acid from a burette, just a little at a time. Swirl the flask carefully, to help the acid and alkali mix.

**3** The indicator suddenly turns colourless. So the alkali has all been used up. The solution is now neutral. Add no more acid!

**4** Find how much acid you added, using the scale on the burette. This tells you how much acid is needed to neutralise 25 cm$^3$ of the alkali.

**5** Now repeat *without* the indicator. (It would be an impurity.) Put 25 cm$^3$ of alkali in the flask. Add the correct amount of acid to neutralise it.

**6** Finally, heat the solution from the flask to evaporate the water. White crystals of sodium chloride will be left behind.

You could use the same method for making potassium salts from potassium hydroxide, and ammonium salts from ammonia solution.

**Q**

1 What will you start with, to make the salt zinc chloride?
2 You would not make lead salts by reacting lead with acids.
   **a** Why not?   **b** Suggest a way to make lead nitrate.
3 Look at step 2 at the top of page 158. The zinc was *in excess*. What does that mean? (Check the glossary?)
4 What is the purpose of a titration?
5 For carrying out a titration, a burette and pipette are used rather than measuring cylinders. Why?
6 You are asked to make the salt ammonium nitrate. Which reactants will you use?

159

ACIDS AND BASES

## 11.7 Making insoluble salts by precipitation

### Not all salts are soluble
The salts we looked at so far have all been soluble. You could obtain them as crystals, by evaporating solutions. But not all salts are soluble.
This table shows the 'rules' for the solubility of salts:

| Soluble | Insoluble |
|---|---|
| *All* sodium, potassium, and ammonium salts | |
| *All* nitrates | |
| Chlorides . . . *except* | silver and lead chloride |
| Sulfates . . . *except* | calcium, barium and lead sulfate |
| Sodium, potassium, and ammonium carbonates . . . | but all other carbonates are insoluble |

### Making insoluble salts by precipitation
Insoluble salts can be made by **precipitation**.
Barium sulfate is an insoluble salt. You can make it by mixing solutions of barium chloride and magnesium sulfate:

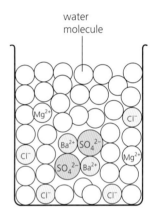

A solution of barium chloride, $BaCl_2$, contains barium ions and chloride ions, as shown here.

A solution of magnesium sulfate, $MgSO_4$, contains magnesium ions and sulfate ions.

When you mix the two solutions, the barium and sulfate ions bond together. Barium sulfate forms as a precipitate.

The equation for the reaction is:

$$BaCl_2\,(aq) + MgSO_4\,(aq) \longrightarrow BaSO_4\,(s) + MgCl_2\,(aq)$$

The ionic equation is:

$$Ba^{2+}\,(aq) + SO_4^{2-}\,(aq) \longrightarrow BaSO_4\,(s)$$

This does not show the magnesium and chloride ions, because they are **spectator ions**. They are present, but do not take part in the reaction.

### The steps in making barium sulfate
1. Make up solutions of barium chloride and magnesium sulfate.
2. Mix them. A white precipitate of barium sulfate forms at once.
3. Filter the mixture. The precipitate is trapped in the filter paper.
4. Rinse the precipitate by running distilled water through it.
5. Then place it in a warm oven to dry.

▲ The precipitation of barium sulfate.

# ACIDS AND BASES

## Choosing the starting compounds

Barium sulfate can also be made from barium nitrate and sodium sulfate, since both of these are soluble. As long as barium ions and sulfate ions are present, barium sulfate will precipitate.

**To precipitate an insoluble salt, you must mix a solution that contains its positive ions with one that contains its negative ions.**

## Some uses of precipitation

Precipitation has some important uses in industry. For example:

- It is used to make coloured pigments for paint.
- It is used in some places to remove harmful substances dissolved in water, when cleaning up waste water.
- It is used in making film, for photography. For this, solutions of silver nitrate and potassium bromide are mixed with gelatine. A precipitate of tiny crystals of insoluble silver bromide forms. The mixture is then coated onto clear film, giving **photographic film**.

Later, when light strikes the film, the silver bromide will break down:

$$2AgBr\,(s) \longrightarrow 2Ag\,(s) + Br_2\,(l)$$

You can find out more about the photographic process on page 145.

▲ The paint we use for home decoration contains insoluble pigments like these – usually made by precipitation.

▲ Putting film in a camera. Most of the film is inside the yellow cartridge, at the top, protected from light.

> **Digital cameras**
>
> Today digital cameras are more popular than cameras that use film.
>
> In a digital camera, the light strikes a surface that generates a current. This is converted to an image by a little computer inside the camera.

▲ Steady on! Most movies are shot on film, which is coated with silver halides in gelatine, just like camera film. Chemicals mixed with the halides provide the colour.

---

**Q**

1. Explain what *precipitation* means, in your own words.
2. Name four salts you could not make by precipitation.
3. Choose two starting compounds you could use to make these insoluble salts:
    a calcium sulfate    b magnesium carbonate
    c zinc carbonate    d lead chloride
4. Write a balanced equation for each reaction in **3**.
5. a  What is a spectator ion?
    b  Identify the spectator ions for your reactions in **3**.
6. Write the ionic equations for the reactions in **3**.
7. Why is precipitation necessary, in making photographic film?

ACIDS AND BASES

## 11.8 Finding concentrations by titration

### How to find a concentration by titration

On page 159, the volume of acid needed to neutralise an alkali was found by adding the acid a little at a time, until the indicator showed that the reaction was complete. This method is called **titration**.

You can find the *concentration* of an acid using the same method. You use a solution of alkali of known concentration (a **standard solution**) and titrate the acid against it.

### An example

You are asked to find the concentration of a solution of hydrochloric acid, using a 1 M solution of sodium carbonate as the standard solution.

**First, titrate the acid against your standard solution.**

- Measure 25 cm³ of the sodium carbonate solution into a conical flask, using a pipette. Add a few drops of methyl orange indicator. The indicator goes yellow.
- Pour the acid into a 50 cm³ burette. Record the level.
- Drip the acid slowly into the conical flask. Keep swirling the flask. Stop adding acid when a single drop finally turns the indicator red. Record the new level of acid in the burette.
- Calculate the volume of acid used. For example:
  Starting level:     1.0 cm³
  Final level:       28.8 cm³
  Volume used:      27.8 cm³

So 27.8 cm³ of the acid neutralised 25 cm³ of the alkaline solution.

You can now calculate the concentration of the acid.

**Step 1** Calculate the number of moles of sodium carbonate used.

1000 cm³ of 1 M solution contains 1 mole so

25 cm³ contains $\frac{25}{1000} \times 1$ mole or 0.025 mole.

**Step 2** From the equation, find the molar ratio of acid to alkali.

$2HCl\ (aq) + Na_2CO_3\ (aq) \longrightarrow 2NaCl\ (aq) + H_2O\ (l) + CO_2\ (g)$
2 moles         1 mole
The ratio is 2 moles of acid to 1 of alkali.

**Step 3** Work out the number of moles of acid neutralised.
1 mole of alkali neutralises 2 moles of acid so
0.025 mole of alkali neutralises 2 × 0.025 moles of acid.
0.05 moles of acid were neutralised.

**Step 4** Calculate the concentration of the acid.

The volume of acid used was 27.8 cm³ or 0.0278 dm³.

$$\text{concentration} = \frac{\text{number of moles}}{\text{volume in dm}^3} = \frac{0.05}{0.0278} = 1.8\ mol/dm^3$$

So the concentration of the hydrochloric acid is **1.8 M.**

> **Remember!**
> - Concentration is usually given as moles per dm³ or mol/dm³
> - 1000 cm³ = 1 dm³
> - To convert cm³ to dm³ move the decimal point 3 places left. So 250 cm³ = 0.25 dm³

**Use the calculation triangle**

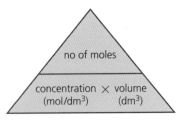

▲ Cover 'concentration' with your finger to see how to calculate it.

# ACIDS AND BASES

▲ You can find how much alkali is needed to neutralise acid by doing a titration using indicator, as here …

▲ … or you could use a pH meter, to measure the pH of the solution. How will you know when neutralisation is complete?

## Another sample calculation

Vinegar is mainly a solution of the weak acid ethanoic acid. 25 cm³ of vinegar were neutralised by 20 cm³ of 1 M sodium hydroxide solution. What is the concentration of ethanoic acid in the vinegar?

**Step 1** Calculate the number of moles of sodium hydroxide used.

1000 cm³ of 1 M solution contains 1 mole so

20 cm³ contains $\frac{20}{1000} \times 1$ mole or 0.02 mole.

**Step 2** From the equation, find the molar ratio of acid to alkali.

$CH_3COOH\ (aq) + NaOH\ (aq) \longrightarrow CH_3COONa\ (aq) + H_2O\ (l)$
1 mole          1 mole

The ratio is 1 mole of acid to 1 mole of alkali.

**Step 3** Work out the number of moles of acid neutralised.

1 mole of alkali neutralises 1 mole of acid so
0.02 mole of alkali neutralise 0.02 mole of acid.

**Step 4** Calculate the concentration of the acid. (25 cm³ = 0.025 dm³)

$$\text{concentration} = \frac{\text{number of moles}}{\text{volume in dm}^3} = \frac{0.02}{0.025} = 0.8\ \text{mol/dm}^3$$

So the concentration of ethanoic acid in the vinegar is 0.8 M.

Note: ethanoic acid is only partly dissociated into ions at any given time. (It is a weak acid.) But as the neutralisation proceeds, it continues to dissociate until it has all reacted.

▲ The ethanoic acid in vinegar – the bottle on the left – gives salad dressing its tasty tang.

## Q

1. What is a *standard solution*?
2. What volume of 2 M hydrochloric acid will neutralise 25 cm³ of 2 M sodium carbonate?
3. 20 cm³ of 1 M sulfuric acid were neutralised by 25 cm³ of ammonia solution. Calculate the concentration of the ammonia solution. (See the equation on page 229.)

ACIDS AND BASES

# Checkup on Chapter 11

## Revision checklist

### Core curriculum
Make sure you can …
- ☐ name the common laboratory acids and alkalis, and give their formulae
- ☐ describe the effect of acids and alkalis on litmus
- ☐ explain what the pH scale is, and what pH numbers tell you
- ☐ describe what *universal indicator* is, and how its colour changes across the pH range
- ☐ define a *base*, and say that alkalis are soluble bases
- ☐ say what is formed when acids react with:
  - metals    bases    carbonates
- ☐ explain what a *neutralisation* reaction is, and identify one from its equation
- ☐ say what gas is given off when strong bases are heated with ammonium compounds
- ☐ say why it is important to control acidity in soil, and how this is done
- ☐ explain what *basic oxides* and *acidic oxides* are, and give examples
- ☐ choose suitable reactants for making a salt
- ☐ describe methods for preparing a solid salt, starting with:
  - a metal or insoluble base
  - an alkaline solution
- ☐ explain how and why an indicator is used, in a titration

### Extended curriculum
Make sure you can also …
- ☐ define *strong acids* and *weak acids*, with examples
- ☐ define *strong alkalis* and *weak alkalis*, with examples
- ☐ explain why the reaction between an acid and a metal is a redox reaction
- ☐ explain what happens in a neutralisation reaction, and give the ionic equation
- ☐ give a definition for acids and bases using the idea of *proton transfer*
- ☐ say what *amphoteric oxides* and *neutral oxides* are, and give examples
- ☐ choose suitable reactants for making an insoluble salt by precipitation
- ☐ say what *spectator ions* are, and identify the spectator ions in a precipitation reaction
- ☐ calculate the concentration of a solution of acid or alkali, using data from a titration

## Questions

### Core curriculum

1  Rewrite the following, choosing the correct word from each pair in brackets.
   Acids are compounds that dissolve in water giving hydrogen ions. Sulfuric acid is an example. It can be neutralised by (acids/bases) to form salts called (nitrates/sulfates).
   Many (metals/non-metals) react with acids to give (hydrogen/carbon dioxide). Acids react with (chlorides/carbonates) to give (hydrogen/carbon dioxide).
   Since they contain ions, solutions of acids are (good/poor) conductors of electricity. They also affect indicators. Litmus turns (red/blue) in acids while phenolphthalein turns (pink/colourless). The level of acidity of an acid is shown by its (concentration/pH number). The (higher/lower) the number, the more acidic the solution.

2  **A** and **B** are white powders. **A** is insoluble in water, but **B** dissolves. Its solution has a pH of 3.
   A mixture of **A** and **B** bubbles or effervesces in water, giving off a gas. A clear solution forms.
   a  Which of the two powders is an acid?
   b  The other powder is a carbonate. Which gas bubbles off in the reaction?
   c  Although **A** is insoluble in water, a clear solution forms when the mixture of **A** and **B** is added to water. Explain why.

3  Oxygen reacts with other elements to form oxides. Three examples are: calcium oxide, phosphorus pentoxide, and copper(II) oxide.
   a  Which of these is:
      i   an insoluble base?
      ii  a soluble base?
      iii an acidic oxide?
   b  When the soluble base is dissolved in water, the solution changes the colour of litmus paper. What colour change will you see?
   c  Name the gas given off when the soluble base is heated with ammonium chloride.
   d  i   Write a word equation for the reaction between the insoluble base and sulfuric acid.
      ii  What is this type of reaction called?
   e  Name another acidic oxide.

| Method of preparation | Reactants | Salt formed | Other products |
|---|---|---|---|
| a  acid + alkali | calcium hydroxide and nitric acid | calcium nitrate | water |
| b  acid + metal | zinc and hydrochloric acid | ................ | ................ |
| c  acid + alkali | ................ and potassium hydroxide | potassium sulfate | water only |
| d  acid + carbonate | ................ and ................ | sodium chloride | water and ................ |
| e  acid + metal | ................ and ................ | iron(II) sulfate | ................ |
| f  acid + ................ | nitric acid and sodium hydroxide | ................ | ................ |
| g  acid + insoluble base | ................ and copper(II) oxide | copper(II) sulfate | ................ |
| h  acid + ................ | ................ and ................ | copper(II) sulfate | carbon dioxide and ................ |

**4** The table above is about the preparation of salts.
  **i** Copy it and fill in the missing details.
  **ii** Write balanced equations for the eight reactions.

**5** The drawings show the preparation of copper(II) ethanoate, a salt of ethanoic acid.

  **a** Which gas is given off in stage **ii**?
  **b i** Write a word equation for the reaction in ii.
   **ii** How can you tell when it is over?
  **c** Which reactant above is:
   **i** present in excess? What is your evidence?
   **ii** completely used up in the reaction?
  **d** Copper(II) carbonate is used in powder form, rather than as lumps. Suggest a reason.
  **e** Name the residue in stage **iv**.
  **f** Write a list of instructions for carrying out this preparation in the laboratory.
  **g** Suggest another copper compound to use instead of copper(II) carbonate, to make the salt.

### Extended curriculum

**6** Magnesium sulfate ($MgSO_4$) is the chemical name for Epsom salts. It can be made in the laboratory by neutralising the base magnesium oxide (MgO).
  **a** Which acid should be used to make Epsom salts?
  **b** Write a balanced equation for the reaction.
  **c i** The acid is fully dissociated in water. Which term describes this type of acid?
   **ii** Which ion causes the 'acidity' of the acid?
  **d i** What is a base?
   **ii** Write an ionic equation that shows the oxide ion ($O^{2-}$) acting as a base.

**7 a i** From the list on page 160, write down two starting compounds that could be used to make the insoluble compound silver chloride.
   **ii** What is this type of reaction called?
  **b i** Write the ionic equation for the reaction.
   **ii** List the spectator ions for the reaction.

**8** **Washing soda** is crystals of hydrated sodium carbonate, $Na_2CO_3.xH_2O$.
The value of $x$ can be found by titration.
In the experiment, 2 g of hydrated sodium carbonate neutralised 14 $cm^3$ of a standard 1 M solution of hydrochloric acid.
  **a** What does *hydrated* mean?
  **b** Write a balanced equation for the reaction that took place during the titration.
  **c** How many moles of HCl were neutralised?
  **d** How many moles of sodium carbonate, $Na_2CO_3$, were in 2 g of the hydrated salt?
  **e** What mass of sodium carbonate, $Na_2CO_3$, is this? ($M_r$: Na = 23, C = 12, O = 16)
  **f** What mass of the hydrated sodium carbonate was water?
  **g** How many moles of water is this?
  **h** How many moles of water are there in 1 mole of $Na_2CO_3.xH_2O$?
  **i** Write the full formula for washing soda.

THE PERIODIC TABLE

# 12.1 An overview of the Periodic Table

## What is the Periodic Table?

You met the Periodic Table briefly in Chapter 3. Let's review its key points.

- The Periodic Table is a way of classifying the elements.
- It shows them in order of their proton number.
  Lithium has 3 protons, beryllium has 4, boron has 5, and so on.
  (The proton number is the *lower* number beside each symbol.)
- When arranged by proton number, the elements show **periodicity**: elements with similar properties appear at regular intervals. The similar elements are arranged in columns.
- Look at the columns numbered 0 to VII. The elements in these form families called **groups**. Look where Group 0 is.
- The rows are called **periods**. They are numbered 0 to 7.
- The heavy zig-zag line above separates **metals** from **non-metals**, with the non-metals to the right (except for hydrogen).

## More about the groups

- The group number is the same as the number of outer-shell electrons in the atoms, except for Group 0. In Group I the atoms have one outer-shell electron, in Group II they have two, and so on.
- The outer-shell electrons are also called **valency electrons**. And they are very important: they dictate how an element behaves.
- So all the elements in a group have similar reactions, because they have the same number of valency electrons.
- The atoms of the Group 0 elements have a very stable arrangement of electrons in their outer shells. This makes them **unreactive**.

**The small numbers**
The two numbers beside a symbol tell you about the particles in the nucleus of its atoms:
  nucleon number
          symbol
  proton number
- The nucleon number is the total number of particles in the nucleus (protons + neutrons)
- The proton number is the number of protons.
These numbers are for the main isotope of each element.

**Groups with special names**
Group I: **the alkali metals**
Group II: **the alkaline earth metals**
Group VII: **the halogens**
Group 0: **the noble gases**

# THE PERIODIC TABLE

## More about the periods
The period number tell you the number of electron shells in the atoms. So in the elements of Period 2, the atoms have two electron shells. In Period 3 they have three, and so on.

## The metals and non-metals
Look again at the table. The metals are to the left of the zig-zag line. There are far more metals than non-metals. In fact over 80% of the elements are metals. Metals and non-metals have very different properties. See Unit 3.5 for more.

## Hydrogen
Find hydrogen in the table. It sits alone. That is because it has one outer electron, and forms a positive ion ($H^+$) like the Group I metals – but unlike them it is a gas, and usually reacts like a non-metal.

## The transition elements
The **transition elements**, in the block in the middle of the Periodic Table, are all metals. There is more about these in Unit 12.5.

## Artificial elements
Some of the elements in the Periodic Table are artificial: they have been created in the lab. Most of these are in the lowest block. They include neptunium (Np) to lawrencium (Lr) in the bottom row. These artificial elements are radioactive, and their atoms break down very quickly. (That is why they are not found in nature.)

## Patterns and trends in the Periodic Table
As you saw, the elements in a group behave in a similar way. But they also show **trends**. For example as you go down Group I, the elements become *more* reactive. Down Group VII, they become *less* reactive.

Across a period there is another trend: a change from metal to non-metal. For example in Period 2, only sodium, magnesium, and aluminium are metals. The rest are non-metals.

So if you know where an element is, in the Periodic Table, you can use the patterns and trends to predict how it will behave.

▲ A world-famous structure, made from iron. Find iron in the Periodic Table. Which block is it in?

▲ Aluminium is used for drinks cans. How many valency electrons?

1 Use the Periodic Table to find the names of:
   a three metals in common use around you
   b two non-metals that you breathe in.
2 Using *only* the Periodic Table to help you, write down everything you can about:  a nitrogen  b magnesium
3 Only two groups in the table are completely non-metal. Which two?
4 Name three elements that are likely to react in a similar way to:  a sodium  b fluorine
5 Which is likely to be more reactive, oxygen or krypton? Why?
6 Which element is named after:
   a Europe?  b Dmitri Mendeleev?  c America?
7 Chemists consider the Periodic Table very useful. Why?

THE PERIODIC TABLE

# 12.2 Group I: the alkali metals

## What are they?
The **alkali metals** are in Group I in the Periodic Table: lithium, sodium, potassium, rubidium, caesium and francium. Only the first three of these are safe to keep in the school lab. The rest are violently reactive.

## Their physical properties
The alkali metals are *not* typical metals.
- Like all metals, they are good conductors of heat and electricity.
- But they are softer than most other metals. You can cut them with a knife.
- They are 'lighter' than most other metals – they have **low density**. So they float on water – while reacting with it.
- They have low melting and boiling points, compared with most metals.

▲ A piece of sodium, cut with a knife.

## The trends in their physical properties
Like any family, the alkali metals are all a little different. Look at this table:

| Metal | This metal is silvery and … | | Density in g/cm³ | | Melts at /°C | |
|---|---|---|---|---|---|---|
| lithium, Li | soft | | 0.53 | | 181 | |
| sodium, Na | a little softer | softness increases | 0.97 | density increases | 98 | melting points decrease |
| potassium, K | softer still | | 0.86 | | 63 | |
| rubidium, Rb | even softer | | 1.53 | | 39 | |
| caesium, Cs | the softest | | 1.88 | | 29 | |

So there is an overall increase or decrease for each property, as you go down the table. This kind of pattern is called a trend.

## Their chemical properties
Let's compare the reactions of lithium, sodium, and potassium, in the lab.

### 1 Reaction with water
All three react violently with water, giving hydrogen and a hydroxide.

| Experiment | What you see | |
|---|---|---|
| *trough of water and indicator, with metal* | • lithium floats and fizzes<br>• sodium shoots across the water<br>• potassium melts with the heat of the reaction, and the hydrogen catches fire | increasing reactivity |

Note the trend in reactivity. For sodium the reaction is:

sodium + water → sodium hydroxide + hydrogen

Sodium hydroxide is an alkali, so the indicator changes colour.

**The alkali metals react vigorously with water. Hydrogen bubbles off, leaving solutions of their hydroxides, which are alkalis.**

## 2 Reaction with chlorine

If you heat the three metals, and plunge them into gas jars of chlorine, they burst into flame. They burn brightly, forming chlorides. For example:

sodium + chlorine → sodium chloride

## 3 Reaction with oxygen

The three metals also burst into flame when you heat them and plunge them into gas jars of oxygen. They burn fiercely to form **oxides**. These dissolve in water to give alkaline solutions.

The same trend in reactivity is shown in all three reactions. Each time, lithium is the least reactive of the three elements, and potassium the most:

**Reactivity increases as you go down Group I.**

## Why do they react in a similar way?

All the alkali metals react in a similar way. Why? Because they have the same number of valency (outer-shell) electrons:

**Atoms with the same number of valency electrons react in a similar way.**

## Why are they so reactive?

**The alkali metals are the most reactive of all the metals.**

Why? Because they need to lose only one electron, to gain a stable outer shell. So they have a strong drive to react with other elements and compounds, in order to give up this electron. They become **ions**. The compounds they form are **ionic**. For example sodium chloride is made up of the ions $Na^+$ and $Cl^-$.

**The alkali metals form ionic compounds, in which the metal ion has a charge of 1+. The compounds are white solids. They dissolve in water to give colourless solutions.**

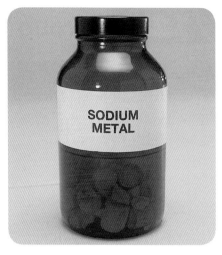

▲ Lithium, sodium, and potassium are stored under oil in the lab, to prevent reaction with oxygen and water.

### Why does reactivity *increase* down Group I?

In reactions, the Group I atoms lose their outer electron, to gain a stable outer shell.

The more shells there are, the further the outer electron is from the positive nucleus – so the easier to lose.

And the easier it is to lose an electron, the more reactive the metal will be!

### And the winner is …

Lithium is the lightest of all metals.

---

**Q**

**1 a** What is the other name for the Group I elements?
 **b** Why are they called that?
**2** Which best describes the Group I metals:
 **a** soft or hard? **b** reactive or unreactive?
**3** The Group I metals show a *trend* in melting points.
 **a** What does that mean?
 **b** Describe two other physical trends for the group.
 **c** One measurement in the table on page 168 does *not* fit the trend. See if you can spot it.

**4 a** What forms when potassium reacts with chlorine?
 **b** What colour is this compound?
 **c** What will you *see* when you dissolve it in water?
 **d** Will the solution conduct electricity? Explain.
**5** Which holds its outer electron more strongly: a lithium atom, or a sodium atom? Explain why you think so.
**6** Rubidium is below potassium, in Group I. Predict how it will react with: **a** water **b** chlorine
 and describe the products that form.

THE PERIODIC TABLE

## 12.3 Group VII: the halogens

### A non-metal group
Group VII is a group of non-metal elements. It includes fluorine, chlorine, bromine, and iodine. These are usually called the **halogens**. They all:
- form coloured gases. Fluorine is a pale yellow gas and chlorine is a green gas. Bromine forms a red vapour, and iodine a purple vapour.
- are poisonous.
- form diatomic molecules (containing two atoms). For example, $Cl_2$.

### Trends in their physical properties
As usual, the group shows trends in physical properties. Look at these:

| Halogen | At room temperature the element is … | | | Boiling point /°C | |
|---|---|---|---|---|---|
| fluorine, $F_2$ | a yellow gas | | | −188 | |
| chlorine, $Cl_2$ | a green gas | colour gets deeper | density increases | −35 | boiling points increase |
| bromine, $Br_2$ | a red liquid | | | 59 | |
| iodine, $I_2$ | a black solid | | | 184 | |

### Trends in their chemical properties
The halogens are among the most reactive elements in the Periodic Table.
They react with metals to form compounds called **halides**. For example:

| Halogen | Reaction with iron wool | | The product | Its appearance |
|---|---|---|---|---|
| fluorine | Iron wool bursts into flame as fluorine passes over it – without any heating! | | iron(III) fluoride, $FeF_3$ | pale green solid |
| chlorine | Hot iron wool glows brightly when chlorine passes over it. | reactivity decreases | iron(III) chloride, $FeCl_3$ | yellow solid |
| bromine | Hot iron wool glows, but less brightly, when bromine vapour passes over it. | | iron(III) bromide, $FeBr_3$ | red-brown solid |
| iodine | Hot iron wool shows a faint red glow when iodine vapour passes over it. | | iron(III) iodide, $FeI_3$ | black solid |

So they all react in a similar way. But note the trend in reactivity:
**Reactivity decreases as you go down Group VII.**

### Why do they react in a similar way?
The halogens react in a similar way because their atoms all have 7 valency (outer-shell) electrons. Compare the fluorine and chlorine atoms:

**Atoms with the same number of valency electrons react in a similar way.**

▲ Iodine is a disinfectant. His skin is being wiped with a solution of iodine in ethanol, before he gives blood.

# THE PERIODIC TABLE

## Why are they so reactive?

The halogen atoms need just one more electron to reach a stable outer shell of 8 electrons. So they have a strong drive to react with other elements or compounds, to gain this electron. That is why they are so reactive.

When halogen atoms react with metal atoms they accept electrons, forming halide ions. So the products are ionic. For example the reaction between iron and chlorine gives iron(III) chloride, made up of $Fe^{3+}$ and $Cl^-$ ions.

But with non-metal atoms such as hydrogen and carbon, they share electrons, forming molecules with covalent bonds. For example hydrogen and chlorine atoms share electrons, to form molecules of hydrogen chloride, HCl.

> **Why does reactivity *decrease* down Group VII?**
>
> Halogen atoms react to gain or share an electron. The positive nucleus of the atom attracts the extra electron.
>
> The more shells there are, the further the outer shell is from the nucleus. So attracting an electron becomes more difficult. So reactivity falls.

## How the halogens react with halides

1 When chlorine water (a solution of chlorine) is added to a colourless solution of potassium bromide, the solution turns orange, as shown in the photo. This reaction is taking place:

$Cl_2$ (*aq*) + 2KBr (*aq*) → 2KCl (*aq*) + $Br_2$ (*aq*)
colourless                                                       orange

Bromine has been pushed out of its compound, or **displaced**.

2 And when chlorine water is added to a colourless solution of potassium iodide, the solution turns red-brown, because of this reaction:

$Cl_2$ (*aq*) + 2KI (*aq*) → 2KCl (*aq*) + $I_2$ (*aq*)
colourless                                                       red-brown

This time iodine has been displaced.

But what happens if you use bromine or iodine instead of chlorine? This table gives the results:

▲ Chlorine displacing bromine from aqueous potassium bromide.

| If the solution contains ... | when you add chlorine ... | when you add bromine ... | when you add iodine ... |
|---|---|---|---|
| chloride ions ($Cl^-$) |  | there is no change | there is no change |
| bromide ions ($Br^-$) | bromine is displaced |  | there is no change |
| iodide ions ($I^-$) | iodine is displaced | iodine is displaced |  |

You know already that chlorine is more reactive than bromine, and bromine is more reactive than iodine. So now you can see that:

**A halogen will displace a less reactive halogen from a solution of its halide.**

## Q

1 What do the halogens look like? Describe them.
2 a Describe the trend in reactivity in Group VII.
  b Is this trend the same as for Group I? (Check back!)
3 a Describe any similarities you notice in the *products* that form when the halogens react with iron wool.
  b Which type of bonding do they have?
4 What makes the halogens so reactive?
5 a Write a word equation for the reaction of bromine with potassium iodide. What do you expect to *see*?
  b Now explain *why* the reaction in **a** occurs.
6 The fifth element in Group VII is called astatine. It is a very rare element. Do you expect it to be:
  a a gas, a liquid, or a solid? Give your reason.
  b coloured or colourless?   c harmful or harmless?

THE PERIODIC TABLE

## 12.4 Group 0: the noble gases

### The noble gases
This group of non-metals contains the elements helium, neon, argon, krypton and xenon. These elements are all:

- non-metals
- colourless gases, which occur naturally in air
- **monatomic** – they exist as single atoms
- unreactive. This is their most striking property. They do not normally react with anything. That is why they are called **noble**.

### Why are they unreactive?
As you have seen, atoms react in order to gain a stable outer shell of electrons. But the atoms of the noble gases already have a stable outer shell – with 8 electrons, except for helium which has 2 (since it has only one shell):

| Noble gas | % in air |
|---|---|
| helium | tiny traces |
| argon | just under 1 % |
| neon | 0.002 % |
| krypton | 0.0001 % |
| xenon | less than 0.0001 % |

Helium is the second most abundant element in the universe, after hydrogen. But it is so light that it escapes from our atmosphere.

a helium atom   a neon atom   an argon atom

So the atoms have no need to react in order to gain or lose electrons. **The noble gases are unreactive, and monatomic, because their atoms already have a stable outer electron shell.**

### Where we get them
We obtain helium from natural gas, in which it is an impurity. We get the other noble gases from the air, in the fractional distillation of liquid air (page 212).

### Trends in their physical properties
Like all groups, the Group 0 elements do show trends. Look at this table.

| Noble gas | Its atoms | | A balloon full of this gas ... | | Boiling point / °C | |
|---|---|---|---|---|---|---|
| helium | $^4_2$He | | rises quickly into the air | | −269 | |
| neon | $^{20}_{10}$Ne | the atoms increase in size and mass | rises slowly | the density of the gases increases | −246 | the boiling points increase |
| argon | $^{40}_{18}$Ar | | falls slowly | | −186 | |
| krypton | $^{84}_{36}$Kr | | falls quickly | | −152 | |
| xenon | $^{131}_{54}$Xe | | falls very quickly | | −107 | |

The gases grow denser (or 'heavier') down the group, because the mass of the atoms increases. The increase in boiling points is a sign of increasing attraction between atoms. It gets harder to separate them to form a gas.

Compare these physical trends with those for the Group VII non-metals on page 170. What do you notice?

# THE PERIODIC TABLE

▲ Colourful signs in Tokyo, thanks to neon.

▲ Cool blue headlamps, thanks to xenon.

## Uses of the noble gases

The noble gases are unreactive or inert, which makes them safe to use. They also glow when a current is passed through them at low pressure. These properties lead to many uses.

- Helium is used to fill balloons and airships, because it is much lighter than air – and will not catch fire.
- Argon is used to provide an inert atmosphere. For example it is used:
  - as a filler in tungsten light bulbs. (If air were used, the oxygen in it would make the tungsten filament burn away.)
  - to protect metals that are being welded. It won't react with the hot metals (unlike the oxygen in air).
- Neon is used in advertising signs. It glows red, but the colour can be changed by mixing it with other gases.
- Krypton is used in lasers – for example for eye surgery – and in car headlamps.
- Xenon gives a light like bright daylight, but with a blue tinge. It is used in lighthouse lamps, lights for hospital operating rooms, and car headlamps.

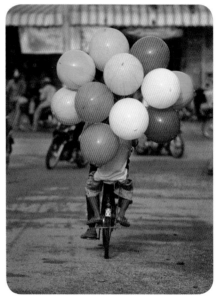

▲ There is an easy way to blow up balloons: buy a canister of helium.

## Q

1. Why do the members of Group 0 have similar properties?
2. Explain why the noble gases are unreactive.
3. **a** What are the trends in density and boiling point for the noble gases?
   **b** Are these trends the same as for:
      **i** Group I?   **ii** Group VII?   (Check back!)
4. The noble gases are widely used. Explain why, and give one use for each.
5. The sixth element in Group 0 is radon (Rn). Would you expect it to be:
   **a** a gas, a liquid, or a solid, at room temperature?
   **b** heavier, or lighter, than xenon?
   **c** chemically reactive?

THE PERIODIC TABLE

## 12.5 The transition elements

### What are they?
The transition elements are the block of 30 elements in the middle of the Periodic Table. They are all metals, and include most of the metals we use every day – such as iron, tin, copper, and silver.

### Their physical properties
Here are three of the transition elements:

Iron: the most widely used metal; grey with a metallic lustre (shine).

Copper: reddish with a metallic lustre.

Nickel: silvery with a metallic lustre.

Here is some data for them, with sodium for comparison:

| Element | Symbol | Density in g/cm³ | Melting point/°C |
|---|---|---|---|
| iron | Fe | 7.9 | 1535 |
| copper | Cu | 8.9 | 1083 |
| nickel | Ni | 8.9 | 1455 |
| sodium | Na | 0.97 | 98 |

**Some transition elements**

iron  copper
nickel  zinc
silver  gold
platinum  mercury
chromium  cadmium

The transition elements share these physical properties:
- **hard, tough and strong**. They are not soft like the Group I metals.
- **high melting points**. Look at the values in the table. But mercury is an exception. It is a liquid at room temperature. (It melts at −39 °C.)
- **malleable** (can be hammered into different shapes) and **ductile** (can be drawn out into wires).
- **good conductors of heat and electricity**. Of all the metals, silver is the best conductor of electricity, and copper is next.
- **high density**. They are heavy. 1 cm³ cube of iron weighs 7.9 grams – over 8 times more than 1 cm³ cube of sodium.

### Their chemical properties
1 **They are much less reactive than the Group I metals.**
   For example copper and nickel do not react with water, or catch fire in air – unlike sodium. In general, the transition elements do not corrode readily in the atmosphere. But iron is an exception – it rusts easily. We spend a fortune every year on rust prevention.
2 **They show no clear trend in reactivity, unlike the Group I metals.**
   But those next to each other in the Periodic Table do tend to be similar.

▲ Because they are coloured, compounds of the transition elements are used in pottery glazes.

THE PERIODIC TABLE

3 **Most transition elements form coloured compounds.** In contrast, the Group I metals form white compounds.

4 **Most can form ions with different charges.** Compare these:

| Metal | Charge on ions | Examples |
|---|---|---|
| Group I metals | always 1+ | sodium: $Na^+$ |
| Group II metals | always 2+ | magnesium: $Mg^{2+}$ |
| Group III metals | always 3+ | aluminium: $Al^{3+}$ |
| Transition elements | variable | copper: $Cu^+$, $Cu^{2+}$ <br> iron: $Fe^{2+}$, $Fe^{3+}$ |

So we say the transition elements show **variable valency**.

5 **They can form more than one compound with another element.** This is because of their variable valency. For example:

copper(I) oxide, $Cu_2O$     copper(II) oxide, $CuO$
iron(II) oxide, $FeO$     iron(III) oxide, $Fe_2O_3$

The Roman numeral in brackets tells you how many electrons the metal atom has lost. This number is called its **oxidation state**.

6 **Most transition elements can form complex ions.** For example, if you add ammonia to a solution containing copper(II) ions, a pale blue precipitate of copper(II) hydroxide forms. It dissolves again if you add more ammonia, giving a deep blue solution.

It dissolves because each copper ion attracts four ammonia molecules and two water molecules, forming a large soluble **complex ion**. This ion gives the solution its deep blue colour.

## Uses of the transition elements

- The hard, strong transition elements are used in structures such as bridges, buildings, and cars. Iron is the most widely used – usually in the form of **alloys** called **steels**. (In alloys, small amounts of other substances are mixed with a metal, to improve its properties.)

- Many transition elements are used in making alloys. For example, chromium and nickel are mixed with iron to make **stainless steel**.

- Transition elements are used as conductors of heat and electricity. For example, steel is used for radiators, and copper for electric wiring.

- Many transition elements and their compounds acts as **catalysts**. Catalysts speed up reactions, while remaining unchanged themselves. For example, iron is used as a catalyst in making ammonia (page 127).

**Salts of transition elements**
- The oxides and hydroxides of *all* metals are bases; they react with acids to form salts.
- So you can make salts of the transition elements by starting with their oxides or hydroxides, and reacting these with acids.

**Testing for copper ions**
The reaction in point 6 is used in the test for copper(II) ions (page 286.)
The formula of the complex ion is $[Cu(H_2O)_2(NH_3)_4]^{2+}$.

▲ Iron rods give the building strength.

1 Name five transition elements.
2 Which best describes the transition elements, overall:
   a soft or hard?    b high density or low density?
   c high melting point or low melting point?
   d reactive or unreactive, with water?
3 What is unusual about mercury?
4 Most paints contain compounds of transition elements. Why do you think this is?
5 Suggest reasons why copper is used in hot water pipes, while iron is not.

THE PERIODIC TABLE

## 12.6 Across the Periodic Table

### Trends across Period 3

As you saw, there are trends within groups in the Periodic Table. There are also trends across a period. Look at this table for Period 3:

| Group | I | II | III | IV | V | VI | VII | 0 |
|---|---|---|---|---|---|---|---|---|
| Element | sodium | magnesium | aluminium | silicon | phosphorus | sulfur | chlorine | argon |
| Valency electrons | 1 | 2 | 3 | 4 | 5 | 6 | 7 | 8 |
| Element is a... | metal | metal | metal | metalloid | non-metal | non-metal | non-metal | non-metal |
| Reactivity | high → | | | low → | | | high | unreactive |
| Melting point/°C | 98 | 649 | 660 | 1410 | 590 | 119 | −101 | −189 |
| Boiling point/°C | 883 | 1107 | 2467 | 2355 | (ignites) | 445 | −35 | −186 |
| Oxide is... | basic | | amphoteric | | | | acidic | – |
| Typical compound | NaCl | $MgCl_2$ | $AlCl_3$ | $SiCl_4$ | $PH_3$ | $H_2S$ | HCl | – |
| Valency shown in that compound | 1 | 2 | 3 | 4 | 3 | 2 | 1 | – |

Notice these trends across the period:

1 The number of valency (outer-shell) electrons increases by 1 each time. It is the same as the group number, for Groups I to VII.

2 The elements go from metal to non-metal. Silicon is in between. It is like a metal in some ways and a non-metal in others. It is called a **metalloid**.

3 Melting and boiling points rise to the middle of the period, then fall to very low values on the right. (Only chlorine and argon are gases at room temperature.)

4 The oxides of the metals are **basic** – they react with acids to form salts. Those of the non-metals are **acidic** – they react with alkalis to form salts. But aluminium oxide is in between: it reacts with both acids and alkalis to form salts. So it is called an **amphoteric oxide**. (See page 157 for more.)

The elements in Period 2 show similar trends.

### The change from metal to non-metal

The change from metal to non-metal is not clear-cut. Silicon is called a metalloid because it is like metal in some ways, and a non-metal in others.

In fact there are metalloids in all the periods of the table. They lie along the zig-zag line that separates metals from non-metals. Look on the right.

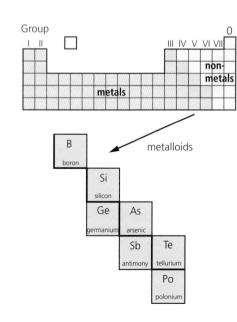

# THE PERIODIC TABLE

Metals conduct electricity. Metalloids can too, under certain conditions. So they are called **semi-conductors**. This leads to their use in computer chips and PV cells for solar power. Silicon is used the most.

## Valency

Look at the last two rows in the table. One shows a typical compound of each element. The other shows the **valency** of the element in that compound.

**The valency of an element is the number of electrons its atoms lose, gain or share, to form a compound.**

Sodium always loses 1 electron to form a compound. So it has a valency of 1. Chlorine shares or gains 1, so it also has a valency of 1. Valency rises to 4 in the middle of the period, then falls again. It is zero for the noble gases.

Note that *valency* is not the same as *the number of valency electrons*. But:
- the valency does match the number of valency electrons, up to Group IV
- the valency matches the charge on the ion, where an element forms ions.

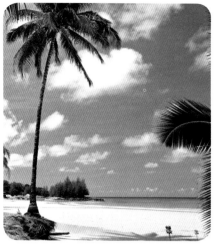

▲ Silicon occurs naturally in sand as silica (silicon dioxide). To extract it the silica is heated with carbon (coke).

## What about reactivity?

As you know, metal atoms lose their outer electrons when they react, while non-metal atoms accept or share electrons.

Reactivity across Period 3 changes roughly like this:

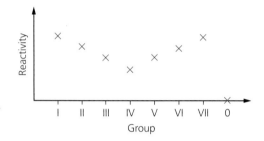

Note that:
- reactivity *decreases* across the metals. Aluminium is a lot less reactive than sodium, for example. Why? Because the more electrons a metal atom needs to lose, the more difficult it is. (The electrons must have enough energy to overcome the pull of the nucleus.)
- reactivity *increases* across the non-metals (apart from Group 0). So chlorine is more reactive than sulfur. Why? Because the fewer electrons a non-metal atom needs to gain, the easier it is to attract them.

▲ Silicon is the main element used in solar cells, to generate electricity from sunlight. It has to be 99.9999% pure!

**Q**

1. **a** Describe how the number of valency electrons changes with group number, across the Periodic Table.
   **b** Describe the change in character from metal to non-metal, across Period 3.
2. How does the reactivity of the metals change as you move across a period? Why?
3. What does *valency of an element* mean? Give two examples.
4. What is a *metalloid*? Give three examples.
5. What is a *semi-conductor*? Name one.
6. **a** A challenge! Make a table like the one opposite, but for *Period 2*. For each element the table should show:
   i   the group number
   ii  the name of the element
   iii the number of valency electrons it has
   iv  a typical compound
   v   the valency shown in that compound.
   **b** Now try to predict melting and boiling points for the elements in the period. (The earlier units may help!)

THE PERIODIC TABLE

# How the Periodic Table developed

### Life before the Periodic Table
Imagine you find a box of jigsaw pieces. You really want to build that jigsaw. But the lid has only scraps of the picture. Many of the pieces are missing. And the image on some pieces is not complete. How frustrating!

That's how chemists felt, about 150 years ago. They had found more and more new elements. For example 24 metals, including lithium, sodium, potassium, calcium, and magnesium, were discovered between 1800 and 1845. They could tell that these fitted a pattern of some kind. They could see fragments of the pattern – but could not work out what the overall pattern was.

And then the Periodic Table was published in 1869, and everything began to make sense.

### A really clever summary
The Periodic Table is *the* summary of chemistry. It names the elements that make up our world. It shows the families they belong to, and how these relate to each other. It even tells you about the numbers of protons, electrons, and electron shells in their atoms.

Today we take the Periodic Table for granted. But it took hundreds of years, and hard work by hundreds of chemists, to develop. There were some good tries along the way, like the 'Law of Octaves'.

### The Law of Octaves
By 1863, 56 elements were known. John Newlands, an English chemist, noted that there were many pairs of similar elements. In each pair, the **atomic weights** (or relative atomic masses) differed by a multiple of 8. So he produced a table with the elements in order of increasing atomic weight, and put forward the **Law of Octaves**: *an element behaves like the eighth one following it in the table.*

This was the first table to show a repeating or **periodic** pattern of properties. But it had many inconsistencies. For example it had copper and sodium in the same group – even though they behave very differently. So it was rejected by other chemists.

▲ 250 years ago, nobody knew of aluminium. Today, planes are about 80% aluminium by mass.

▲ Lithium was discovered in 1817. Lithium batteries are used in pacemakers, to keep the heartbeat steady.

**Examples of octaves**

| element | atomic weight |
|---|---|
| potassium | 39 |
| sodium | − 23 |
| | $\overline{16}$ or 2 × 8 |
| calcium | 40 |
| magnesium | − 24 |
| | $\overline{16}$ or 2 × 8 |

Now we use **relative atomic mass** instead of **atomic weight**.

◀ Newlands knew of all these, in 1865. How many of them can you name? Find Di (for didymium). This 'element' was later found to be a mixture.

Newland's Table of Octaves, presented to the Chemical Society in London in 1865

| H | Li | Be | B | C | N | O |
|---|---|---|---|---|---|---|
| F | Na | Mg | Al | Si | P | S |
| Cl | K | Ca | Cr | Ti | Mn | Fe |
| Co, Ni | Cu | Zn | Y | In | As | Se |
| Br | Rb | Sr | Ce, La | Zr | Di, Mo | Ro, Ru |
| Pd | Ag | Cd | U | Sn | Sb | Te |
| I | Cs | Ba, V | Ta | W | Nb | Au |
| Pt, Ir | Tl | Pb | Th | Hg | Bi | Os |

# THE PERIODIC TABLE

▲ Dmitri Mendeleev (1834–1907). Element 101 in the Periodic Table – the artificial element Mendelevium (Md) – is named after him. So is a crater on the moon.

▲ New elements are still being added to the Periodic Table. This is the team that created the artificial element 112, which was officially named copernicium (Cn) in 2010.

## The Periodic Table arrives

Dmitri Ivanovich Mendeleev was born in Russia in 1834, the youngest of at least 14 children. By the age of 32, he was a Professor of Chemistry.

Mendeleev had gathered a huge amount of data about the elements. He wanted to find a pattern that made sense of it, to help his students. So he made a card for each of the known elements (by then 63). He played around with the cards on a table, first putting the elements in order of atomic weight, and then into groups with similar behaviour. The result was the Periodic Table. It was published in 1869.

Mendeleev took a big risk: he left gaps for elements not yet discovered. He even named three: **eka-aluminium**, **eka-boron**, and **eka-silicon**, and predicted their properties. And soon three new elements were found, that matched his predictions – gallium, scandium and germanium. This helped to convince other chemists, and his table was accepted.

## Atomic structure and the Periodic Table

Mendeleev had put the elements in order of *atomic weight*. But he was puzzled, because he then had to swop some to get them into the right groups. For example potassium ($A_r = 39$) is lighter than argon ($A_r = 40$), so should come before argon. But a reactive metal like potassium clearly belongs to Group I, not Group 0. So he switched those two around.

In 1911 the proton was discovered. It soon became clear that the **proton number** was the key factor in deciding an element's position in the table. So Mendeleev was right to swop those elements. But it was still not clear why the groups were so different. Then scientists discovered that:

- the number of electrons equals the number of protons, in an atom
- the electrons are arranged in shells
- the outer-shell electrons dictate reactions. So elements with the same number of outer-shell electrons react in the same way.

By 1932, 63 years after it appeared, Mendeleev's table finally made sense. Today's Periodic Table contains many more elements. But his table, nearly 150 years old, is still the blueprint for it.

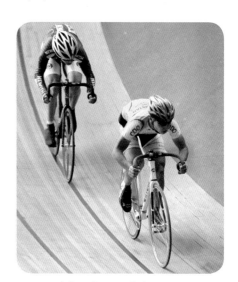

▲ Mendeleev knew of aluminium, titanium, and molybdenum, which are all used in today's racing bikes.

▲ Mendeleev would recognise all the elements in these health tablets too.

# THE PERIODIC TABLE

# Checkup on Chapter 12

## Revision checklist

### Core curriculum
Make sure you can ...
- [ ] state the link between the Periodic Table and proton number
- [ ] point out where in the Periodic Table these are:
  *Group I*   *Group VII*   *Group 0*
  *hydrogen*   *the transition elements*
- [ ] define *valency electrons*
- [ ] state the link between:
  – group number and the number of valency electrons
  – period number and the number of electron shells
- [ ] describe the change from metal to non-metal, across a period
- [ ] say why elements in a group react in a similar way
- [ ] give the other name for Group I, and name at least three elements in this group
- [ ] describe the trends in softness, melting point, density, and reactivity, for the Group I elements
- [ ] give at least two typical reactions for Group I elements, and describe the products
- [ ] explain why the Group I elements are so reactive
- [ ] give the other name for Group VII
- [ ] name at least four Group VII elements and say what they look like at room temperature
- [ ] describe the trend in reactivity for Group VII
- [ ] explain why the Group VII elements are so reactive
- [ ] describe how halogens react with solutions of other halides, and explain the pattern
- [ ] give the other name for Group 0, and name five elements in this group
- [ ] explain why the Group 0 elements are unreactive
- [ ] give one use for each Group 0 element you name
- [ ] give three physical properties and three chemical properties of the transition elements
- [ ] explain why compounds of transition elements often have Roman numerals in their names
- [ ] give some uses of transition elements, including as catalysts

### Extended curriculum
Make sure you can also ...
- [ ] give more detail about the trends across a period, including the change from metal to non-metal

## Questions

### Core curriculum

**1** This extract from the Periodic Table shows the symbols for the first 20 elements.

| H | | | | | | | | | | | | | | | | | He |
|---|---|---|---|---|---|---|---|---|---|---|---|---|---|---|---|---|---|
| Li | Be | | | | | | | | | | | B | C | N | O | F | Ne |
| Na | Mg | | | | | | | | | | | Al | Si | P | S | Cl | Ar |
| K | Ca | | | | | | | | | | | | | | | | |

Look at the row from lithium (Li) to neon (Ne).
**a** What is this row of the Periodic Table called?
**b** Which element in it is the least reactive? Why?
Look at the column of elements from lithium (Li) to potassium (K).
**c** What is this column of the table called?
**d** Of the three elements shown in this column, which one is the most reactive?

**2** Rubidium is an alkali metal. It lies below potassium in Group I. Here is data for Group I:

| Element | Proton number | Melting point/°C | Boiling point/°C | Chemical reactivity |
|---|---|---|---|---|
| lithium | 3 | 180 | 1330 | quite reactive |
| sodium | 11 | 98 | 890 | reactive |
| potassium | 19 | 64 | 760 | very reactive |
| rubidium | 37 | ? | ? | ? |
| caesium | 55 | 29 | 690 | violently reactive |

**a** Describe the trends in melting point, boiling point, and reactivity, as you go down the group.
**b** Now predict the missing data for rubidium.
**c** In a rubidium atom:
  **i** how many electron shells are there?
  **ii** how many electrons are there?
  **iii** how many valency electrons are there?

**3** Identify these non-metal elements:
**a** a colourless gas, used in balloons and airships
**b** a poisonous green gas
**c** a colourless gas that glows with a red light in advertising signs
**d** a red liquid
**e** a yellow gas which is so reactive that it is not allowed in school labs
**f** a black solid that forms a purple vapour when you heat it gently.

4 This diagram shows some of the elements in Group VII of the Periodic Table.

a What are the elements in this group called?
b Chlorine reacts explosively with hydrogen. The word equation for the reaction is:
   hydrogen + chlorine → hydrogen chloride
The reaction requires sunlight, but not heat.
  i How would you expect fluorine to react with hydrogen?
  ii Write the word equation for the reaction.
c i How might bromine react with hydrogen?
  ii Write the word equation for that reaction.

5 The Periodic Table is the result of hard work by many scientists, in many countries, over hundreds of years. They helped to develop it by discovering, and investigating, new elements.

The Russian chemist Mendeleev was the first person to produce a table like the one we use today. He put all the elements he knew of into his table. But he realized that gaps should be left for elements not yet discovered. He even predicted the properties of some of these.

Mendeleev published his Periodic Table in 1869. The extract on the right below shows Groups I and VII from his table. Use the modern Periodic Table (page 314) to help you answer these questions.

a What does Period 2 mean?
b i How does Group I in the modern Periodic Table differ from Group I in Mendeleev's table?
  ii The arrangement in the modern table is more appropriate for Group I. Explain why.
  iii What do we call the Group I elements today?
c i What do we call the Group VII elements?
  ii The element with the symbol Mn is out of place in Group VII. Why?
  iii Where is the element Mn in today's table?
d Mendeleev left gaps in several places in his table. Why did he do this?
e There was no group to the right of Group VII, in Mendeleev's table. Suggest a reason for this omission.

**Extended curriculum**

6 This question is about **elements** from these families: alkali metals, alkaline earth metals (Group II), transition elements, halogens, noble gases.
  A is a soft, silvery metal that reacts violently with water.
  B is a gas at room temperature. It reacts violently with other elements, without heating.
  C is an unreactive gas that sinks in air.
  D is a hard solid at room temperature, and forms coloured compounds.
  E conducts electricity, and reacts slowly with water. Its atoms each give up two electrons.
  F is a reactive liquid; it does not conduct electricity; it shows a valency of 1 in its compounds.
  G is a hard solid that conducts electricity, can be beaten into shape, and rusts easily.
  a For each element above, say which of the listed families it belongs to.
  b i Comment on the position of elements **A**, **B**, and **C** within their families.
    ii Describe the valence (outer) shell of electrons for each of the elements **A**, **B**, and **C**.
  c Explain why the arrangement of electrons in their atoms makes some elements very reactive, and others unreactive.
  d Name elements that fit descriptions **A** to **G**.
  e Which of **A** to **G** may be useful as catalysts?

7 The elements of Group 0 are called the noble gases. They are all monatomic gases.
  a Name four of the noble gases.
  b i What is meant by *monatomic*?
    ii Explain *why* the noble gases, unlike all other gaseous elements, are monatomic.
  *When elements react, they become like noble gases.*
  c i Explain what the above statement means.
    ii What can you conclude about the reactivity of Group VII *ions*?

*An extract from Mendeleev's Periodic Table*

|  | Group I | Group VII |
|---|---|---|
| Period 1 | H |  |
| Period 2 | Li | F |
| Period 3 | Na | Cl |
| Period 4 | K | Mn |
|  | Cu | Br |
| Period 5 | Rb |  |
|  | Ag | I |

THE BEHAVIOUR OF METALS

## 13.1 Metals: a review

### So far …
You have met quite a lot of information about metals already. We review it in this unit, before going on to look more closely at their reactivity.

### Metals and the Periodic Table
The metals lie to the left of the zig-zag line in the Periodic Table. There are far more metals than non-metals. In fact over 80% of the elements are metals. Here is a reminder of some of them:

**Group I – the alkali metals** (lithium, sodium, potassium …)

**Group III** has aluminium, the most abundant metal in the Earth's crust, in this position …

**Group II – the alkaline earth metals** (beryllium, magnesium, calcium …)

**the transition elements** – all are metals, and they include most of the metals in everyday use, like iron, copper, tin, zinc, lead, silver, gold …

### The physical properties of metals
Metals *usually* have these physical properties.

1. They are **strong**. If you press on them, or drop them, or try to tear them, they won't break – and it is hard to cut them.
2. They are **malleable**. They can be hammered into shape without breaking.
3. They are **ductile**: they can be drawn out into wires.
4. They are **sonorous**: they make a ringing noise when you strike them.
5. They are shiny when polished.
6. They are good conductors of electricity and heat.
7. They have high melting and boiling points. (They are all solid at room temperature, except mercury.)
8. They have high **density** – they feel 'heavy'. (Look at the blue panel.)

### The chemical properties of metals
1. They react with oxygen to form **oxides**. For example, magnesium burns in air to form magnesium oxide.
2. Metal oxides are **bases**: they neutralise acids, forming salts and water.
3. Metals form positive ions when they react. For example, magnesium forms magnesium ions ($Mg^{2+}$) when it reacts with oxygen.
4. For the metals in the numbered groups, the charge on the ion is the same as the group number. But the transition elements have **variable valency**: they can form ions with different charges. For example $Cu^+$ and $Cu^{2+}$.

**What is density?**
It tells you 'how heavy'.

$$\text{density} = \frac{\text{mass (in grams)}}{\text{volume (in cm}^3\text{)}}$$

Compare these:

1 cm³ of iron, mass 7.86 g
density 7.86 g/cm³

1 cm³ of lead, mass 11.34 g
density 11.34 g/cm³

▲ Dropping anchor … helped by the density of the iron.

# THE BEHAVIOUR OF METALS

▲ We are over 96% non-metal (mainly oxygen, carbon, nitrogen, hydrogen) but we also contain metals: calcium, potassium, sodium, magnesium, copper, zinc, iron and more ...

▲ Trucks at a copper mine in the USA. Metals are big business. World trade in metals is worth over 2000 billion US dollars a year, and the metals industry employs around 70 million people.

## All metals are different!

The properties on the opposite page are typical of metals. But all metals are different. They do not share *all* of those properties. For example, all *do* conduct electricity, and their oxides act as bases. But compare these:

Iron is malleable and strong. Good for gates like these! But it rusts easily in damp air. And unlike most other metals, it is **magnetic**. It melts at 1530 °C.

Sodium is so soft that you can cut it with a knife. It floats on water – and reacts immediately with it, forming a solution. So no good for gates. It melts at only 98 °C.

Gold is unreactive. It is malleable, ductile, and looks attractive. It is also quite rare. So it is used for jewellery and precious objects. It melts at 1064 °C.

So of those three metals, sodium is clearly the most reactive, and gold the least. In the next two units we will look at reactions you can do in the lab, to compare the reactivities of metals.

**Q**

1. Not all metals share the typical metal properties. See if you can name a metal (not shown in the photos) that is *not*:
   a hard and strong   b malleable at room temperature
2. 10 cm³ of aluminium weighs 28 g.
   10 cm³ of tin weighs 74 g.
   a Which is more dense, aluminium or tin?
   b How many times more dense is it than the other metal?
3. Suggest reasons for this use of a metal:
   a silver for jewellery   b copper for electrical wiring
4. For some uses, a highly *sonorous* metal is needed. See if you can give two examples.
5. Try to think of *two* reasons why:
   a mercury is used in thermometers
   b aluminium is used for soft-drinks cans.

THE BEHAVIOUR OF METALS

# 13.2 Comparing metals for reactivity

## What does *reactive* mean?
A *reactive* element has a strong drive to become a compound, so that its atoms gain stable outer shells. So the metal reacts readily with other elements and compounds. Compare the reactions below.

### 1 The reaction of metals with water

Sodium reacts violently with cold water, whizzing over the surface. Hydrogen gas and a clear solution of sodium hydroxide are formed.

The reaction between calcium and cold water is slower. Hydrogen bubbles off, and a cloudy solution of calcium hydroxide forms.

Magnesium reacts very slowly with cold water, but vigorously on heating in steam: it glows brightly. Hydrogen and solid magnesium oxide form.

You can tell from their behaviour that sodium is the most reactive of the three metals, and magnesium the least.

Compare the equations for the three reactions, below.
What pattern do you notice?

$$2Na\ (s) + 2H_2O\ (l) \longrightarrow 2NaOH\ (aq) + H_2\ (g)$$
$$Ca\ (s) + 2H_2O\ (l) \longrightarrow Ca(OH)_2\ (aq) + H_2\ (g)$$
$$Mg\ (s) + H_2O\ (g) \longrightarrow MgO\ (s) + H_2\ (g)$$

Now compare the reactions of those metals with the others in this table:

| Metal | Reaction | Order of reactivity | Products |
|---|---|---|---|
| potassium | very violent with cold water; catches fire | most reactive | hydrogen and a solution of potassium hydroxide, KOH |
| sodium | violent with cold water | | hydrogen and a solution of sodium hydroxide, NaOH |
| calcium | less violent with cold water | | hydrogen and calcium hydroxide, $Ca(OH)_2$, which is only slightly soluble |
| magnesium | very slow with cold water, but vigorous with steam | | hydrogen and solid magnesium oxide, MgO |
| zinc | quite slow with steam | | hydrogen and solid zinc oxide, ZnO |
| iron | slow with steam | | hydrogen and solid iron oxide, $Fe_3O_4$ |
| copper | no reaction | | |
| silver | | | |
| gold | | least reactive | |

Note the order of reactivity, based on the reaction with water.
And note that only the first three metals in the list produce hydroxides.
The others produce insoluble oxides, if they react at all.

## 2 The reaction of metals with hydrochloric acid

It is not safe to add sodium or potassium to acid in the lab, because the reactions are explosively fast. But other metals can be tested safely. Compare these reactions with hydrochloric acid:

| Metal | Reaction with hydrochloric acid | Order of reactivity | Products |
|---|---|---|---|
| magnesium | vigorous | most reactive ↑ | hydrogen and a solution of magnesium chloride, $MgCl_2$ |
| zinc | quite slow | | hydrogen and a solution of zinc chloride, $ZnCl_2$ |
| iron | slow | | hydrogen and a solution of iron(II) chloride, $FeCl_2$ |
| lead | slow, and only if the acid is concentrated | | hydrogen and a solution of lead(II) chloride, $PbCl_2$ |
| copper | no reaction, even with concentrated acid | | |
| silver | | | |
| gold | | least reactive | |

The equation for the reaction with magnesium this time is:

$$Mg\ (s)\ +\ 2HCl\ (aq)\ \longrightarrow\ MgCl_2\ (aq)\ +\ H_2\ (g)$$

Now compare the order of the metals in the two tables, and the equations for the reactions. What patterns can you see?

### Hydrogen is displaced

When a metal *does* react with water or hydrochloric acid, it drives hydrogen out, and takes its place. This shows that the metal is *more reactive* than hydrogen. It has a stronger drive to form a compound.

But copper and silver do not react with water or acid. So they are *less reactive* than hydrogen.

### It is a redox reaction

The displacement of hydrogen is a **redox reaction**. When magnesium reacts with hydrochloric acid, its atoms lose electrons. The hydrogen ions from the acid gain electrons. The half-equations are:

magnesium: $\quad Mg\ (s) \longrightarrow Mg^{2+}\ (aq) + 2e^-\quad$ (oxidation)
hydrogen ions: $\ 2H^+\ (aq) + 2e^- \longrightarrow H_2\ (g)\quad$ (reduction)

▲ Magnesium displacing hydrogen from hydrochloric acid.

Remember **OIL RIG!**
**O**xidation **Is L**oss of electrons
**R**eduction **Is G**ain of electrons.

**Q**

1. Write a balanced equation for the reaction of potassium with water.
2. Which is more reactive? And what is your evidence?
   a  potassium or sodium?   b  copper or zinc?
3. Which gas is always produced if a metal reacts with water, or dilute acid?
4. Explain why the reaction of iron with hydrochloric acid is a redox reaction.

THE BEHAVIOUR OF METALS

## 13.3 Metals in competition

### When metals compete
You saw that metals can be put in order of reactivity, using their reactions with water and hydrochloric acid. Now let's see what happens when they compete with each other, and with carbon, to form a compound.

### 1 Competing with carbon

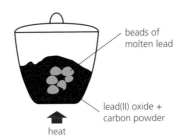

| calcium | more reactive than carbon |
| aluminium | |
| carbon | |
| zinc | less reactive than carbon |
| iron | |
| copper | |

Magnesium oxide is mixed with powdered carbon and heated. No reaction! So magnesium must be more reactive than carbon.

But when lead(II) oxide is used instead, it turns to molten lead, and carbon dioxide gas forms. So carbon is more reactive than lead.

The oxides of the metals above were also tested. Two were found to be more reactive than carbon. The other three were less reactive than carbon.

The equation for the reaction with lead(II) oxide is:

$$2PbO\,(s) + C\,(s) \rightarrow 2Pb\,(s) + CO_2\,(g)$$
$$\text{lead(II) oxide} + \text{carbon} \rightarrow \text{lead} + \text{carbon dioxide}$$

The lead has lost oxygen: it has been reduced. Carbon is the reducing agent. The reaction is a redox reaction.

**Carbon is more reactive than some metals. It will reduce their oxides to form the metal.**

### 2 Competing with other metals, for oxygen

Some powdered iron is heated with copper(II) oxide, CuO. Can the iron grab the oxygen from the copper(II) oxide?

The reaction gives out heat, once it gets going. The mixture glows. Iron(II) oxide and copper are formed. The iron has won.

Other metals were compared in the same way. This shows their order of reactivity. It is the same as in the table on page 185.

The tests confirm that iron, zinc, and lead are all more reactive than copper. The equation for the reaction with iron is:

$$Fe\,(s) + CuO\,(s) \rightarrow FeO\,(s) + Cu\,(s)$$
$$\text{iron} + \text{copper(II) oxide} \rightarrow \text{iron(II) oxide} + \text{copper}$$

The iron is acting as a **reducing agent**, removing oxygen.

**A metal will reduce the oxide of a less reactive metal. The reduction always gives out heat – it is exothermic.**

# 3 Competing to form ions in solution

Copper(II) sulfate solution contains blue copper(II) ions and sulfate ions. An iron nail is placed in it. Will anything happen?

Yes! Copper soon coats the nail. The solution turns green, which indicates iron(II) ions. Iron has pushed copper out of solution.

Other metals and solutions were tested too, with the results above. What do you notice about the order of the metals in this list?

Once again, iron wins against copper. It **displaces** the copper from the copper(II) sulfate solution:

$$Fe\ (s)\ +\ CuSO_4\ (aq)\ \longrightarrow\ FeSO_4\ (aq)\ +\ Cu\ (s)$$
iron + copper(II) sulfate ⟶ iron(II) sulfate + copper
        (blue)                   (green)

Other metals displace less reactive metals in the same way.

**A metal displaces a less reactive metal from solutions of its compounds.**

## They are all redox reactions

All the reactions in this unit are redox reactions: electron transfer takes place in them all. Compare the competitions between iron and copper:

|  | **Competing for oxygen** | **Competing to form ions in solution** |
|---|---|---|
| **Equation** | $Fe\ (s)\ +\ CuO\ (s)\ \longrightarrow\ FeO\ (s)\ +\ Cu\ (s)$ | $Fe\ (s)\ +\ CuSO_4\ (aq)\ \longrightarrow\ FeSO_4\ (aq)\ +\ Cu\ (s)$ |
| **The half-equations** for electron loss  for electron gain | $Fe \longrightarrow Fe^{2+} + 2e^-$  $Cu^{2+} + 2e^- \longrightarrow Cu$ | $Fe \longrightarrow Fe^{2+} + 2e^-$  $Cu^{2+} + 2e^- \longrightarrow Cu$ |
| **The ionic equation** (add the half-equations and cancel the electrons) | $Fe + Cu^{2+} \longrightarrow Fe^{2+} + Cu$ | $Fe + Cu^{2+} \longrightarrow Fe^{2+} + Cu$ |
| **Conclusion** | In each case the iron has given up electrons to form positive ions. The copper has accepted electrons, to form copper. **The more reactive metal forms positive ions more readily.** | |

## Q

1. In the reaction between carbon and lead(II) oxide, which substance is oxidised?
2. **a** What do you expect to happen when carbon powder is heated with: **i** calcium oxide? **ii** zinc oxide?
   **b** Give a word equation for any reaction that occurs in **a**.
3. When chromium(III) oxide is heated with powdered aluminium, chromium and aluminium oxide form. Which is more reactive, chromium or aluminium?
4. Iron displaces copper from copper(II) sulfate solution. Explain what *displaces* means, in your own words.
5. When copper wire is put into a colourless solution of silver nitrate, crystals of silver form on the wire, and the solution goes blue. Explain these changes.
6. For the reaction described in **5**:
   **a** write the half equations, to show the electron transfer
   **b** give the ionic equation for the reaction.

## THE BEHAVIOUR OF METALS

# 13.4 The reactivity series

### Pulling it all together: the reactivity series

We can use the results of the experiments in the last two units to put the metals in final order, with the most reactive one first. The list is called **the reactivity series**. Here it is.

| The reactivity series | | |
|---|---|---|
| potassium, K | most reactive | |
| sodium, Na | | |
| calcium, Ca | | |
| magnesium, Mg | | metals above the blue line: |
| aluminium, Al | | carbon can't reduce their oxides |
| *carbon* | | |
| zinc, Zn | increasing | |
| iron, Fe | reactivity | metals above the red line: they |
| lead, Pb | | displace hydrogen from acids |
| *hydrogen* | | |
| copper, Cu | | |
| silver, Ag | | |
| gold, Au | least reactive | |

The non-metals carbon and hydrogen are included for reference.
The list is not complete, of course. You could test many other metals, for example tin, and nickel, and platinum, and add them in the right place.

### Things to remember about the reactivity series

1 The reactivity series is really a list of the metals in order of their drive to form positive ions, with stable outer shells. The more easily its atoms can give up electrons, the more reactive the metal will be.

2 So a metal will react with a compound of a less reactive metal (for example an oxide, or a salt in solution) by pushing the less reactive metal out of the compound and taking its place.

3 The more reactive the metal, the more **stable** its compounds are. They do not break down easily.

4 The more reactive the metal, the more difficult it is to extract from its ores, since these are stable compounds. For the most reactive metals you need the toughest method of extraction: electrolysis.

5 The less reactive the metal, the less it likes to form compounds. That is why copper, silver and gold are found as elements in the Earth's crust. The other metals are *always* found as compounds.

> **Metals we had to wait for ...**
> - Because they are easy to obtain from their ores, the less reactive metals have been known and used for thousands of years. For example copper has been in wide use for 6000 years, and iron for 3500 years.
> - But the more reactive metals, such as sodium and potassium, had to wait until the invention of electrolysis, in 1800, for their discovery.

▲ Copper is used for roofing, since it is unreactive. But over time it does form a coat of blue-green copper(II) carbonate.

▲ A metal's position in the reactivity series will give you clues about its uses. Only unreactive metals are used in coins.

## Comparing the stability of some metal compounds

Many compounds break down easily on heating. In other words, they undergo **thermal decomposition**.

But reactive metals have more stable compounds. Will they break down easily? Let's compare some compounds of sodium and copper:

| Compound | Effect of heat on the sodium compound | Effect of heat on the copper compound |
|---|---|---|
| carbonate | There is no change in this white compound. | This blue-green compound readily breaks down to black copper(II) oxide and carbon dioxide:<br><br>$CuCO_3 (s) \longrightarrow CuO (s) + CO_2 (g)$ |
| hydroxide | There is no change in this white compound. | This pale blue compound readily breaks down to copper(II) oxide and water:<br><br>$Cu(OH)_2 (s) \longrightarrow CuO (s) + H_2O (l)$ |
| nitrate | This white compound partly decomposes to the nitrite and oxygen:<br><br>$2NaNO_3 (s) \longrightarrow 2NaNO_2 (s) + O_2 (g)$<br>sodium nitrite | This bright blue compound readily breaks down to copper(II) oxide and the brown gas nitrogen dioxide:<br><br>$2Cu(NO_3)_2 (s) \longrightarrow 2CuO (s) + 4NO_2 (g) + O_2 (g)$ |

So the compounds of copper, the less reactive metal, break down easily. The compounds of sodium do not.

## The general rules for thermal decomposition

These are the general rules:

- The lower a metal is in the reactivity series, the more readily its compounds decompose when heated.
- Carbonates, *except* those of sodium and potassium, decompose to the oxide and carbon dioxide.
- Hydroxides, *except* those of sodium and potassium, decompose to the oxide and water.
- Nitrates, *except* those of sodium and potassium, decompose to the oxide, nitrogen dioxide, and oxygen. The nitrates of sodium and potassium form nitrites and oxygen.

▲ Limestone (calcium carbonate) being heated in a lime kiln to give calcium oxide (called lime, or quicklime). The lime might be used to make limewash for buildings, or mixed with sand to make lime mortar.

1  **a** List the metals of the reactivity series, in order.
   **b** Beside each, say where it occurs in the Periodic Table.
   **c** To which group in the Periodic Table do the most reactive metals belong?
   **d** Where in the table are the least reactive ones found?
2  Why is magnesium never found as the element, in nature?
3  Gold has been known and used for thousands of years longer than aluminium. Explain why.
4  Which will break down more easily on heating, magnesium nitrate or silver nitrate? Why?
5  Write a balanced equation for the thermal decomposition of lead(II) nitrate.

THE BEHAVIOUR OF METALS

# 13.5 Making use of the reactivity series

### Those differences in reactivity are useful!
We make clever use of the differences in reactivity of metals. Here are four examples.

### 1 The thermite process
This is used to repair rail and tram lines. Powdered aluminium and iron(III) oxide are put in a container over the damaged rail. When the mixture is lit, the aluminium reduces the iron(III) oxide to molten iron, in a very vigorous reaction. The iron runs into the cracks and gaps in the rail, and hardens:

$$Fe_2O_3(s) + 2Al(s) \longrightarrow 2Fe(l) + Al_2O_3(s)$$

### 2 Making simple cells
The diagram on the right shows a simple cell – two metal strips standing in an electrolyte. (You may have met one on page 120.) The bulb is lit, so a current must be flowing. Hydrogen is forming at the copper strip. The magnesium strip is dissolving. Why is all this happening?

1 Magnesium is more reactive than copper: it has a stronger drive to form ions. So when it is connected to the copper strip, it gives up electrons and goes into solution as ions:

$$Mg(s) \longrightarrow Mg^{2+}(aq) + 2e^- \quad \text{(oxidation)}$$

2 Electrons flow along the wire to the copper strip, as a current. The bulb lights up as the current flows through it.

3 The solution contains $Na^+$ and $Cl^-$ ions from sodium chloride, and some $H^+$ and $OH^-$ ions from water. Hydrogen is less reactive than sodium, so the $H^+$ ions accept electrons from the copper strip:

$$2H^+(aq) + 2e^- \longrightarrow H_2(g) \quad \text{(reduction)}$$

So the difference in reactivity has caused a redox reaction, that gives out energy in the form of electricity.

**A simple cell consists of two different metals in an electrolyte. Electrons flow from the more reactive metal, so it is called the negative pole. The other metal is the positive pole.**

▲ The thermite process being used to join new tram lines.

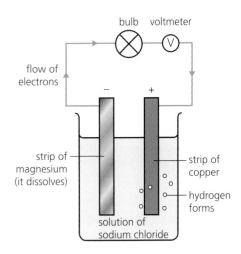

**The poles in cells ...**
... are sometimes called **electrodes**. Don't confuse them with the rods in electrolysis!

### Using other metals in simple cells
You can use other metals in place of copper and magnesium, in a simple cell.

A voltmeter measures the 'push' or voltage that makes electrons flow. This chart shows the voltage for different pairs of metals. For example 2.7 V for copper/magnesium, and 0.47 V for copper/lead. **The further apart the metals are in reactivity, the higher the voltage will be.**

Notice how the voltages in the chart add up: 0.47 V for copper/lead, 0.31 for lead/iron, and 0.78 V (0.47 + 0.31) for copper/iron.

## 3 The sacrificial protection of iron

Iron is used in big structures such as oil rigs and ships. But it has one big drawback: it reacts with oxygen and water, forming iron(III) oxide or **rust**.

To prevent this, the iron can be teamed up with a more reactive metal like zinc or magnesium. For example a block of zinc may be welded to the side of a ship. Zinc is more reactive than iron – so the zinc dissolves:

$2Zn\ (s) \longrightarrow 2Zn^{2+}\ (aq) + 4e^-$         (oxidation)

The electrons flow to the iron, which passes them on, in this reaction:

$O_2\ (g) + 2H_2O\ (l) + 4e^- \longrightarrow 4OH^-\ (aq)$     (reduction)

The overall equation for the reaction is:

$2Zn\ (s) + O_2\ (g) + 2H_2O\ (l) \longrightarrow 2Zn(OH)_2\ (aq)$

So the zinc is oxidised instead of the iron. This is called **sacrificial protection**. The zinc block must be replaced before it all dissolves away.

## 4 Galvanising

This is another way of using zinc to protect iron. It is used for the steel in car bodies, and the corrugated iron for roofing.

- In **galvanising,** the iron or steel is coated with zinc. For car bodies, this is carried out by a form of electrolysis. For roofing, the iron is dipped in a bath of molten zinc.
- The zinc coating keeps air and moisture away. But if the coating gets damaged, the zinc will still protect the iron, by sacrificial protection.

## A note about the reactivity of aluminium

Aluminium is more reactive than iron. But we can use it for things like TV aerials, and satellite dishes, and ladders, without protecting it. Why?

Because aluminium protects itself! It reacts rapidly with oxygen, forming a thin coat of aluminium oxide – so thin you cannot see it.

This sticks tight to the metal, acting as a barrier to further corrosion. So the aluminium behaves as if it were unreactive.

(You saw on page 190 that it reacts very vigorously with iron(III) oxide in the thermite process. But for this, powdered aluminium is used, and a very hot flame to start the reaction off.)

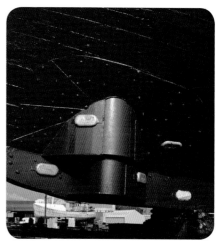

▲ Here blocks of magnesium have been welded to a ship's hull, to prevent the steel (an alloy of iron) from corroding.

▼ The aluminium ladder is protected from corrosion by its oxide layer.

**Q**

1. A copper rod and an iron rod stand in an electrolyte. If you connect a bulb between them, it will light dimly.
   a Why does the current flow?
   b Which acts as the positive electrode: copper or iron?
   c Suggest two metals you could use to get a brighter light.
2. From the chart on page 190, see if you can work out the voltage for a cell that uses magnesium and zinc.
3. a Steel for cars is *galvanised*. What does that mean?
   b Explain how this protects the steel.
4. Aluminium is more reactive than iron. But unlike iron, we do not need to protect it from corrosion. Why not?
5. a Write a word equation for the thermite reaction.
   b See if you can give *two* reasons why the aluminium is powdered, for this reaction.

# THE BEHAVIOUR OF METALS

## Checkup on Chapter 13

### Revision checklist

**Core curriculum**

Make sure you can ...
☐ explain these terms used about metals:
  *malleable    ductile    sonorous
  high density    conductors*
☐ give at least five physical properties of metals
☐ give four chemical properties of metals
☐ explain what a *reactive* element is
☐ explain what the reactivity series is, and list the metals in it, in the correct order
☐ describe how the metals in the series react with
  – water
  – dilute acids
  and give word equations where reactions occur
☐ explain what *displacement of hydrogen* means
☐ explain why hydrogen and carbon are often shown in the reactivity series, and say where they fit in
☐ predict the products, when carbon is heated with the oxide of a metal below it in the series

**Extended curriculum**

Make sure you can also ...
☐ state what the products will be, when:
  – a metal is heated with the oxide of a less reactive metal
  – a metal is placed in the solution of a compound of a less reactive metal
☐ explain why those reactions are redox reactions
☐ define *thermal decomposition*
☐ give the 'rules' for the effect of heat on:
  – metal carbonates
  – metal hydroxides
  – metal nitrates
  and give word equations where reactions occur
☐ explain what simple cells are, and
  – say why a current is produced
  – predict which metal will be the positive pole
  – decide which pair of metals will give the largest voltage, and why
☐ explain what these are for, and why they work, and name the metals used:
  – sacrificial protection
  – galvanising

### Questions

**Core curriculum**

1.

| Metal | Density in g/cm³ |
|---|---|
| aluminium | 2.7 |
| calcium | 1.6 |
| copper | 8.9 |
| gold | 19.3 |
| iron | 7.9 |
| lead | 11.4 |
| magnesium | 1.7 |
| sodium | 0.97 |

a List the metals given in the table above in order of increasing density.
b i What is meant by *density*?
  ii A block of metal has a volume of 20 cm³ and a mass of 158 g. Which metal is it?
c Now list the metals in order of reactivity.
d i The most reactive metal in the list has a density of .....?
  ii The least reactive one has a density of .....?
  iii Does there appear to be a link between density and reactivity? If yes, what?
e Using low-density metals for vehicles saves money on fuel and road repairs. Explain why.
f Which of the low-density metals above is the most suitable for vehicles? Why? Give three reasons.

2. This shows metals in order of reactivity:

  sodium                              (most reactive)
  calcium
  magnesium
  zinc
  iron
  lead
  copper
  silver                              (least reactive)

a Which element is stored in oil?
b Which elements will react with cold water?
c Choose one metal that will react with steam but *not* cold water. Draw a diagram of suitable apparatus for this reaction. (You must show how the steam is generated.)
d i Name the gas given off in **b** and **c**.
  ii Name another reagent that reacts with many metals to give the same gas.

3 For each description below, choose one metal that fits the description. Name the metal. Then write a word equation for the reaction that takes place.
   a  A metal that displaces copper from copper(II) sulfate solution.
   b  A metal that reacts gently with dilute hydrochloric acid.
   c  A metal that floats on water and reacts vigorously with it.
   d  A metal that reacts quickly with steam but very slowly with cold water.

4 Look again at the list of metals in **2**. Carbon can be placed between zinc and aluminium.
   a  Which two of these will react?
      i   carbon + aluminium oxide
      ii  carbon + copper(II) oxide
      iii magnesium + carbon dioxide
   b  Write a word equation for the two reactions, and underline the substance that is reduced.

**Extended curriculum**

5 When magnesium powder is added to copper(II) sulfate solution, a displacement reaction occurs and solid copper forms.

   a  Write a word equation for the reaction.
   b  Why does the displacement reaction occur?
   c  i   Write a half-equation to show what happens to the magnesium atoms.
      ii  Which type of reaction is this?
   d  i   Write a half-equation to show what happens to the copper ions.
      ii  Which type of reaction is this?
      iii Which metal shows the greater tendency to form a positive ion?
   e  i   Write the ionic equation for the displacement reaction, by adding the half-equations.
      ii  Which type of reaction is it?
   f  Use the reactivity series of metals to decide whether these will react together:
      i   iron + copper(II) sulfate solution
      ii  silver + calcium nitrate solution
      iii zinc + lead(II) nitrate solution
   g  For those that react:
      i   describe what you would *see*
      ii  write the ionic equations for the reactions.

6 When magnesium and copper(II) oxide are heated together, this redox reaction occurs:
   $Mg\,(s) + CuO\,(s) \longrightarrow MgO\,(s) + Cu\,(s)$
   a  What does the word *redox* stand for?
   b  For the above reaction, name:
      i  the reducing agent   ii  the oxidising agent
   c  Describe the electron transfer in the reaction.
   d  Explain as fully as you can why the *reverse* reaction does not occur.
   e  i   Name one metal that would remove the oxygen from magnesium oxide.
      ii  Does this metal *gain* electrons, or *lose* them, more easily than magnesium does?

7 When the pale blue compound copper(II) hydroxide is heated, thermal decomposition occurs and steam is given off.
   a  i   What does *thermal decomposition* mean?
      ii  Write the chemical equation for the reaction.
      iii What colour change would you observe?
   b  Name a hydroxide that does not decompose when heated.
   c  In further experiments, nitrates of copper and sodium are heated.
      i   Which gas is released in both experiments?
      ii  One of the nitrates also releases the brown gas nitrogen dioxide. Which one?
      iii Write the equation for this reaction.
   d  Relate the observations in **c** to the positions of copper and sodium in the reactivity series.

8

   Look at the three cells above.
   a  How can you tell that the three unknown metals are all more reactive than copper?
   b  Place the metals in order, most reactive first.
   c  What voltage will be obtained in a cell using:
      i   metals A and B?   ii  metals B and C?
   d  For each cell in c, state which metal is the negative terminal.

9 In simple cells, chemical reactions give electricity.
   a  Which other set-up also involves electricity and chemical reactions?
   b  What is the key difference between it and the simple cell?

MAKING USE OF METALS

# 14.1 Metals in the Earth's crust

## The composition of the Earth's crust

We get some metals from the sea, but most from the Earth's **crust** – the Earth's hard outer layer.

The crust is mostly made of **compounds**. But it also contains some **elements** such as copper, silver, mercury, platinum, and gold. These occur **native,** or uncombined, because they are unreactive.

If you could break all the crust down to elements, you would find it is almost half oxygen! This shows its composition:

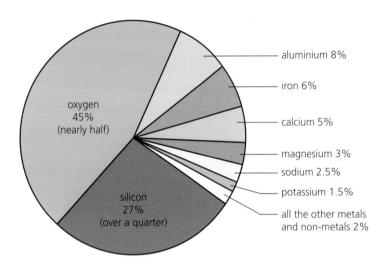

Note that:
- two non-metals, oxygen and silicon, make up nearly three-quarters of the crust. They occur together in compounds such as silicon dioxide (**silica** or **sand**). Oxygen is also found in compounds such as iron(III) oxide, aluminium oxide, and calcium carbonate.
- just six metals – aluminium to potassium in the pie chart – make up over one-quarter of the crust. Aluminium is the most abundant of these, and iron next. All six occur as compounds, because they are reactive.

## Scarce, and precious …

All the other metals *together* make up less than 2% of the Earth's crust. Many, including lead, zinc, and copper, are considered scarce.

Gold, silver, platinum, and palladium are called **precious metals** because they are scarce, expensive, and often kept as a store of wealth.

The car industry uses a lot of metal. Cars are mainly steel, plus over 5% aluminium. But the steel is coated with zinc, and the bumpers with nickel and chromium. The electrics depend on copper, the battery uses lead, and modern exhausts contain palladium, platinum, and rhodium as catalysts.

▲ Light metals such as aluminium and titanium are used in the International Space Station, 360 km above us.

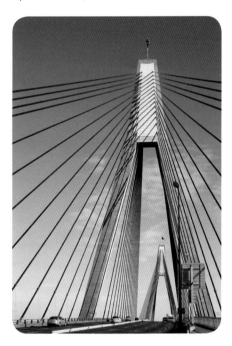

▲ We use about nine times more iron than all the other metals put together.

▼ Metals on wheels.

## Metal ores

The rocks in the Earth's crust are a mixture of substances. Some contain a large amount of one metal compound, or one metal, so it may be worth digging them up to extract the metal. Rocks from which metals are obtained are called ores. For example:

This is a chunk of **rock salt**, the main ore of sodium. It is mostly **sodium chloride**.

This is a piece of **bauxite**, the main ore of aluminium. It is mostly **aluminium oxide**.

Since gold is unreactive, it occurs native (uncombined). This sample is almost pure gold.

## To mine or not to mine?

Before starting to mine an ore, the mining company must decide whether it is economical. It must find answers to questions like these:

1 How much ore is there?
2 How much metal will we get from it?
3 Are there any special problems about getting the ore out?
4 How much will it cost to mine the ore and extract the metal from it? (The cost will include roads, buildings, mining equipment, the extraction plant, transport, fuel, chemicals, and wages.)
5 How much will we be able to sell the metal for?
6 So will we make a profit if we go ahead?

The answers to these questions will change from year to year. For example if the selling price of a metal rises, even a low-quality or **low-grad**e ore may become worth mining.

The local people may worry that the area will be spoiled, and the air and rivers polluted. So they may object to plans for a new mine. On the positive side, they may welcome the new jobs that mining will bring.

▲ The world's biggest man-made hole: the Bingham Canyon copper mine in Utah, USA. Started in 1906, it is now over 1 km deep and 4 km wide.

**Q**

1 Which is the main *element* in the Earth's crust?
2 Which is the most common *metal* in the Earth's crust? Which is the second?
3 Gold occurs *native* in the Earth's crust. Explain.
4 Is it true that the most reactive metals are quite plentiful in the Earth's crust?
5 Some metals are called *precious*. Why? Name four.
6 One metal is used more than all the others put together. Which one? Why is it so popular?
7 What is a *metal ore*?
8 Name the main ore of: **a** sodium **b** aluminium What is the main compound in each ore?

MAKING USE OF METALS

## 14.2 Extracting metals from their ores

### Extraction

After mining an ore, the next step is to remove or **extract** the metal from it. How you do this? It depends on the metal's **reactivity**.

- The most unreactive metals – such as silver and gold – occur in their ores as elements. All you need to do is separate the metal from sand and other impurities. This is like removing stones from soil. It does not involve chemical reactions.
- The ores of all the other metals contain the metals as compounds. These have to be reduced, to give the metal:

  metal compound $\xrightarrow{\text{reduction}}$ metal

- The compounds of the more reactive metals are very stable, and need electrolysis to reduce them. This is a powerful method, but it costs a lot because it uses a lot of electricity.
- The compounds of the less reactive metals are less stable, and can be reduced using a suitable **reducing agent**.

> **Reduction of metal ores**
> Remember, you can define reduction as:
> - loss of oxygen
>   $Fe_2O_3 \rightarrow Fe$
> - or gain of electrons
>   $Fe^{3+} + 3e^- \rightarrow Fe$
>
> Either way, the ore is reduced to the metal.

### Extraction and the reactivity series

So the method of extraction is strongly linked to the **reactivity series**, as shown below. Carbon is included for reference.

| Metal | | | Method of extraction from ore | | |
|---|---|---|---|---|---|
| potassium | | | | | |
| sodium | | | | | |
| calcium | metals more reactive | ores more difficult to decompose | electrolysis | method of extraction more powerful | method of extraction more expensive |
| magnesium | | | | | |
| aluminium | | | | | |
| *carbon* | | | | | |
| zinc | | | heating with a reducing agent – carbon or carbon monoxide | | |
| iron | | | | | |
| lead | | | | | |
| silver | | | occur naturally as elements so no chemical reactions needed | | |
| gold | | | | | |

### Carbon as a reducing agent

As you saw on page 186, carbon will react with oxides of metals less reactive than itself, reducing them to the metal.

Luckily, many ores are oxides, or compounds easily converted to oxides.

The table shows that carbon can be used to extract zinc, iron, and lead. It is used in the form of coke (made from coal), which is heated with the metal oxide in a furnace. But in the process, carbon may react with a limited supply of oxygen, giving carbon monoxide gas (CO). In that case, the carbon monoxide brings about the actual reduction.

▲ No need to reduce gold …

# MAKING USE OF METALS

## Three examples of ore extraction

**1 Iron ore** This is mainly iron(III) oxide. It is reduced like this:

iron(III) oxide + carbon monoxide ⟶ iron + carbon dioxide
$Fe_2O_3 (s)$ + $3CO (g)$ ⟶ $2Fe (l)$ + $3CO_2 (g)$

We will look more closely at this extraction in Unit 14.3.

**2 Aluminium ore** This is mainly aluminium oxide. Aluminium is more reactive than carbon, so electrolysis is needed for this reduction:

aluminium oxide ⟶ aluminium + oxygen
$2Al_2O_3 (l)$ ⟶ $4Al (l)$ + $3O_2 (g)$

We will look more closely at this extraction in Unit 14.4.

**3 Zinc blende** This is mainly zinc sulfide, ZnS. First it is roasted in air, giving zinc oxide and sulfur dioxide:

zinc sulfide + oxygen ⟶ zinc oxide + sulfur dioxide
$2ZnS (s) + 3O_2 (g)$ ⟶ $2ZnO (s) + 2SO_2 (g)$

Then the oxide is reduced in one of these two ways:

**i Using carbon monoxide** This is carried out in a furnace:

zinc oxide + carbon monoxide ⟶ zinc + carbon dioxide
$ZnO (s)$ + $CO (g)$ ⟶ $Zn (s)$ + $CO_2 (g)$

The final mixture contains zinc and a slag of impurities. The zinc is separated by fractional distillation. (It boils at 907 °C.)

**ii Using electrolysis** For this, a compound must be melted, or in solution. But zinc oxide has a very high melting point (1975 °C), and is insoluble in water!

Instead, it is dissolved in dilute sulfuric acid (made from the sulfur dioxide produced in the roasting stage). Zinc oxide is a base, so it neutralises the acid, giving a solution of zinc sulfate. This undergoes electrolysis, and zinc is deposited at the cathode:

$Zn^{2+} (aq) + 2e^- \longrightarrow Zn (s)$     (reduction)

The zinc is scraped off the cathode, and melted into bars to sell.

In fact most zinc is extracted by electrolysis, because this gives zinc of very high purity. Cadmium and lead occur as impurities in the zinc blende, and these metals are recovered and sold too.

▲ After extraction, some aluminium is made into rolls like these, ready for sale.

**Zinc metal is used ...**
- to **galvanise** iron – coat it to stop it rusting (page 191)
- in the **sacrificial protection** of iron structures (page 191)
- to make **alloys** such as brass and bronze (page 203)
- to make batteries (page 122)

▲ An iron bucket, galvanised with zinc.

---

**Q**

1 Why is no chemical reaction needed to get gold?
2 Lead is extracted by heating its oxide with carbon:
   lead oxide + carbon ⟶ lead + carbon monoxide
   a Why can carbon be used for this reaction?
   b One substance is *reduced*. Which one?
   c Which substance is the reducing agent?
3 The reaction in question **2** is a *redox* reaction. Why?
4 Sodium is extracted from rock salt (sodium chloride).
   a Electrolysis is needed for this. Explain why.
   b Write a balanced equation for the reaction.
5 Zinc blende is an ore of zinc. It is mainly ... ?
6 Describe the extraction of zinc by electrolysis.

MAKING USE OF METALS

## 14.3 Extracting iron

### The blast furnace

This diagram shows the **blast furnace** used for extracting iron from its ore. It is an oven shaped like a chimney, at least 30 metres tall.

▲ Blast furnaces run non-stop 24 hours a day.

A mixture called the **charge**, containing the iron ore, is added through the top of the furnace. Hot air is blasted in through the bottom. After a series of reactions, liquid iron collects at the bottom of the furnace.

### What's in the charge?

The charge contains three things:

1 **Iron ore.** The chief ore of iron is called **hematite**. It is mainly iron(III) oxide, $Fe_2O_3$, mixed with sand and some other compounds.
2 **Limestone.** This common rock is mainly calcium carbonate, $CaCO_3$.
3 **Coke.** This is made from coal, and is almost pure carbon.

▲ Mining hematite.

MAKING USE OF METALS

## The reactions in the blast furnace

| Reactions, products, and waste gases | Comments |
|---|---|
| **Stage 1: The coke burns, giving off heat**<br>The blast of hot air starts the coke burning.<br>It reacts with the oxygen in the air, giving carbon dioxide:<br>carbon + oxygen → carbon dioxide<br>$C\,(s)\ +\ O_2\,(g)\ \rightarrow\ CO_2\,(g)$ | This, like all combustion, is a **redox reaction**.<br>The carbon is **oxidised** to carbon dioxide.<br>The blast of air provides the oxygen for the reaction.<br>The reaction is **exothermic** – it gives off heat, which helps to heat the furnace. |
| **Stage 2: Carbon monoxide is made**<br>The carbon dioxide reacts with more coke, like this:<br>carbon + carbon dioxide → carbon monoxide<br>$C\,(s)\ +\ CO_2\,(g)\ \rightarrow\ 2CO\,(g)$ | In this redox reaction, the carbon dioxide loses oxygen.<br>It is **reduced**.<br>The reaction is **endothermic** – it takes in heat from the furnace.<br>That's good: stage 3 needs a lower temperature. |
| **Stage 3: The iron(III) oxide is reduced**<br>This is where the actual extraction occurs.<br>Carbon monoxide reacts with the iron ore, giving liquid iron:<br>iron(III) oxide + carbon monoxide → iron + carbon dioxide<br>$Fe_2O_3\,(s)\ +\ 3CO\,(g)\ \rightarrow\ 2Fe\,(l)\ +\ 3CO_2\,(g)$<br>The iron trickles to the bottom of the furnace. | In this redox reaction, carbon monoxide is the **reducing agent**.<br>It reduces the iron(III) oxide to the metal.<br>The carbon monoxide is **oxidised** to carbon dioxide. |
| **What is the limestone for?**<br>The limestone breaks down in the heat of the furnace:<br>$CaCO_3 \rightarrow CaO\,(s) + CO_2\,(g)$<br>The calcium oxide that forms reacts with the sand, which is mainly silicon dioxide or **silica**:<br>calcium oxide + silica → calcium silicate<br>$CaO\,(s)\ +\ SiO_2\,(s)\ \rightarrow\ CaSiO_3\,(s)$<br>The calcium silicate forms a **slag** which runs down the furnace and floats on the iron. | The purpose of this reaction is to produce calcium oxide, which will remove the sand that was present in the ore.<br><br>Calcium oxide is a basic oxide. Silica is an acidic oxide.<br>Calcium silicate is a salt.<br><br>The molten slag is drained off. When it solidifies it is sold, mostly for road building. |
| **The waste gases:** hot **carbon dioxide** and **nitrogen** come out from the top of the furnace. The heat is transferred from them to heat the incoming blast of air. | The carbon dioxide is from the reaction in stage **3**.<br>The nitrogen is from the air blast. It has not taken part in the reactions so has not been changed. |

## Where next?

The iron from the blast furnace is called **pig iron**. It is impure. Carbon and sand are the main impurities.

Some is run into moulds to make **cast iron**. This is hard but brittle, because of its high carbon content – so it is used only for things like canisters for bottled gas (page 252) and drain covers.

But most of the iron is turned into **steels**. You can find how this is done in Unit 14.6.

▶ A cast-iron drain cover.

1 Write the equation for the redox reaction that gives iron.
2 What is the 'blast' of the blast furnace?
3 Name the waste gases from the blast furnace.
4 The calcium carbonate in the blast furnace helps to purify the iron. Explain how, with an equation.
5 The slag and waste gases are both useful. How?

MAKING USE OF METALS

## 14.4 Extracting aluminium

### From rocks to rockets

Aluminium is the most abundant metal in the Earth's crust. Its main ore is **bauxite**, which is aluminium oxide mixed with impurities such as sand and iron oxide. The impurities make it reddish brown.

These are the steps in obtaining aluminium:

**1** First, geologists test rocks and analyse the results, to find out how much bauxite there is. If the tests are satisfactory, mining begins.

**2** Bauxite is red-brown in colour. It is usually near the surface, so is easy to dig up. This is a bauxite mine in Jamaica.

**3** The ore is taken to a bauxite plant, where impurities are removed. The result is white **aluminium oxide**, or **alumina**.

**4** The alumina is taken to another plant for electrolysis. It may even be sent to another country where electricity is cheaper.

**5** The extracted metal is made into sheets and blocks, and sold to other industries. It has a great many uses. For example …

**6** … it is used to make drinks cans, food cartons, cooking foil, bikes, TV aerials, electricity cables, ships, trains, and space rockets.

### The electrolysis

The electrolysis is carried out in a large steel tank. (See next page.) This is lined with carbon, which acts as the cathode (−). Big blocks of carbon hang in the middle of the tank, and act as the anode (+).

Alumina melts at 2045 °C. It would be impossible to keep the tank that hot. Instead, the alumina is dissolved in molten **cryolite**, or sodium aluminium fluoride, which has a much lower melting point.

## The electrolysis tank
This is the tank for the electrolysis of aluminium:

## The reactions at the electrodes
Once the alumina dissolves, its aluminium and oxide ions are free to move. They move to the electrode of opposite charge.

**At the cathode**  The aluminium ions gain electrons:

$$4Al^{3+} (l) + 12e^- \longrightarrow 4Al (l) \qquad \text{(reduction)}$$

The aluminium drops to the bottom of the cell as molten metal. This is run off at intervals. Some will be mixed with other metals to make **alloys**. Some is run into moulds, to harden into blocks.

**At the anode**  The oxygen ions lose electrons:

$$6O^{2-} (l) \longrightarrow 3O_2 (g) + 12e^- \qquad \text{(oxidation)}$$

The oxygen gas bubbles off, and reacts with the anode:

$$C (s) + O_2 (g) \rightarrow CO_2 (g) \qquad \text{(oxidation of carbon)}$$

So the carbon blocks get eaten away, and need to be replaced.

**The overall reaction**  The alumina is broken down, giving aluminium:

aluminium oxide $\longrightarrow$ aluminium + oxygen
$$2Al_2O_3 (l) \longrightarrow 4Al (l) + 3O_2 (g)$$

## Some properties of aluminium
1  It is a bluish-silver, shiny metal.
2  It has a low density – it is 'light'. Iron is three times heavier.
3  It is a good conductor of heat and electricity.
4  It is malleable and ductile.
5  It resists corrosion. This is because a fine coat of aluminium oxide forms on its surface, and acts as a barrier. (See page 191.)
6  It is not very strong when pure, but it can be made much stronger by mixing it with other metals to form **alloys**. (See page 203.)
7  It is generally **non-toxic** (not harmful to health). But taking in large quantities could harm you.

▲ Moltem aluminium from the tank was run into these moulds, to make blocks.

▲ Electricity cables: aluminium (light) with a steel core (strong).

**Q**

1  Which compounds are used in the extraction of aluminium? Say what role each plays.
2  a  Sketch the electrolysis cell for extracting aluminium.
   b  Why do the aluminium ions move to the cathode?
   c  What happens at the cathode? Give an equation.
   d  The anode is replaced regularly. Why?
3  These terms all describe properties of aluminium. Say what each term means.
   a  malleable   b  ductile   c  non-toxic
   d  low density   e  resistant to corrosion
4  List six uses of aluminium. For each, say which properties of the metal make it suitable.

MAKING USE OF METALS

# 14.5 Making use of metals and alloys

## Properties dictate uses

Think of all the solid things you own, or use, or see around you. Some are probably made of wood, or plastic, or stone, or cloth. But which are made of metal, or contain metal?

Metals share some properties. Each also has its own special properties. How we use the metals depends on their properties. You would not use poisonous metals for food containers, for example.

Here are some examples of uses:

Aluminium foil is used for food cartons because it is non-toxic, resists corrosion, and can be rolled into thin sheets.

Copper is used for electrical wiring in homes, because it is an excellent conductor, and easily drawn into wires.

Zinc is coated onto steel car bodies, before they are painted, because zinc protects the steel from rusting, by sacrificial protection.

## A summary of their uses

The three metals above have other uses too. Look at this table:

| Metal | Used for ... | Properties that make it suitable |
|---|---|---|
| aluminium | overhead electricity cables (with a steel core for strength) | a good conductor of electricity (not as good as copper, but cheaper and much lighter); ductile, resists corrosion |
| | cooking foil and food cartons | non-toxic, resistant to corrosion, can be rolled into thin sheets |
| | drinks cans | light, non-toxic, resistant to corrosion |
| | coating CDs and DVDs | can be deposited as a thin film; shiny surface reflects laser beam |
| copper | electrical wiring | one of the best conductors of electricity, ductile |
| | saucepans and saucepan bases | malleable, conducts heat well, unreactive, tough |
| zinc | protecting steel from rusting | offers sacrificial protection to the iron in steel |
| | coating or **galvanising** iron and steel | resists corrosion, but offers sacrificial protection if coating cracks |
| | for torch batteries | gives a current when connected to a carbon pole, packed into a paste of electrolyte |

## Alloys: making metals more useful

The uses on page 202 are for the pure metals. But often a metal is more useful when mixed with another substance. The mixture is called an **alloy**.

For example, mixing molten zinc and copper gives the gold-coloured alloy called **brass**. When this solidifies, it is hard, strong, and shiny. It is used for door locks, keys, knobs, and musical instruments such as trumpets.

**Turning a metal into an alloy changes its properties, and makes it more useful.**

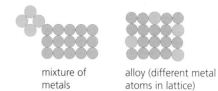

▲ In an alloy, the atoms of the second metal must enter the lattice. So you need to melt the metals first, then mix them.

## Why an alloy has different properties

This shows the atoms in a pure metal. They are arranged in a regular lattice. (In fact they are metal ions in a sea of electrons, as you saw on page 62.)

When pressure is applied, for example by hammering the metal, the layers can slide over each other easily. That is why a metal is malleable and ductile.

But when the metal is turned into an alloy, new atoms enter the lattice. The layers can no longer slide easily. So the alloy is stronger than the original metal.

It is not only strength that changes: other properties will change too. For example the alloy may be more resistant to corrosion than the original metal was.

You can add more than one substance. You can try out different amounts of different substances, to design an alloy with exactly the properties that you need.

## Some examples of alloys

There are *thousands* of different alloys. Here are a couple of examples, for the metals on the opposite page.

▲ Blow your own (brass) trumpet.

| Alloy | Made from | Special properties | Uses |
| --- | --- | --- | --- |
| brass | 70% copper<br>30% zinc | harder than copper,<br>does not corrode | musical instruments, ornaments, door knobs and other fittings |
| aluminium alloy 7075 TF | 90.25% aluminium<br>6% zinc<br>2.5% magnesium<br>1.25% copper | light but very strong<br>does not corrode | aircraft |

Look at the aluminium alloy. Aircraft need materials that are light but very strong, and will not corrode. Pure aluminium is not strong enough. So *hundreds* of aluminium alloys have been developed, for aircraft parts.

1. See if you can give two *new* examples of a use for a metal, that depends on the metal being:
   a a good conductor   b non-toxic
   c strong   d resistant to corrosion

2. Bronze is 95% copper and 5% tin.
   a What is a mixture like this called?
   b What can you say about its properties?
   c See if you can give an example of a use for bronze.

MAKING USE OF METALS

## 14.6 Steels and steel-making

### Steels: alloys of iron

Iron is very widely used – but almost never on its own.

Pure iron is no good for building things, because it is too soft, and stretches quite easily, as you can see from this photo. Even worse, it rusts very easily.

But when it has a very small amount of carbon mixed with it, the result is a **mild steel**. This is hard and strong. It is used for buildings, ships, car bodies, and machinery.

When nickel and chromium are mixed with iron, the result is **stainless steel**. This is hard and rustproof. It is used for cutlery, and equipment in chemical factories.

So mild steel and stainless steel are **alloys** of iron. Some typical mixtures are:

mild steel – 99.7% iron and 0.3% carbon
stainless steel – 70% iron, 20% chromium, and 10% nickel

There are many other types of steel too, all with different properties.

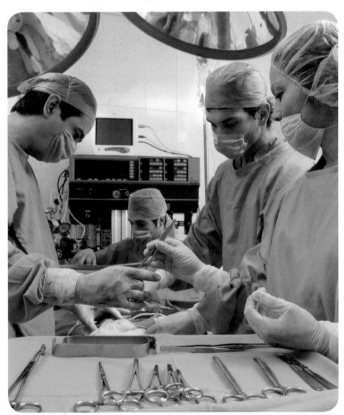

▲ Scalpel, please! Stainless steel saves lives. It is easy to sterilise, which is very important in surgery.

▲ Mild steel is used for washing machines and fridges, as well as for cars.

# MAKING USE OF METALS

▲ Very often, scrap iron is added to the molten pig iron in the oxygen furnace. So the iron gets recycled as new steel.

▲ Molten steel being poured out, in a steel works. Look how it glows. Temperature: around 1400 °C.

## Making steels

Steels are made using the molten iron from the blast furnace.

As you saw, this molten iron is impure. It contains about 5% carbon, from the coke used in the furnace, plus sand (which is mainly silicon dioxide) and phosphorus and sulfur compounds, from the iron ore.

1 **First, unwanted impurities are removed from the iron.**
   - The molten iron from the blast furnace is poured into an **oxygen furnace**, and a jet of oxygen is turned on.
   - The oxygen reacts with the carbon, forming carbon monoxide and carbon dioxide gases, which are carried off. It reacts with other impurities to form acidic oxides.
   - Then calcium oxide is added. It is a basic oxide. It reacts with silicon dioxide and the other acidic oxides, to give a **slag** that is skimmed off.

   For some steels, *all* impurities are removed. But many steels are just iron plus a little carbon. Carbon makes steel stronger – but too much makes it brittle, and hard to shape. So the carbon content is checked continually. When it is correct, the oxygen is turned off.

2 **Then other elements may be added.**
   These are measured out carefully, to give steels such as stainless steel, with the required properties.

> **About calcium oxide**
> Calcium oxide is also called **lime** and **quicklime**.

> **Recycling iron and steel**
> Iron and steel are the most recycled metals in the world.
> - It costs far less to recycle them than to mine new ore and extract the iron – and it is better for the environment.
> - It is easy to separate them from other scrap: use a big magnet.
> - It is easy to recyle them: just add them to the oxygen furnace.
>
> Old cars, fridges and washing machines ... all get recycled.

1 What is the main metal, in steels?
2 a Why is oxygen blown through the molten iron?
   b See if you can write a word equation for two reactions which occur at this stage, that remove carbon.
3 What is special about calcium oxide, that makes it useful in steel-making?
4 Mild steel contains a very small % of carbon.
   a Draw diagrams to show clearly the difference between pure iron and mild steel. (Show atoms!)
   b Explain why mild steel is stronger than iron.
5 Alloys are a type of solution. Explain why.
6 Name the metals in stainless steel.

MAKING USE OF METALS

# Metals, civilisation, and you

## No you without metals

Without metals you probably would not exist. You certainly would not be reading this book. The world would not have nearly 7 billion people.

Our human history has been shaped by the discovery of metals. That is why two of our eras are known as the Bronze Age and the Iron Age.

> **Time past and present**
> - CE stands for *common era*. It means the period since **1 AD**.
> - BCE stands for *before common era*. It means the same as BC.

## Metals and civilisation

| | |
|---|---|
| | 200 000 years ago |
| farming starts around here | 8000 BCE |
| gold known | 6000 BCE |
| copper and silver known | 4000 BCE |
| tin and lead known | 3500 BCE |
| iron began to be used | 1500 BCE |
| just 7 metals known by 1 CE: gold, copper, silver, lead, tin, iron, and mercury | 1 CE |
| Industrial Revolution starts around here. 24 metals known | 1750 CE / 1800 CE |
| 65 metals known | 1900 CE |
| over 90 metals known | 2000 CE |

### The Stone Age
The early humans were hunters and gatherers. We had to kill and chop to get meat and fruit and firewood. We used stone and bone as tools and weapons.

Then, about 10 000 years ago, we began to farm. Farming started in the Middle East. As it spread, great civilisations grew – for example in the valley of the River Indus, in Asia, around 9000 years ago.

### The Bronze Age
By 3500 BCE (over 5000 years ago) copper and tin were known – but not much used, since they are quite soft. Then someone made a discovery: mixing molten copper and tin gives a strong hard metal that can be hammered into different shapes. It was the alloy **bronze**. The Bronze Age was here!

Now a whole new range of tools could be made, for both farming and fighting.

### The Iron Age
Our ancestors had no use for iron – until one day, around 2500 years ago, some got heated up with charcoal (carbon). Perhaps by accident, in a hot fire. A soggy mess of impure iron collected at the bottom of the fire.

The iron was hammered into shape with a stone. The result was a metal with vast potential. In time it led to the Industrial Revolution. It is still the most widely used metal on the Earth.

### The Digital Age
We still depend on iron. But computers now touch every aspect of life. There would be no computing – and no satellites, or TV, or mobile phones – without hi-tech metals like selenium and titanium, and old favourites like aluminium and copper. Many new hi-tech uses are being found for these metals and their alloys.

**Where next?**

# MAKING USE OF METALS

▲ The lead ore galena (lead sulfide) was used as eye make-up in ancient Egypt. Lead was probably produced by heating galena in an open fire.

▲ A modern use of metals: this x-ray shows a replacement hip joint made from titanium. Titanium is suitable since it is strong, light, non-toxic, and does not corrode.

## The earliest metals

The first metals found were the unreactive ones that exist as elements: gold, copper, and silver. But these were too soft on their own to have many uses, other than for ornaments, jewellery, and coins.

Tin, lead, and iron occur naturally as compounds, so have to be extracted by chemical reactions. It could have happened by accident, at the start. Some ore could have fallen into a fire where charcoal was burning. Or the molten metals could have appeared from clay being baked in pottery kilns, where the enclosed fires are very hot.

## The invention of electrolysis

By 250 years ago, only 24 metals were known: those found naturally as elements, plus others that could be extracted easily in a furnace, using carbon. Nobody had set eyes on sodium or magnesium, for example.

Then in 1800, the first ever electrolysis was carried out (of water). The scientist Humphry Davy heard about it, and tried it out on molten compounds – with amazing results! He discovered potassium and sodium in 1807, and magnesium, calcium, strontium, and barium in 1808.

## The discovery of aluminium

Aluminium is the most common metal in the Earth's crust. But it was not extracted until 1825, when aluminium chloride and potassium were heated together. (Potassium, being more reactive, displaces aluminium from its compounds.)

Only small amounts of aluminium could be extracted this way. So it became more valuable than gold! Then in 1886, the way to extract it by electrolysis, using cryolite, was developed. Aluminium had arrived.

## Any more metals to find?

We know about all the metals in the Earth's crust by now. However, new metals are still being discovered. But they are artificial elements, made in labs, and radioactive. They usually break down very very fast.

▲ This gold collar is around 3000 years old. It was found in Ireland.

▲ Humphry Davy died at 52, of an illness probably caused by harmful vapours from electrolysis. Beware!

MAKING USE OF METALS

# Checkup on Chapter 14

## Revision checklist

### Core curriculum
Make sure you can ...
- ☐ name the two most common metals in the Earth's crust
- ☐ explain what an *ore* is, and name the main ores of aluminium and iron
- ☐ explain what *extracting a metal* means
- ☐ say how the method used to extract a metal depends on the reactivity of the metal
- ☐ explain why electrolysis is needed to extract some metals, and name at least two of them
- ☐ for the extraction of iron in the blast furnace:
  - name the raw materials and explain the purpose of each
  - draw a labelled sketch of the blast furnace
  - give word equations for the reactions that take place
  - give uses for the waste products that form
  - name two impurities present in the molten iron
- ☐ give at least two uses of aluminium, and state the properties that make it suitable for those uses
- ☐ explain what alloys are, and draw a diagram to show the structure of an alloy
- ☐ explain why an alloy is usually harder and stronger than the original metal
- ☐ explain what brass is, and say what it is used for
- ☐ describe how iron from the blast furnace is turned into steels
- ☐ say what is in this alloy, and what it is used for:
  - mild steel
  - stainless steel

### Extended curriculum
Make sure you can also ...
- ☐ describe the extraction of zinc from zinc blende
- ☐ describe the extraction of aluminium from aluminium oxide, with the help of cryolite (you will not be asked to draw the electrolysis cell)
- ☐ give uses for copper and zinc, and state the properties that lead to those uses

## Questions

### Core curriculum

1 This table gives information about the extraction of three different metals from their main ores:

| metal | formula of main compound in ore | method of extraction |
|---|---|---|
| iron | $Fe_2O_3$ | heating with carbon |
| aluminium | $Al_2O_3$ | electrolysis |
| sodium | NaCl | electrolysis |

a Give:
  i the chemical name of each compound shown
  ii the common names for the three ores
b Arrange the three metals in order of reactivity.
c i How are the two more reactive metals extracted from their ores?
  ii Explain why this is a reduction of the ores.
d i How is the least reactive metal extracted from its ore?
  ii Explain why this is a reduction of the ore.
  iii Why can this method not be used for the more reactive metals?
e Which of the methods would you use to extract:
  i potassium? ii lead? iii magnesium?
f Gold is found native in the Earth's crust. Explain what *native* means.
g Where should gold go, in your list for **b**?
h Name another metal that occurs native.

2 a Draw a diagram of the blast furnace. Show clearly on your diagram:
  i where air is blasted into the furnace
  ii where the molten iron is removed
  iii where a second liquid is removed
b i Name the three raw materials used.
  ii What is the purpose of each material?
c i Name the second liquid that is removed.
  ii When it solidifies, does it have any uses? If so, name one.
d i Name a waste gas from the furnace.
  ii Does this gas have a use? If so, what?
e i Write an equation for the chemical reaction that produces the iron.
  ii Explain why this is a reduction of the iron compound.
  iii What acts as the reducing agent?

# MAKING USE OF METALS

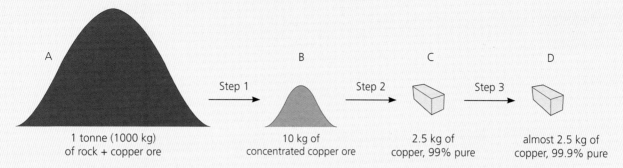

| A | | B | | C | | D |
|---|---|---|---|---|---|---|
| 1 tonne (1000 kg) of rock + copper ore | →Step 1→ | 10 kg of concentrated copper ore | →Step 2→ | 2.5 kg of copper, 99% pure | →Step 3→ | almost 2.5 kg of copper, 99.9% pure |

## Extended curriculum

**3** The diagram above shows stages in obtaining copper from a low-grade ore. The ore contains copper(II) sulfide, CuS. It may also contain small amounts of silver, gold, platinum, iron, cadmium, zinc, and arsenic.
  **a** What is an *ore*?
  **b** What is a *low-grade* ore?
  **c i** How much waste rock is removed per tonne in **step 1**?
   **ii** What % of the ore in B is finally extracted as pure copper?
  **d** Why might it be economic to extract copper from a low-grade ore like this?
  **e i** Which type of reaction occurs in **step 2**?
   **ii** With what could the copper ore be reacted, to obtain the metal?
  **f i** Which process is carried out at **step 3**, to purify the metal?
   **ii** What will the main cost in this process be?
   **iii** As well as pure copper, this process may produce other valuable substances. Explain why, and where they will be found.
  **g** List some of the environmental problems that may arise in going from A to D.

**4** Zinc and lead are obtained from ores that contain only the metal and sulfur, in the molar ratio 1:1.
  **a** Name the compounds in these ores.
  **b** Write down their formulae.
  In the extraction of the metal, the compounds are roasted in air to obtain the oxide of the metal. The sulfur forms sulfur dioxide.
  **c i** Write equations for the roasting of the ores.
   **ii** Which type of reaction is this?
   **iii** Care must be taken in disposing of the sulfur dioxide produced. Explain why.
  Then the oxide can be heated with coke (carbon) to obtain the metal and carbon monoxide.
  **d i** Write equations for the reactions with carbon.
   **ii** Which substances are reduced, in the reactions?

**5** Aluminium is the most abundant metal in the Earth's crust. Iron is next.
Iron and aluminium are extracted from their ores in large quantities. The table below summarises the two extraction processes.

| Extraction | Iron | Aluminium |
|---|---|---|
| Chief ore | hematite | bauxite |
| Formula of main compound in ore | $Fe_2O_3$ | $Al_2O_3$ |
| Energy source used | burning of coke in air (exothermic reaction) | electricity |
| Other substances used in extraction | limestone | carbon (graphite) cryolite |
| Temperature at hottest part of reactor/°C | 1900 | 1000 |
| How the metal separates from the reaction mixture | melts and collects at the bottom | melts and collects at the bottom |
| Other products | carbon dioxide sulfur dioxide slag | carbon dioxide |

  **a** In each extraction, is the metal oxide *oxidised* or *reduced*, when it is converted to the metal?
  **b** Explain *why* each of these substances is used:
   **i** limestone in the extraction of iron
   **ii** carbon in the extraction of aluminium
   **iii** cryolite in the extraction of aluminium.
  **c** Describe any two similarities in the extraction processes for aluminium and iron.
  **d** Give a use for the slag that is produced as a by-product in the extraction of iron.
  **e** Aluminium costs over three times as much per tonne as iron. Suggest two reasons why aluminium is more expensive than iron, even though it is more abundant in the Earth's crust.
  **f** Most of the iron produced is converted into steels. **i** Why? **ii** How is this carried out?
  **g** Both steel and aluminium are recycled. Suggest reasons why it is important to recycle these metals.

AIR AND WATER

## 15.1 What is air?

### The Earth's atmosphere

The **atmosphere** is the blanket of gas around the Earth.

It is about 700 km thick. We live in the lowest layer, the **troposphere**. (Look at the diagram.) The gas is at its most dense here, thanks to gravity. As you go up, it quickly thins out. In fact 90% of the mass of the atmosphere is in the lowest 16 km.

Here in the troposphere, we usually call the atmosphere air.

### What is in air?

This pie chart shows the gases that make up clean air:

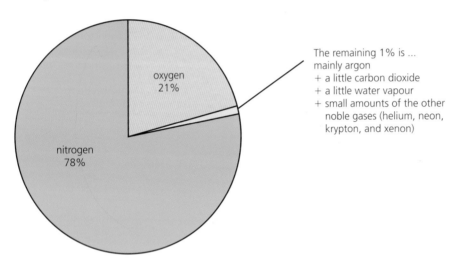

The remaining 1% is ...
mainly argon
+ a little carbon dioxide
+ a little water vapour
+ small amounts of the other noble gases (helium, neon, krypton, and xenon)

▲ The Earth's atmosphere.

The composition of the air changes very slightly from day to day, and place to place. For example:

- there is more water vapour in the air around you on a damp day.
- pollutants such as carbon monoxide and sulfur dioxide are likely to be given out from busy cities and industrial areas. But since air is continually on the move, the pollutants get spread around too.

▲ We cannot live without oxygen. So deep-sea divers have to bring some with them ...

◀ ... and so do astronauts.

## Oxygen: the gas we need most

Most of the gases in air are essential to us. For example we depend on plants for food, and they depend on carbon dioxide. And without nitrogen to dilute the oxygen, fuels would burn too fast for us.

But the gas we depend on most is oxygen. Without it, we would quickly die. We need it for the process called **respiration**, that goes on in all our cells:

glucose + oxygen ⟶ carbon dioxide + water + energy

The energy from respiration keeps us warm, and allows us to move, and enables hundreds of different reactions to go on in our bodies. (And note that respiration, in some form, takes place in the cells of all living things, not only humans.)

▲ Oxygen enters through your mouth and nose, passes into your lungs, and from there diffuses into your blood.

## Measuring the percentage of oxygen in air

**The apparatus** A tube of hard glass is connected to two gas syringes A and B. The tube is packed with small pieces of copper wire. At the start, syringe A contains 100 cm³ of air. B is empty.

**The method** These are the steps:

1. Heat the tube containing copper using a Bunsen burner. Then push in A's plunger, as shown above. This forces the air into B. When A is empty, push in B's plunger, forcing the air back to A. Repeat several times. As the air is pushed to and fro, the oxygen in it reacts with the hot copper, turning it black.

2. Stop heating the tube after about 3 minutes, and allow the apparatus to cool. Then push all the gas into one syringe and measure its volume. (It is now less than 100 cm³.)

3. Repeat steps 1 and 2 until the volume of the gas remains steady. This means all the oxygen has been used up. Note the final volume.

**The results** Starting volume of air: 100 cm³. Final volume of air: 79 cm³. So the volume of oxygen in 100 cm³ air is 21 cm³.
The percentage of oxygen in air is therefore $\frac{21}{100} \times 100 =$ **21%**.

▲ Fish take in the oxygen dissolved in water, through their gills.

---

**Q**

1. What percentage of air is made up of:
   a nitrogen?   b oxygen?   c nitrogen + oxygen?
2. About how much more nitrogen is there than oxygen in air, by volume?
3. What is the combined percentage of all the other gases in air?
4. Mount Everest is over 8.8 km high. Climbers carry oxygen when attempting to reach its summit. Explain why.
5. Which do you think is the most *reactive* gas in air? Why?
6. a Write down the name and formula of the black substance that forms in the experiment above.
   b Suggest a way to turn it back into copper. (Page 92!)

# AIR AND WATER

## 15.2 Making use of air

### Separating gases from the air
As you saw, air is a mixture of gases. Most of them are useful to us. But first, we must separate them from each other.

How can we separate gases? There is a very clever way. First the air is cooled until it turns into a liquid. Then the liquid mixture is separated using a method you met in Chapter 2: **fractional distillation**.

### The fractional distillation of liquid air
This method works because the gases in air have different boiling points. (Look at the table.) So when liquid air is warmed up, the gases boil at different temperatures, and can be collected one by one.

| The boiling points of the gases in air (°C) | |
|---|---|
| carbon dioxide | −32 |
| xenon | −108 |
| krypton | −153 |
| oxygen | −183 |
| argon | −186 |
| nitrogen | −196 |
| neon | −246 |
| helium | −269 |

increasing

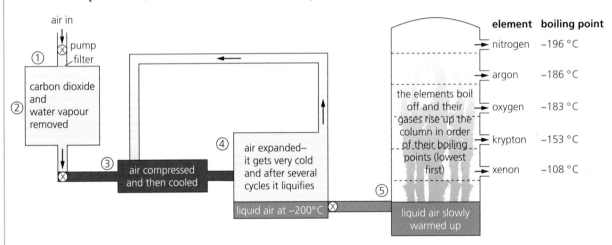

**The steps** The diagram shows the steps.

1 Air is pumped into the plant, and filtered to remove dust particles.

2 Next, water vapour, carbon dioxide, and pollutants are removed (since these would freeze later and block the pipes). Like this:
   - First the air is cooled until the water vapour condenses to water.
   - Then it is passed over beds of adsorbent beads to trap the carbon dioxide, and any pollutants in it.

3 Now the air is forced into a small space, or **compressed**. That makes it hot. It is cooled down again by recycling cold air, as the diagram shows.

4 The cold, compressed air is passed through a jet, into a larger space. It expands rapidly, and this makes it very cold.

Steps 3 and 4 are repeated several times. The air gets colder each time. By −200 °C, it is liquid, except for neon and helium. These gases are removed. They can be separated from each other by adsorption on charcoal.

5 The liquid air is pumped into the fractionating column. There it is slowly warmed up. The gases boil off one by one, and are collected in tanks or cylinders. Nitrogen boils off first. Why?

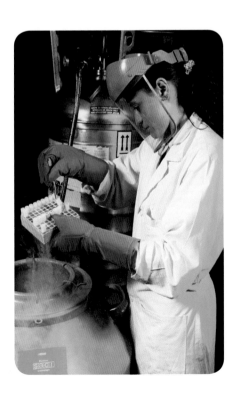

▶ Liquid nitrogen is used in medical research, to keep tissue samples frozen.

▲ An infant in an oxygen tent, to help it breathe.

▲ Easy: slicing through steel with an oxy-acetylene flame.

## Some uses of oxygen
- Planes carry oxygen supplies. So do divers and astronauts.
- In hospitals, patients with breathing problems are given oxygen through an **oxygen mask**, or in an **oxygen tent**. This is a plastic tent that fits over the bed. Oxygen-rich air is pumped into it.
- In steel works, oxygen is used in converting the impure iron from the blast furnace into steels. See page 205.
- A mixture of oxygen and the gas **acetylene** ($C_2H_2$) is used as the fuel in **oxy-acetylene torches** for cutting and welding metal. When this gas mixture burns, the flame can reach 6000 °C. Steel melts at around 3150 °C, so the flame cuts through it by melting it.

## Some uses of nitrogen
- Liquid nitrogen is very cold. (It boils at −196 °C.) So it is used to quick-freeze food in food factories, and to freeze liquid in cracked pipes before repairing them. It is also used in hospitals to store tissue samples.
- Nitrogen is unreactive. So it is flushed through food packaging to remove oxygen and keep the food fresh. (Oxygen helps decay.)

## Some uses of the noble gases
The noble gases are unreactive or **inert**. This leads to many uses.
- Argon provides the inert atmosphere in ordinary tungsten light bulbs. (In air, the tungsten filament would quickly burn away.)
- Neon is used in advertising signs because it glows red when a current is passed through it.
- Helium is used to fill balloons, since it is very light, and safe.

For futher examples of their uses, see page 173.

▲ Frozen food on sale. In the frozen-food factory, food is dipped into liquid nitrogen to quick-freeze it.

1 In the separation of air into its gases:
   a why is the air compressed and then expanded?
   b why is argon obtained *before* oxygen?
   c what do you think is the biggest expense? Explain.
2 Give two uses of oxygen gas.
3 A mixture of oxygen and acetylene burns with a much hotter flame than a mixture of air and acetylene. Why?
4 Nitrogen is used to keep food frozen during transportation. Which properties make it suitable for this?
5 Give three uses for noble gases. (Check page 173 too.)

AIR AND WATER

## 15.3 Pollution alert!

### The air: a dump for waste gases
Everyone likes clean fresh air. But every year we pump billions of tonnes of harmful gases into the air. Most come from burning **fossil fuels**.

### The fossil fuels
These are **coal**, **petroleum** (or **crude oil**) and **natural gas**. Natural gas is mainly methane, $CH_4$. Coal and petroleum are mixtures of many compounds. Most are **hydrocarbons** – they contain only carbon and hydrogen. But some contain other elements, such as sulfur.

Fossil fuels provide us with energy for heating, and transport, and generating electricity. But there is a drawback: burning them produces harmful compounds. Look at the table below.

▲ Don't breathe in!

### The main air pollutants
This table shows the main pollutants found in air, and the harm they do:

| Pollutant | How is it formed? | What harm does it do? |
|---|---|---|
| **Carbon monoxide, CO** colourless gas, insoluble, no smell | Forms when the carbon compounds in fossil fuels burn in too little air. For example, inside car engines and furnaces. | Poisonous even in low concentrations. It reacts with the **haemoglobin** in blood, and prevents it from carrying oxygen around the body – so you die from oxygen starvation. |
| **Sulfur dioxide, $SO_2$** an acidic gas with a sharp smell | Forms when sulfur compounds in the fossil fuels burn. Power stations are the main source of this pollutant. | Irritates the eyes and throat, and causes **respiratory** (breathing) problems. Dissolves in rain to form **acid rain**. Acid rain attacks stonework in buildings, especially limestone and marble – they are calcium carbonate. It lowers the pH in rivers and lakes, killing fish and other river life. It also kills trees and insects. |
| **Nitrogen oxides, NO and $NO_2$** acidic gases | Form when the nitrogen and oxygen in air react together, inside hot car engines and hot furnaces. | Cause respiratory problems, and dissolve in rain to give acid rain. |
| **Lead compounds** | A compound called tetra-ethyl lead used to be added to petrol, to help it burn smoothly in car engines. It is still added in some countries. On burning, it produces particles of other lead compounds. | Lead damages children's brains. It also damages the kidneys and nervous system in adults. |

◀ Power stations are a major source of pollution – and especially those that burn coal. It can contain a lot of sulfur, and it also forms soot.

# AIR AND WATER

## Reducing air pollution
These are some steps being taken to cut down air pollution.
- In modern power stations, the waste gas is treated with slaked lime (calcium hydroxide). This removes sulfur dioxide by reacting with it to give calcium sulfate. The process is called **flue gas desulfurisation**. See page 241 for more.
- Most countries have now banned lead in petrol. So lead pollution is much less of a problem. But it can still arise from plants where lead is extracted, and from battery factories.
- The exhausts of new cars are fitted with **catalytic converters**, in which harmful gases are converted to harmless ones. See below.

## Catalytic converters for car exhausts
When petrol burns in a car engine, harmful gases are produced, including:
  oxides of nitrogen
  carbon monoxide, CO
  unburnt hydrocarbons from the petrol; these can cause cancer.

To tackle the problem, modern car exhausts contain a **catalytic converter**. In this, the harmful gases are adsorbed onto the surface of catalysts, where they react to form harmless gases. The catalysts speed up the reaction.

▲ Some days the pollution is so bad that she has to wear a mask.

The converter usually has two compartments, marked A and B below:

catalyst compartments

> In **A**, harmful compounds are **reduced**. For example:
> $2NO\,(g) \longrightarrow N_2\,(g) + O_2\,(g)$
> The nitrogen and oxygen from this reaction then flow into **B**.

The catalysts are usually the transition elements platinum, palladium, and rhodium. They are coated onto a ceramic honeycomb, or ceramic beads, to give a large surface area for adsorbing the harmful gases. The harmless products flow out the exhaust pipe.

> In **B**, harmful compounds are **oxidised**, using the oxygen from **A**. For example:
> $2CO\,(g) + O_2\,(g) \longrightarrow 2CO_2\,(g)$
> The harmless products then flow out the exhaust pipe.

---

**Q**

1. Look at the pollutants in the table on page 214.
   a Which come from petrol burned in car engines?
   b Which come from air? How and why do these form?
2. Natural gas or methane is a fossil fuel. In a plentiful supply of air, it burns to give carbon dioxide and water. Write a balanced equation to show this.
3. If methane burns in a poor supply of air it will give carbon monoxide and water instead. See if you can write a balanced equation to show this.
4. Catalytic converters can remove carbon monoxide.
   a Give a word equation for the reaction that takes place.
   b What is the purpose of the transition elements?

AIR AND WATER

## 15.4 The rusting problem

### What is rusting?
This car was once new and shiny. But it has been **corroded** – broken down by reaction with something in the atmosphere. In time, it will all be dust.

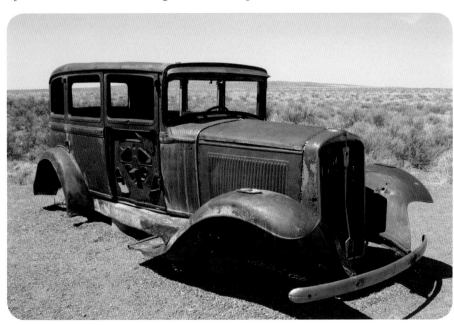

The corrosion of iron and steel has a special name: **rusting**. The red-brown substance that forms is called **rust**.

### An experiment to investigate rusting
1 Stand three identical nails in three test-tubes.
2 Now prepare the test-tubes as below, so that:
 – test-tube 1 contains dry air
 – test-tube 2 contains water but no air
 – test-tube 3 has both air and water.
3 Leave the test-tubes to one side for several days.

**Result** After several days, the nails in test-tubes 1 and 2 show no signs of rusting. But the nail in test-tube 3 has rust on it. This is because:
**Rusting requires oxygen and water.**

In fact the iron is oxidised, in this reaction:

$$4Fe\,(s) + 3O_2\,(g) + 4H_2O\,(l) \longrightarrow 2Fe_2O_3.2H_2O\,(s)$$
$$\text{iron} + \text{oxygen} + \text{water} \longrightarrow \text{hydrated iron(III) oxide (rust)}$$

▲ Rust forms as flakes. Oxygen and moisture can get behind them - so in time the iron rusts right through.

> **Rusting and salt**
> - Iron rusts faster in salty water. (Salt speeds up the oxidation.)
> - So this is a problem for ships … and for cars, where salt is sprinkled on roads in winter, to melt ice. It helps the cars to rust!

▲ A rusting anchor chain. Iron rusts faster in salty water.

> **Stainless steel**
> Remember, the alloy **stainless steel** does not rust. But other steels do.

# AIR AND WATER

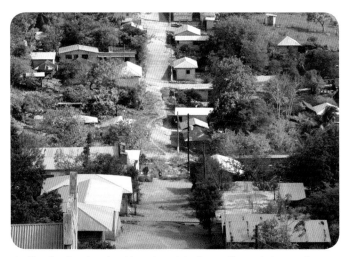

▲ Roofs of galvanised iron in a Mexican village. It is usually called **corrugated iron**.

▲ They are painting the outside of this huge ferry, to help prevent rusting.

## How to prevent rusting

Iron is the most widely used metal in the world – for everything from needles to ships. But rusting destroys things. How can you prevent it? There are two approaches.

### 1 Cover the iron

The aim is to keep out oxygen and water. You could use:

- **paint**. Steel bridges and railings are usually painted.
- **grease**. Tools and machine parts are coated with grease or oil.
- **another metal**. Iron is coated with **zinc**, by dipping it into molten zinc, for roofing. Steel is electroplated with zinc, for car bodies. Coating with zinc has a special name: **galvanising**.
  For food tins, steel is coated with **tin** by electroplating.

### 2 Let another metal corrode instead

During rusting, iron is oxidised: it loses electrons. Magnesium is more reactive than iron, which means it has a stronger drive to lose electrons. So when a bar of magnesium is attached to the side of a steel ship, or the leg of an oil rig, it will corrode instead of the iron.

Without magnesium: $Fe \longrightarrow Fe^{2+} + 2e^-$
With magnesium: $Mg \longrightarrow Mg^{2+} + 2e^-$

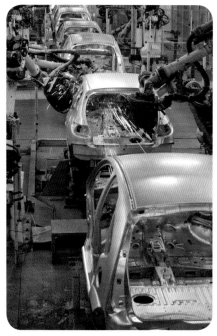

▲ Electroplating is used to galvanise car bodies. The anode is a zinc rod, and the cathode is the steel car body.

The magnesium dissolves. It has been sacrificed to protect the iron. This is called **sacrificial protection**.
The magnesium bar must be replaced before it all dissolves.
Note that zinc could also be used for this. See page 191 for more.

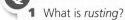

1 What is *rusting*?
2 Which two substances cause iron to rust?
3 See if you can think of a way to *prove* that it is the oxygen in air, not nitrogen, that causes rusting.
4 Iron that is tin-plated does not rust. Why not?
5 You have a new bike. Suggest steps you could take to make sure it does not rust. Give a reason for each one.
6 a What does the *sacrificial protection* of iron mean?
   b Both magnesium and zinc can be used for it. Why?
   c But copper will not work. Explain why.

# AIR AND WATER

## 15.5 Water supply

### Everyone needs water
We all need water.
- **At home** we need it for drinking, cooking, washing things (including ourselves) and flushing toilet waste away.
- **On farms** it is needed as a drink for animals, and to water crops.
- **In industry**, they use it as a solvent, and to wash things, and to keep hot reaction tanks cool. (Cold-water pipes are coiled around the tanks.)
- **In power stations** it is heated to make steam. The steam then drives the turbines that generate electricity.

### So where does the water come from?
Much of the water we use is taken from rivers. But some is pumped up from below ground, where water that has drained down through the soil lies trapped in rocks.

This underground water is called **groundwater**. A large area of rock may hold a lot of groundwater, like a sponge. This rock is called an **aquifer**.

▲ In many places, our water supply is pumped from rivers. The water is cleaned up, the germs are killed, and then it is pumped to homes.

### Is it clean?
River water is not clean – even if it looks it! It will contain particles of mud, and animal waste, and bits of dead vegetation. But worst of all are the **microbes**: bacteria and other tiny organisms that can make us ill.

Over 1 billion people around the world have no access to clean water. They depend on dirty rivers for their drinking water. And over 2 million people, mainly children, die each year from **diarrhoea** and diseases such as **cholera** and **typhoid**, caused by drinking infected water.

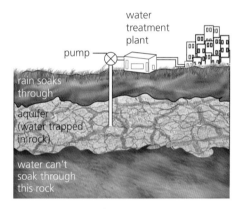

▲ Using an aquifer as a water supply.

### Providing a water supply on tap
No matter where in the world you are, the steps in providing a clean safe water supply, on tap, are the same:

1. Find a clean source – a river or aquifer – to pump water from.

2. Remove as many solid particles from the water as you can.
   - You could make fine particles stick together and skim them off.
   - You could filter the water through clean gravel or sand.

3. Add something to kill the microbes in the water. (Usually chlorine.)

4. Store the water in a clean covered reservoir, ready for pumping to taps.

How well you can clean the water up depends on how dirty it is, and what kind of treatment you can afford!

▲ River water – now safe to drink.

# A modern treatment plant

This diagram shows a modern water treatment plant. Follow the numbers to see how particles are removed and microbes killed.

**2** A **coagulant** is added – a chemical to make small suspended particles stick together. It could be iron(III) sulfate, for example.

**3** Next, air is blown through the water in **flotation tanks**, to make the coagulated particles float to the top. They are skimmed off.

**1** The water is pumped in. A screen traps any big particles, such as twigs.

**4** The water is passed through a bed of fine sand to filter it.

**5** It may go through further filters. For example more sand, or **charcoal** to remove bad tastes and smells.

**6** Chlorine is added to kill the bacteria and other microbes.

**7** A fluoride compound is added in some plants, to help fight tooth decay.

**8** The water is pumped to the storage reservoir, ready for pumping to homes.

This treatment can remove even the tiniest particles. And chlorine can kill all the microbes. But the water may still have harmful substances dissolved in it. For example, nitrates from fertilisers, that can make babies ill.

It is possible to remove dissolved substances, using special membranes. But that is very expensive, and is not usually done. The best solution is to find the cleanest source you can, for your water supply.

**Two tests for water**
If a liquid contains water, it will …
- turn white anhydrous copper(II) sulfate blue
- turn blue cobalt chloride paper pink.

Both colour changes can be reversed by heating.

**If there is only dirty water to drink …**
- Leave it to sit in a container for a while, to let particles settle.
- Scoop out the clearer water from the top of the container, and boil it for several minutes to kill the microbes.
- If you are not able to boil it, leave it sitting in a clear plastic container in the sun for several hours. This will kill most microbes!

**The test for pure water**
If a liquid is *pure* water, it will also …
- boil at 100 °C, and
- freeze at 0 °C.

**Q**

1 What is: **a** groundwater? **b** an aquifer? **c** a microbe? (Check the glossary?)
2 What is a *coagulant* used for, in water treatment plants?
3 Why is chlorine such an important part of the treatment?
4 A fluoride compound may be added to water. Why?
5 Some water can be harmful even after treatment. Explain.
6 You need a drink of water – but there is only dirty river water. What will you do to clean it?

AIR AND WATER

# Living in space

◀ The International Space Station. The 'wings' carry solar panels.

▲ Just hanging about in the lab.

### The International Space Station

Right now, about 350 km above you, a large satellite is orbiting the Earth. On board are scientists: at least three. They could be asleep as you read this, or listening to music, or taking exercise. But most of the time they are doing experiments.

The satellite is the International Space Station (ISS). It is a floating lab, where scientists from different countries carry out a range of experiments. These can be more exciting than usual, since everything is weightless!

### Where do they get their oxygen?

Inside the space station, the air is like that on the Earth. The main gas is nitrogen, which does not get used up. But the oxygen does – and the scientists would die without a steady supply. There is no oxygen outside the space station. So where do they get it?

- **From the electrolysis of water**   A special polymer is used as the electrolyte. The overall reaction is:

  $2H_2O \, (l) \longrightarrow 2H_2(g) + O_2 \, (g)$

  The scientists breathe the oxygen. The hydrogen is vented to space.

- **From 'oxygen candles'**   These are a back-up. They contain sodium chlorate ($NaClO_3$) mixed with iron powder. When the mixture is lit, the sodium chlorate starts breaking down to sodium chloride and oxygen. Some of the oxygen in turn reacts with the iron, giving iron oxide – and this reaction gives out the heat needed to keep the main reaction going.

- **From oxygen cylinders**   These are for emergencies only! (It costs too much to deliver oxygen cylinders from the Earth.)

The carbon dioxide that accumulates in the air is collected, and vented to space. (In future it may be kept to grow plants.)

> **What is the ISS for?**
> It is a place to study the effects of low gravity – for example on:
> - the human body (and mind)
> - how liquids behave
> - the rate of chemical reactions
>
> They expect to learn things that will be useful on Earth – and on long journeys to other planets.

▲ Getting to grips with the project.

# AIR AND WATER

> **Who owns the ISS?**
> - The ISS is jointly owned by the space agencies of the USA, Russia, Japan, and Canada, and the European Space Agency. They all send scientists to it.
> - Other countries may join in before the project ends (around 2020 or later).

## What about water?

The scientists would die without water too. So where does the water come from, to drink and for electrolyis? Mostly from their own bodies!

All urine is collected. So is the water vapour in the air (from the scientists' breath) and any waste water from washing. It is filtered through many kinds of filter, to remove dissolved substances, and treated to kill bacteria.

In fact this water ends up much purer than our drinking water down here on the Earth! As a back-up, some containers of water are stored on board.

Note that the scientists use very little water for washing, since it is so precious. They usually have a wipe with a damp cloth.

▲ Time to grab a snack?

> **What about sleep?**
> - The scientists sleep in sleeping bags – tied to a wall or seat to stop them floating away.
> - They wear eye masks to keep the sunlight out.

## Keeping the lights on

The other essential is electricity. Easy! Solar panels on the 'wings' of the space station convert sunlight to electricity. Some of this is used to charge batteries, for the hours when the sun is hidden. (The space station orbits the Earth once every 90 minutes, so the scientists enjoy 16 sunrises and sunsets each day!)

The electricity is used for electrolysis, and lighting, and laptops, and music players. Cooking does not take much. The food is mostly dried ready-made meals, sent up from the Earth. Add water, and warm in the oven!

## A self-contained unit

This diagram sums up the life support systems, in the space station:

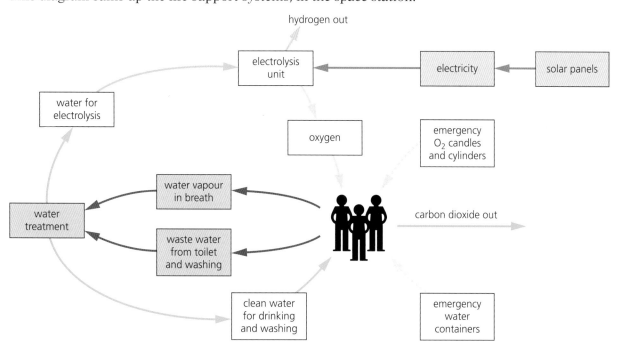

The systems developed for the space station could be very useful one day, if more of us have to move into space!

# Checkup on Chapter 15

## Revision checklist

### Core curriculum
Make sure you can ...
- [ ] name the gases in clean air, and give the percentages for the two main gases present
- [ ] explain why oxygen is so important to us
- [ ] describe an experiment to find the % of oxygen in air
- [ ] name the fossil fuels and give examples of their use as fuels
- [ ] – name four common pollutants in air
  – give the source, for each
  – describe the harm they do
- [ ] explain what rusting is
- [ ] say which substances must be present for rusting to occur
- [ ] describe an experiment to investigate the conditions needed for rusting
- [ ] give examples of ways to prevent rusting, by keeping oxygen and moisture away
- [ ] give examples of how we use water in homes, on farms, in industry, and in power stations
- [ ] describe the steps in the treatment of water, to give a clean safe water supply

### Extended curriculum
Make sure you can also ...
- [ ] describe the separation of gases from the air, using fractional distillation
- [ ] list the harmful gases produced in car engines
- [ ] – explain what catalytic converters are
  – name the metals they usually use as catalysts
  – explain how nitrogen oxides and carbon monoxide are converted to harmless gases in them

## Questions

### Core curriculum

1. Copy and complete:
   Air is a ............of different gases. 99% of it consists of the two elements ......... and .......... . Some of the remaining 1% consists of two compounds, ......... and ......... . The rest is made up of elements called the ......... . These all belong to Group .... of the Periodic Table. The gas we depend on most is ......... This gas combines with glucose in our body cells, releasing energy. The process is called ......... . This gas is also used in combination with ......... , in torches for welding and cutting metal.

2. Air is a *mixture* of different gases.
   a Which gas makes up about 78% of the air?
   b Only one gas in the mixture will allow things to burn in it. Which gas is this?
   c Which noble gas is present in the greatest quantity, in air?
   d i Which gas containing sulfur is a major cause of air pollution?
     ii What harmful effect does this gas have?
   e Name two other gases that contribute to air pollution, and say what harm they do.

3. The rusting of iron wool gives a way to find the percentage of oxygen in air, using this apparatus:

   After five days the water level had risen by 2.5 cm, and the iron had rusted to iron(III) oxide.
   a Write a balanced equation for the reaction between iron wool and oxygen.
   b The iron wool was dampened with water before being put in the tube. Why?
   c Why does the water rise up the tube?
   d What result does the experiment give, for the percentage of oxygen in air?
   e What sort of result would you get if you doubled the amount of iron wool?

4 Oxygen and nitrogen, the two main gases in air, are both slightly soluble in water. A sample of water was boiled, and the gases collected. The water vapour was allowed to condense and the remaining gases were measured. In a 50 cm$^3$ sample of these gases, 18 cm$^3$ were oxygen.
   a  i  What % of the dissolved air was oxygen?
      ii How does this compare to the % of oxygen in the atmosphere?
   b About what % of atmospheric air is nitrogen?
   c Which gas, nitrogen or oxygen, is more soluble in water?

5 This diagram shows one stage in the treatment of water to make it ready for piping to homes:

   a Name the process being carried out here.
   b The water entering this stage has already been treated with a *coagulant*, aluminium sulfate. What does a coagulant do?
   c Which kinds of impurities will the above process: i remove?  ii fail to remove?
   d The next stage in treatment is *chlorination*.
      i What does this term mean?
      ii Why is this process carried out?
   e In some places the water is too acidic to be piped to homes. What could be added to reduce the acidity level?
   f At the end of treatment, another element may be added to water, for dental reasons, in the form of a salt. Which element is this?

## Extended curriculum

6 Nitrogen and oxygen are separated from air by fractional distillation. Oxygen boils at −183 °C and nitrogen at −196 °C.
   a Write the chemical formulae of these two gases.
   b What state must the air be in, before fractional distillation can be carried out?
   c Very low temperatures are required for b. How are these achieved?
   d Explain, using their boiling points, how the nitrogen and oxygen are separated.
   e Name one other gas obtained in the process.

7 Modern cars are fitted with catalytic converters. An outline of one is shown below:

   a  i Where in the car is the catalytic converter?
      ii What is the purpose of a catalytic converter?
      iii Which type of elements are rhodium, platinum, and palladium?
   b Look at the gases that enter the converter.
      i How and where are the oxides of nitrogen formed?
      ii Where do the unburnt hydrocarbons come from?
   c The gases below enter the catalytic converter. Name substances **i – iv**, to show what the gases are converted into.

   | gases in | converted to |
   | --- | --- |
   | oxides of nitrogen | i ............................... |
   | carbon monoxide | ii ............................... |
   | unburnt hydrocarbons | iii ............................... |
   |  | and iv ............................... |

8 In the catalytic converters fitted to modern cars, carbon monoxide and oxides of nitrogen in the exhaust gas are converted to other substances.
   a  i Why is carbon monoxide removed?
      ii Give one harmful effect of nitrogen dioxide.
   b What is meant by a *catalytic* reaction?
   c In one reaction in a catalytic converter, nitrogen monoxide (NO) reacts with carbon monoxide to form nitrogen and carbon dioxide. Write a balanced equation for this reaction.

9 Underwater steel pipelines need to be protected from corrosion. One method is to attach a block of a second metal to the pipeline.
   a  i What is the key factor in choosing the second metal?
      ii Name a suitable metal.
      iii Write a half-equation to show what happens to this metal when it is attached to the pipeline.
   b What name is given to this type of protection?

SOME NON-METALS AND THEIR COMPOUNDS

# 16.1 Hydrogen, nitrogen, and ammonia

### Hydrogen: the lightest element
Hydrogen is the lightest of all the elements. It is so light that there is none in the air: it has escaped from the Earth's atmosphere.

But out in space, it is the most common element in the universe. Inside the sun, hydrogen atoms fuse to form helium atoms. The energy given out provides the Earth with heat and light. We could not live without it.

### Making hydrogen in the lab
Hydrogen is made in the lab by using a metal to drive it out or *displace* it from dilute acid. Zinc and dilute sulfuric acid are the usual choice.

The apparatus is shown on the right, below. The reaction is:

$$Zn\ (s)\ +\ H_2SO_4\ (aq)\ \longrightarrow\ ZnSO_4\ (aq)\ +\ H_2\ (g)$$
zinc    dilute sulfuric acid    zinc sulfate    hydrogen

▲ Sunshine, thanks to hydrogen!

### The properties of hydrogen
1. It is the lightest of all gases – about 20 times lighter than air.
2. It is colourless, with no smell.
3. It combines with oxygen to form water.
   A mixture of the two gases will explode when lit. The reaction is:
   $$2H_2\ (g)\ +\ O_2\ (g)\ \longrightarrow\ 2H_2O\ (l)$$
   It gives out so much energy that it is used to fuel space rockets. The same overall reaction takes place in hydrogen fuel cells, without burning, and gives energy in the form of electricity (page 121).
4. Hydrogen is more reactive than copper, as you can see from the panel. So it will take oxygen from copper(II) oxide, reducing it to copper. The hydrogen is itself oxidised:
   $$CuO\ (s)\ +\ H_2\ (g)\ \longrightarrow\ Cu\ (s)\ +\ H_2O\ (l)$$

▲ One way to make hydrogen in the lab.

### Nitrogen
Nitrogen is a colourless, odourless, unreactive gas, that makes up nearly 80% of the air. You breathe it in – and breathe it out again, unchanged.

But you also take in nitrogen in the **proteins** in your food. Your body uses these to build muscle, bone, skin, hair, blood, and other tissues. In fact you are nearly 3% nitrogen, by mass!

### The properties of nitrogen
1. It is a colourless gas, with no smell.
2. It is only slightly soluble in water.
3. It is very unreactive compared with oxygen.
4. But it will react with hydrogen to form ammonia:
   $$N_2\ (g)\ +\ 3H_2\ (g)\ \rightleftharpoons\ 2NH_3\ (g)$$

   This reversible reaction is the first step in making nitric acid, and nitrogen fertilisers. There is more about it in the next unit.

**The hydrogen displacers**
potassium
sodium
calcium
magnesium
aluminium
zinc
iron
lead
*hydrogen*
copper
silver
gold

increasing reactivity

metals above hydrogen in the reactivity series can displace it from acids

5  Nitrogen also combines with oxygen at high temperatures to form oxides: nitrogen monoxide (NO) and nitrogen dioxide ($NO_2$).

The reactions occur naturally in the air during lightning – and also inside hot car engines, and power station furnaces. The nitrogen oxides are acidic, and cause air pollution, and acid rain. (See page 214.)

## Ammonia

Ammonia is a gas with the formula $NH_3$. It is a very important compound, because it is used to make fertilisers. It is made in industry by reacting nitrogen with hydrogen. The details are in the next unit.

## Making ammonia in the lab

You can make ammonia in the lab by heating *any* ammonium compound with a strong base. The base displaces ammonia from the compound. For example:

$$2NH_4Cl\,(s) \; + \; Ca(OH)_2\,(s) \longrightarrow CaCl_2\,(s) \; + \; 2H_2O\,(l) \; + \; 2NH_3\,(g)$$
ammonium       calcium         calcium       water        ammonia
chloride       hydroxide       chloride

This reaction can be used as a test for ammonium compounds. If an unknown compound gives off ammonia when heated with a strong base, it must be an ammonium compound.

## The properties of ammonia

1  It is a colourless gas with a strong, choking smell.
2  It is less dense than air.
3  It reacts with hydrogen chloride gas to form a white smoke. The smoke is made of tiny particles of solid ammonium chloride:

$$NH_3\,(g) \; + \; HCl\,(g) \longrightarrow NH_4Cl\,(s)$$

This reaction can be used to test whether a gas is ammonia.
4  It is very soluble in water. (It shows the fountain effect.)
5  The solution in water is alkaline – it turns red litmus blue.
6  Since ammonia solution is alkaline, it reacts with acids to form salts. For example with nitric acid it forms ammonium nitrate:

$$NH_3\,(aq) \; + \; HNO_3\,(aq) \longrightarrow NH_4NO_3\,(aq)$$

▲ The **fountain experiment**. The flask contains ammonia. It dissolves in the first drops of water that reach the top of the tube, so a fountain of water rushes up to fill the vacuum. (Any very soluble gas will show this effect.)

---

**Q**

1  Hydrogen can be made using zinc and dilute sulfuric acid. See if you can suggest a different metal and acid, for making it. (Use the reactivity series?)
2  **a** Hydrogen is able to react with copper(II) oxide. Why?
   **b** Which type of reaction is this?
3  Which is more reactive, nitrogen or oxygen?
4  Two examples of *displacement reactions* have been given in this unit. What is a displacement reaction?
5  Write a word equation for the reaction of powdered sodium hydroxide with ammonium sulfate.

# SOME NON-METALS AND THEIR COMPOUNDS

## 16.2 Making ammonia in industry

### It's a key chemical
Ammonia is a very important chemical, because it is needed to make fertilisers – and we depend on fertilisers to grow enough food.
It is made from nitrogen and hydrogen. The reaction is **reversible**.

### The Haber process
The process used to make ammonia is called the **Haber process**.

▲ Ammonia plants are often built close to oil refineries, to make use of the hydrogen from cracking.

1. The reactants are nitrogen and hydrogen. The nitrogen is obtained from air, and the hydrogen by reacting natural gas (methane) with steam, or by cracking hydrocarbons. See the details on the right. The two gases are mixed, and **scrubbed** (cleaned) to remove impurities.
2. The gas mixture is **compressed**. More and more gas is pumped in, until the pressure reaches 200 atmospheres.
3. The compressed gas flows to the **converter** – a round tank with beds of hot iron at 450 °C. The iron catalyses the reversible reaction:

    $$N_2 (g) + 3H_2 (g) \rightleftharpoons 2NH_3 (g)$$

    But only 15% of the mixture leaving the converter is ammonia.
4. The mixture is cooled until the ammonia condenses to a liquid. The nitrogen and hydrogen are recycled to the converter for another chance to react. Steps **3** and **4** are continually repeated.
5. The ammonia is run into tanks, and stored as a liquid under pressure.

### Obtaining the reactants

**Nitrogen**
Air is nearly 80% nitrogen, and 20% oxygen. The oxygen is removed by burning hydrogen:
$$2H_2 (g) + O_2 (g) \longrightarrow 2H_2O (l)$$
That leaves mainly nitrogen, and a small amount of other gases.

**Hydrogen**
- It is usually made by reacting **natural gas** (methane) with steam:
$$CH_4 (g) + 2H_2O (g) \xrightarrow{catalyst} CO_2 (g) + 4H_2 (g)$$
- It can also be made by **cracking hydrocarbons** from petroleum. For example:
$$C_2H_6 (g) \xrightarrow{catalyst} C_2H_4 (g) + H_2 (g)$$
ethane → ethene

## Improving the yield of ammonia

The reaction between nitrogen and hydrogen is reversible, and the forward reaction is exothermic: it gives out heat.

$$N_2(g) + 3H_2(g) \underset{endothermic}{\overset{exothermic}{\rightleftharpoons}} 2NH_3(g)$$

Since the reaction is reversible, a mixture of the two gases will *never* react completely. The yield will never be 100%.

But the yield can be improved by changing the reaction conditions, to shift equilibrium towards the product.

The graph on the right shows how the yield changes with temperature and pressure.

The yield of ammonia at different temperatures and pressures

## The chosen conditions

**The temperature and pressure**  As you can see, the highest yield on the graph is at **X**, at 350 °C and 400 atmospheres.

But the Haber process uses 450 °C and 200 atmospheres, at **Y** on the graph. Why? Because at 350 °C, the reaction is too slow. 450 °C gives a better rate.

And second, a pressure of 400 atmospheres needs very powerful pumps, and very strong and sturdy pipes and tanks, and a lot of electricity. 200 atmospheres is safer, and saves money.

So the conditions inside the converter do not give a high yield. But then the ammonia is removed, so that more will form. And the unreacted gases are recycled, for another chance to react. So the final yield is high.

**The catalyst**  Iron speeds up the reaction. But it does not change the yield!

## More about the raw materials

The panel on the opposite page shows how the raw materials are obtained. Air and water are easy to find. But you need natural gas (methane), or hydrocarbons from petroleum (crude oil), to make hydrogen.

So ammonia plants are usually built close to natural gas terminals or petroleum refineries. In fact many petroleum and gas companies now make ammonia, as a way to increase their profits.

---

**Making ammonia: a summary**

$$N_2 + 3H_2 \underset{endothermic}{\overset{exothermic}{\rightleftharpoons}} 2NH_3$$
4 molecules ⇌ 2 molecules

**To improve the yield:**
- quite high pressure
- remove ammonia

**To get a decent reaction rate:**
- moderate temperature
- use a catalyst

---

1 Ammonia is made from nitrogen and hydrogen.
  a How are the nitrogen and hydrogen obtained?
  b What is the process for making ammonia called?
  c Write an equation for the reaction.
2 Look at the catalyst beds in the diagram on page 226.
  a What is in them?
  b Why are they arranged this way?

3 a Explain *why* high pressure and low temperature help the yield, in making ammonia. (Check Unit 9.6?)
  b 400 atmospheres and 250 °C would give a high yield. Why are these conditions *not* used in the Haber process?
  c What is the % yield of ammonia at 200 atmospheres and 450 °C? (Use the graph.)
  d What happens to the unreacted gases?

SOME NON-METALS AND THEIR COMPOUNDS

## 16.3 Fertilisers

### What plants need

A plant needs carbon dioxide, light, and water. It also needs several different elements. The main ones are **nitrogen**, **potassium**, and **phosphorus**.

Plants need nitrogen for making chlorophyll, and proteins.

Potassium helps them to produce proteins, and to resist disease.

Phosphorus helps roots to grow, and crops to ripen.

Plants obtain these elements from compounds in the soil, which they take in through their roots as solutions. The most important one is nitrogen. Plants take it in as **nitrate** ions and **ammonium** ions.

### Fertilisers

Every crop a farmer grows takes compounds from the soil. Some get replaced naturally. But in the end the soil gets worn out. New crops will not grow well. So the farmer has to add **fertilisers**.

**A fertiliser is any substance added to the soil to make it more fertile.**

Animal manure is a natural fertiliser. **Synthetic fertilisers** are made in factories, and sprinkled or sprayed on fields. Here are some examples.

ammonium nitrate, $NH_4NO_3$     ammonium sulfate, $(NH_4)_2SO_4$

potassium sulfate, $K_2SO_4$     ammonium phosphate, $(NH_4)_3PO_4$

▲ Nutrition for plants: these granules are made of animal manure, a natural fertiliser.

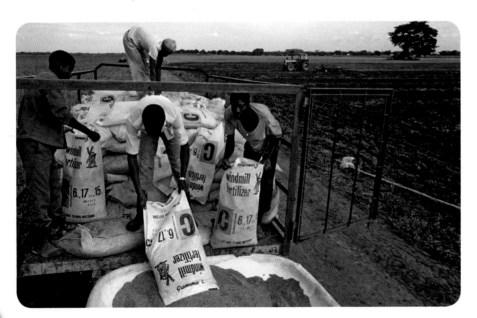

◄ Getting ready to apply fertiliser to fields. (Sometimes spelled *fertilizer*!)

# SOME NON-METALS AND THEIR COMPOUNDS

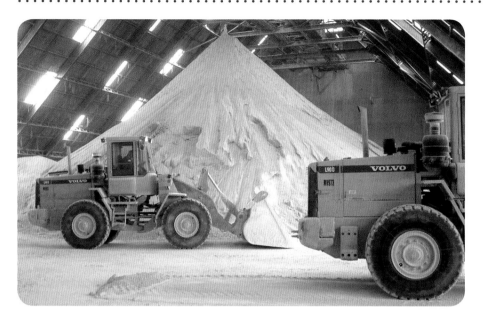

◀ Synthetic fertilisers are made by reactions like those below. Then the solutions are evaporated to give solids. This shows fertiliser in storage.

## Examples of reactions to make synthetic fertilisers

1 Ammonia reacts with nitric acid to give ammonium nitrate. This fertiliser is an excellent source of nitrogen:

$$NH_3\ (aq) \quad HNO_3\ (aq) \quad \rightarrow \quad NH_4NO_3\ (aq)$$
ammonia      nitric acid              ammonium nitrate

2 Ammonia reacts with sulfuric acid to give ammonium sulfate:

$$2NH_3\ (aq) \quad H_2SO_4\ (aq) \quad \rightarrow \quad (NH_4)_2SO_4\ (aq)$$
ammonia      sulfuric acid            ammonium sulfate

**What % of it is nitrogen?**
Ammonium nitrate is rich in nitrogen. What % of it is nitrogen? Find out like this:

Formula: $NH_4NO_3$
$A_r$: N = 14, H = 1, O = 16
$M_r$: $(14 \times 2) + (4 \times 1) + (16 \times 3)$
    $= 80$
% of this that is nitrogen
$= \dfrac{28}{80} \times 100\%$
$= 35\%$

## It's not all good news

Fertilisers help to feed the world. We could not grow enough crops without them. But there are drawbacks – as usual!

**In the river**  Fertilisers can seep into rivers from farmland. In the river, they help tiny water plants called **algae** to grow. These can cover the water like a carpet. When they die, bacteria feed on them, at the same time using up the oxygen dissolved in the water. So fish suffocate.

**In the water supply**  From rivers, the nitrate ions from fertilisers can end up in our water supply. They are converted to nitrite ions in our bodies. These combine with haemoglobin in blood, in place of oxygen, so the blood carries less oxygen around the body. This can cause illness, especially in infants. Their skin may take on a blue tinge.

So farmers should use fertilisers carefully. They should try to keep them away from river banks – and not spread them in wet weather.

▲ Paddling through the algae. Fertiliser from farms helps these plants to grow.

## Q

1 You can buy a mixture of fertilisers called *NPK fertiliser*. It contains elements plants need. Why do you think it is called *NPK*?
2 *Nitrogenous fertilisers* are fertilisers that contain nitrogen. Name three nitrogenous fertilisers.
3 Fertilisers can harm river life. Explain how.
4 The box in the margin above will remind you how to work out % composition.
  a Find the % of nitrogen in ammonium sulfate. ($A_r$: S = 32)
  b Which would provide more nitrogen: 1 kg of ammonium nitrate or 1 kg of ammonium sulfate?
  c Make sure *not* to add it in the rainy season. Why not?

SOME NON-METALS AND THEIR COMPOUNDS

# 16.4 Sulfur and sulfur dioxide

## Where is sulfur found?
Sulfur is a non-metal. It is quite a common element in the Earth's crust.
- It is found, as the *element*, in large underground beds in several countries, including Mexico, Poland and the USA. It is also found around the rims of volcanoes.
- It occurs as a compound in many metal ores. For example in the lead ore **galena**, which is lead(II) sulfide, PbS.
- Sulfur compounds also occur naturally in the fossil fuels: coal, petroleum (crude oil) and natural gas.

▲ A molecule of sulfur. It has 8 atoms – so the molecular formula of sulfur is $S_8$. But it is just called S in equations.

## Extracting the sulfur
**From oil and gas**  Most sulfur is now obtained from the sulfur compounds found in petroleum and natural gas. These compounds are removed to help reduce air pollution.

For example natural gas is mainly methane. But it can have as much as 30% **hydrogen sulfide**. This is separated from the methane. Then it is reacted with oxygen, with the help of a catalyst, to give sulfur:

$$2H_2S\,(g) + O_2\,(g) \longrightarrow 2S\,(s) + 2H_2O\,(l)$$
hydrogen sulfide + oxygen ⟶ sulfur + water

**From sulfur beds**  About 5% of the sulfur we use comes from the underground sulfur beds. Superheated water is pumped down to melt the sulfur and carry it to the surface. (It melts at 115 °C.)

## The properties of sulfur
1. It is a brittle yellow solid.
3. It has two different forms or **allotropes**, as shown on the right.
4. Because it is molecular, it has quite a low melting point.
5. Like other non-metals, it does not conduct electricity.
6. Like most non-metals, it is insoluble in water.
7. It reacts with metals to form sulfides. For example with iron it forms iron(II) sulfide:

    $$Fe\,(s) + S\,(s) \longrightarrow FeS\,(s)$$

    You can see photos of this reaction on page 46.
8. It burns in oxygen to form sulfur dioxide:
    $$S\,(s) + O_2\,(g) \longrightarrow SO_2\,(g)$$

▲ This is a crystal of **rhombic** sulfur, the allotrope that is stable at room temperature.

## Uses of sulfur
- Most sulfur is used to make sulfuric acid.
- It is added to rubber, for example for car tyres, to toughen it. This is called **vulcanizing** the rubber.
- It is used in making drugs, pesticides, dyes, matches, and paper.
- It is used in making cosmetics, shampoos, and body lotions.
- It is added to cement to make **sulfur concrete**. This is not attacked by acid. So it is used for walls and floors in factories that use acid.

▲ If you heat rhombic sulfur slowly to above 96 °C, the molecules rearrange themselves. The result is needle-shaped crystals of **monoclinic sulfur**.

## Sulfur dioxide

Sulfur dioxide ($SO_2$) is a gas. It forms when sulfur burns in air.

1. It is a colourless gas, heavier than air, with a strong, choking smell.
2. Like most non-metal oxides, it is an acidic oxide. It dissolves in water, forming **sulfurous acid**, $H_2SO_3$:

   $$H_2O\ (l) + SO_2\ (g) \longrightarrow H_2SO_3\ (aq)$$

   This breaks down easily again to sulfur dioxide and water.
3. It acts as a bleach when it is damp or in solution. This is because it removes the colour from coloured compounds by **reducing** them.
4. It can kill bacteria.

## Sulfur dioxide as a pollutant

Coal and petroleum contain sulfur compounds – even after petroleum is treated to remove them. Some coals contain a high % of sulfur.

When these fuels are burned in power stations, and factory furnaces, the sulfur compounds are oxidised to sulfur dioxide.

This escapes into the air, where it can cause a great deal of harm. It can attack your lungs, giving breathing problems. It also dissolves in rain to give **acid rain**. This attacks buildings and metal structures, and can kill fish and plants. (See page 214.)

## Uses of sulfur dioxide

- Its main use is in the manufacture of sulfuric acid.
- It is used to bleach wool, silk, and wood pulp for making paper.
- It is used as a sterilizing agent in making soft drinks and jam, and in drying fruit. It stops the growth of bacteria and moulds.

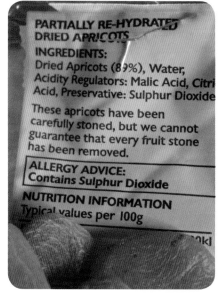

▲ Sulfur dioxide is used to preserve dried fruit (like these apricots) and other foods. (*Sulfur* is often spelled *sulphur*.)

▲ Death by sulfur dioxide: this forest was killed by acid rain.

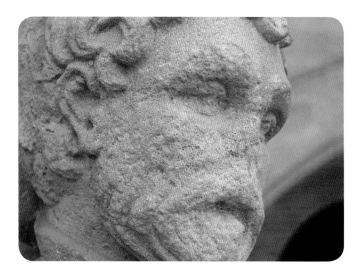

▲ The effect of acid rain on a limestone statue.

**Q**

1. Name three sources of sulfur in the Earth's crust.
2. Sulfur has quite a low melting point. Why is this?
3. Sulfur has two *allotropes*. What does that mean?
4. Sulfur reacts with iron to form iron(II) sulfide. Is this a redox reaction? Explain your answer.
5. **a** Sulfur dioxide is an *acidic* oxide. Explain.
   **b** What problems does this property cause, if sulfur dioxide escapes into the air from power stations?
6. Sulfur dioxide is a heavy gas. Do you think this contributes to air pollution? Explain your answer.

SOME NON-METALS AND THEIR COMPOUNDS

## 16.5 Sulfuric acid

### Making sulfuric acid by the Contact process

More sulfuric acid is made than any other chemical! Most of it is made by the **Contact process**. The raw materials are:

- **sulfur, air, and water** ... or
- **sulfur dioxide, air, and water.**
  The sulfur dioxide is obtained when sulfide ores, such as lead and zinc ores, are roasted in air to extract the metal from them.

Starting with sulfur, the steps in the Contact process are:

sulfur

↓ **1** burned in **air**

sulfur dioxide, $SO_2$
$S(s) + O_2(g) \longrightarrow SO_2(g)$

**2** mixed with more air
**3** passed over four separate beds of catalyst (pellets of vanadium(V) oxide) at 450 °C

sulfur trioxide, $SO_3$
$2SO_2(g) + O_2(g) \rightleftharpoons 2SO_3(g)$

**4** dissolved in concentrated sulfuric acid

thick fuming liquid called **oleum**

**5** mixed carefully with water

concentrated sulfuric acid, $H_2SO_4$
$H_2O(l) + SO_3(g) \longrightarrow H_2SO_4(l)$

▲ No swimming! A lake of natural sulfuric acid, in a volcano in Indonesia.

▲ You will see this sign on tanks of sulfuric acid. What is the message?

### Things to note about the Contact process

- The reaction in step **3** is **reversible**. The sulfur trioxide continually breaks down again. So the mixture is passed over four separate beds of catalyst, to give the reactants further chances to react.
- Sulfur trioxide is removed between the last two beds of catalyst (using step **4**) in order to increase the yield.
- The reaction in step **3** is **exothermic**. So yield rises as temperature falls. But the catalyst will not work below 400 °C, and it works better at higher temperatures. So 450 °C is a compromise.
- To keep the temperature down to 450 °C, heat must be removed from the catalyst beds. So pipes of cold water are coiled around them to carry heat away. The heat makes the water boil. The steam is used to generate electricity for the plant, or for heating buildings.
- In step **4**, the sulfur trioxide is dissolved in concentrated acid instead of water, because with water, a thick, dangerous mist of acid forms.

**Pressure in the Contact process**
In step **3**:
$2SO_2 + O_2 \rightleftharpoons 2SO_3$
3 molecules    2 molecules

- So increasing the pressure will increase the yield of sulfur trioxide. (Page 127 explains why.)
- But in practice, the pressure is increased only a little (to less than two atmospheres).

## Uses of sulfuric acid

Sulfuric acid is one of the world's most important chemicals. It has thousands of uses in industry. Its main uses are in making:

- fertilisers such as ammonium sulfate
- paints, pigments, and dyestuffs
- fibres and plastics
- soaps and detergents.

It is also the acid used in car batteries.

## Dilute sulfuric acid

In the lab, dilute sulfuric acid is made by adding the concentrated acid to water. And never the other way round – because so much heat is produced that the acid could splash out and burn you.

Dilute sulfuric acid shows the usual reactions of acids:

1. acid + metal ⟶ salt + hydrogen
2. acid + metal oxide or hydroxide ⟶ salt + water
3. acid + carbonate ⟶ salt + water + carbon dioxide

Its salts are called **sulfates**. And reactions **2** and **3** are neutralisations: water is produced as well as a salt.

For example dilute sulfuric acid reacts with iron like this:

$H_2SO_4$ (aq) + Fe (s) ⟶ $FeSO_4$ (aq) + $H_2$ (g)
sulfuric acid    iron      iron(II) sulfate    hydrogen

And with copper(II) oxide like this:

$H_2SO_4$ (aq) + CuO (s) ⟶ $CuSO_4$ (aq) + $H_2O$ (l)
sulfuric acid    copper(II) oxide    copper(II) sulfate    water

> **Concentrated sulfuric acid – danger!**
> - Concentrated sulfuric acid is a **dehydrating agent**. It removes water.
> - It 'likes' water so much that it removes hydrogen and oxygen atoms from other substances. For example from sugar (sucrose, $C_{12}H_{22}O_{11}$), leaving just carbon. Look at the photo above.
> - When it is mixed with water, the reaction gives out a great deal of heat.

▲ Concentrated sulfuric acid was added to two teaspoons of sugar – and this is the result. It turned the sugar into carbon. Think what it could do to flesh!

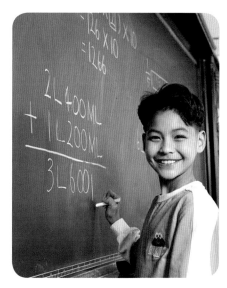

▲ The teacher's sulfate. The white stick they call chalk is not really chalk – it is calcium sulfate.

---

**Q**

Unit 9.6 will help you answer these questions.

1. For making sulfuric acid, name:
    a the process
    b the raw materials
    c the catalyst
2. a The reaction between sulfur dioxide and oxygen is *reversible*. What does that mean?
    b Suggest a reason why a catalyst is needed.
    c At 500 °C, the catalyst makes sulfur trioxide form even faster. Why is this temperature *not* used?
3. Explain how these help to increase the yield of sulfur trioxide, in the Contact process:
    a Several beds of catalyst are used.
    b The sulfur trioxide is removed by dissolving it.
4. Identify two *oxidation* reactions in the manufacture of sulfuric acid.
5. a Write word equations for the reactions of zinc metal, zinc oxide (ZnO) and zinc carbonate ($ZnCO_3$) with dilute sulfuric acid.
    b Now write a balanced equation for each reaction in **a**.

# SOME NON-METALS AND THEIR COMPOUNDS

## 16.6 Carbon and the carbon cycle

### Carbon, the element
Some carbon is found in the Earth's crust as the free element, in two forms: **diamond** and **graphite**. Diamond is a hard, clear solid. Graphite is a dark, greasy solid. So diamond and graphite are **allotropes** (different forms of the same element).

### Carbon compounds
There are *thousands* of carbon compounds in nature: in living things, in the soil, in the oceans, and in the atmosphere (carbon dioxide). You are around 75% water by mass – and around 20% carbon!

### The carbon cycle
Carbon moves between compounds in the atmosphere, living things, the soil, and the ocean, in a non-stop journey called **the carbon cycle**:

▲ **Charcoal**: a form of graphite made by heating coal or wood in a little air.

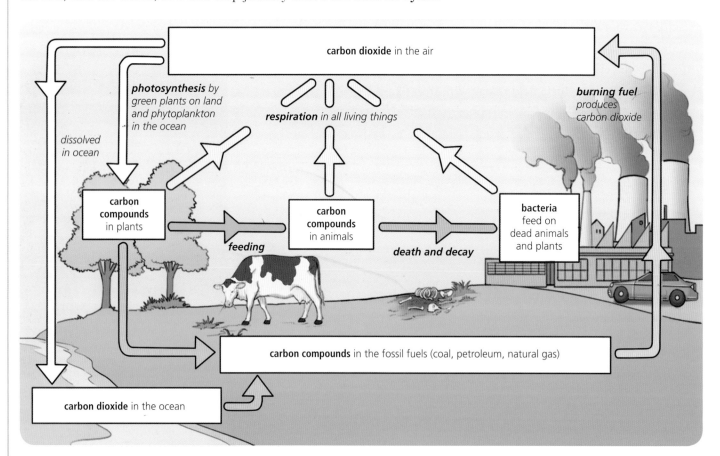

### Note about the carbon cycle
- Carbon moves between the atmosphere, ocean, and living things, in the form of carbon dioxide.
- Carbon dioxide is …
  - removed from the atmosphere by **photosynthesis**, and **dissolving** in the ocean
  - added to it by **respiration**, and the **combustion** (burning) of fuels that contain carbon.

**Two opposite reactions**
**Respiration**, which goes on in your body, is the opposite of **photosynthesis** in plant leaves. You can compare their equations on the next page.

## Removing carbon dioxide from the atmosphere

- **By photosynthesis**  In this process, carbon dioxide and water react in plant leaves, to give glucose and oxygen. Chlorophyll, a green pigment in leaves, is a catalyst for the reaction. Sunlight provides the energy:

$$6CO_2\,(g) + 6H_2O\,(l) \xrightarrow[\text{chlorophyll}]{\text{light}} C_6H_{12}O_6\,(s) + 6O_2\,(g)$$
$$\text{carbon dioxide} \quad \text{water} \qquad\qquad\qquad \text{glucose} \quad\;\; \text{oxygen}$$

The plant uses the glucose to make the other carbon compounds it needs. Then animals eat the plants. So the carbon compounds get passed along the food chain. Many of them end up in your dinner!

Note that photosynthesis also goes on in **phytoplankton**, tiny plants that float in the ocean. These are eaten by fish and other organisms. So carbon is passed along food chains in the ocean too.

- **By dissolving**  Some carbon dioxide from the air dissolves in the ocean. It provides carbonate ions, which shellfish use along with calcium ions from the water, to build their shells. (Shells are made of calcium carbonate.) Fish also use them in building their skeletons.

But only a certain % of carbon dioxide will dissolve. A balance is reached between its concentration in the air and the ocean.

▲ Rice plants: busy with photosynthesis. They will convert the glucose to starch.

## Adding carbon dioxide to the atmosphere

- **By respiration**  This is the process that takes place in our cells (and in the cells of plants and other animals) to provide energy:

$$C_6H_{12}O_6\,(aq) + 6O_2\,(g) \longrightarrow 6CO_2\,(g) + 6H_2O\,(l) + \text{energy}$$
$$\text{glucose} \;\;+\;\; \text{oxygen} \;\longrightarrow\; \text{carbon dioxide} + \text{water} + \text{energy}$$

We get the glucose from food. The energy keeps us warm, and allows us to move, and enables other reactions to go on in our bodies.

- **By the combustion of fuels**  Natural gas or **methane** burns like this:

$$CH_4\,(g) + 2O_2\,(g) \longrightarrow CO_2\,(g) + 2H_2O\,(l) + \text{energy}$$
$$\text{methane} + \text{oxygen} \longrightarrow \text{carbon dioxide} + \text{water} + \text{energy}$$

## The carbon cycle and fossil fuels

In the ocean, the remains of dead organisms fall to the ocean floor, and are buried. Over millions of years their soft parts turn into petroleum (oil) and natural gas. (Hard shells turn into limestone rock.)

Meanwhile, trees and other vegetation get buried in warm swamps. Over millions of years, they turn into coal.

In this way, carbon dioxide from the air ends up in the fossil fuels. And when we burn these, it is released again.

▲ Respiration plus skill ...

**Q**

1. What is the *carbon cycle*?
2. Compare respiration and the combustion of methane.
   a What is similar about the two reactions?
   b What do we use the energy from respiration for?
   c What do we use the energy from burning fuels for?
3. Now compare respiration and photosynthesis. What do you notice about these reactions?
4. See if you can draw a circular flowchart that shows:
   – how carbon dioxide from the air gets locked up as compounds in petroleum and natural gas
   – and how it is released again, millions of years later.
5. One part of the carbon cycle does *not* occur naturally (that is, without help from humans). Which part?
6. What part do *you* play in the carbon cycle?

SOME NON-METALS AND THEIR COMPOUNDS

## 16.7 Some carbon compounds

### Carbon dioxide

The gas carbon dioxide ($CO_2$) occurs naturally in air. It is also a product in these three reactions:

1. The combustion of carbon compounds in plenty of air. For example, when natural gas (methane) burns in plenty of air, the reaction is:

   $$CH_4\,(g) \;+\; 2O_2\,(g) \longrightarrow CO_2\,(g) \;+\; 2H_2O\,(l)$$

2. The reaction between glucose and oxygen, in your body cells:

   $$C_6H_{12}O_6\,(aq) \;+\; 6O_2\,(g) \longrightarrow 6CO_2\,(g) \;+\; 6H_2O\,(l)$$

   This is called **respiration**. You breathe out the carbon dioxide.

3. The reaction between dilute acids and carbonates. For example between hydrochloric acid and marble chips (calcium carbonate):

   $$CaCO_3\,(s) \;+\; 2HCl\,(aq) \longrightarrow CaCl_2\,(aq) \;+\; CO_2\,(g) \;+\; H_2O\,(l)$$

### Properties of carbon dioxide

1. It is a colourless gas, with no smell.
2. It is much heavier than air.
3. Things will not burn in it. We say it **does not support combustion**.
4. It is slightly soluble in water, forming carbonic acid, $H_2CO_3$.

### Carbon monoxide

Carbon monoxide (CO) forms when carbon compounds burn in too little oxygen. For example, when methane burns in insufficient oxygen:

$$2CH_4\,(g) \;+\; 3O_2\,(g) \longrightarrow 2CO\,(g) \;+\; 4H_2O\,(l)$$

It is a deadly poisonous gas. It binds to the haemoglobin in red blood cells, and prevents it from carrying oxygen around the body. So victims die from oxygen starvation.

Carbon monoxide has no smell, which makes it hard to detect. So it is important to have gas heaters and boilers checked regularly, to make sure the air supply is not blocked by soot.

### Carbonates

**Carbonates** are compounds that contain the carbonate ion, $CO_3^{2-}$. One example is calcium carbonate, $CaCO_3$, which occurs naturally as limestone, chalk and marble. These are the main properties of carbonates:

1. They are insoluble in water – *except for* sodium, potassium, and ammonium carbonates, which are soluble.
2. They react with acids to form a salt, water, and carbon dioxide.
3. Most of them break down on heating, to an oxide and carbon dioxide:

   $$CaCO_3\,(s) \;\rightleftharpoons\; CaO\,(s) \;+\; CO_2\,(g)$$
   calcium carbonate    calcium oxide    carbon dioxide
   (limestone)           (lime)

   But sodium and potassium carbonates do not break down, since the compounds of these reactive metals are more stable. (See page 189.)

▲ The 'fizz' in this soft drink is caused by carbon dioxide escaping.

▲ Gas-fuelled water heaters and boilers should be checked regularly. Every year, hundreds of people are killed by carbon monoxide from faulty burners.

## Methane

Methane is the compound $CH_4$.

- It is found in gas deposits in the ocean floor and on land, as **natural gas**. We use natural gas as a fuel.
- It also forms wherever bacteria break down plant material, in the absence of oxygen. For example in paddy fields, and swamps, and landfill sites (rubbish dumps).
- Some animals give out methane as waste gas. They include cattle, sheep, goats, camel, and buffalo. Bacteria in their stomachs help to break down grass and other food, giving methane as one product.

## Organic compounds

Methane is an **organic compound**. Organic compounds *all* contain carbon, and most contain hydrogen. Some contain elements like sulfur and nitrogen too. Many are found in, or derived from, living things.

Methane is the simplest organic compound. There are millions more – far more than all the **inorganic** (non-organic) compounds. They include:

- the proteins, carbohydrates, and fats in your body
- the hundreds of different compounds in petroleum and coal
- the plastics and medical drugs made from the compounds in petroleum.

The study of these carbon compounds is called **organic chemistry**. The next two chapters in this book are about organic chemistry.

▲ This fire extinguisher sprays out carbon dioxide gas. (Things will not burn in an atmosphere of carbon dioxide.)

▲ The Amazon rainforest – packed full of organic compounds.

▲ Built up from organic compounds, and around 20% carbon!

---

**Q**

1. Give the word equation for the combustion of natural gas:
   a. in a gas boiler, when the boiler is working well.
   b. in a gas boiler, when the air inlet is partly blocked with soot.
2. Gas boilers should be checked regularly, to make sure air flows through the burner properly. Why?
3. a. Write an equation to show what happens when lead(II) carbonate is heated.
   b. What is this type of reaction called? (Page 189?)
4. a. Name three sources of methane, $CH_4$.
   b. Which do you think is the main source?
5. Is it organic, or inorganic?  a. sodium chloride  b. water

## 16.8 Greenhouse gases, and global warming

### Carbon dioxide and methane are greenhouse gases

Carbon dioxide and methane are both **greenhouse gases**. That means they absorb heat in the atmosphere, and prevent it from escaping into space.

This is how greenhouse gases work:

1. The sun sends out energy as light and UV rays.
2. These warm the Earth, which reflects some of the energy away again, as heat.
3. Some of this heat escapes from the atmosphere.
4. But some is absorbed by greenhouse gases in the atmosphere. So the air, and Earth, are warmed.

There are several greenhouse gases. Carbon dioxide and methane are the two main ones we add to the atmosphere, through human activity. There is much more carbon dioxide than methane in the atmosphere. But the levels of both are rising:

- The level of carbon dioxide is rising because we burn more fossil fuel each year. The carbon dioxide from this goes into the atmosphere. It cannot escape into space, and the ocean can dissolve only some of it.
- The level of methane is rising because there is an increase in animal farming, and rice farming, around the world – and more and more landfill sites.

We need greenhouse gases. Without them, we would freeze to death at night, when the sun was not shining. But many scientists think the level of greenhouse gases is now so high that it is causing global warming.

▲ Two methane manufacturers.

### Global warming

Measurements show that average temperatures around the world are rising. We call this **global warming**.

Why is it happening? Some scientists say it is a natural change, like similar changes in the past.

However, a panel of scientists from around the world examined all the data, and concluded that greenhouse gases are *almost certainly* the main cause.

They picked out carbon dioxide as the main culprit. The rise in average temperatures over time appears to match the rise in carbon dioxide levels over time. Compare the two graphs on the right.

SOME NON-METALS AND THEIR COMPOUNDS

## Climate change

Air temperature affects rainfall, and cloud cover, and wind patterns. So as average temperatures rise, climates around the world change too.

Scientists try to predict what will happen, using computer models. They cannot make really good predictions yet, because they do not fully understand the links between weather, and clouds, and the ocean.

But they do predict that:

- some places with quite a lot of rain will become very dry, and other places will get much wetter.
- melting land-ice in the Arctic and Antarctica will cause sea levels to rise, so low-lying countries will be at risk of flooding.
- storms, floods, and wildfires will be more frequent and severe.
- species that cannot adapt to the changing climate will die out.
- more drought is likely, which will led to famine – so more people will become refugees.

Most experts agree that climate change is already underway.

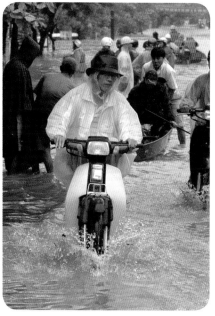

▲ Many countries are already having more severe floods than usual.

## What can we do?

If global warming is a natural change, we can do nothing to stop it. We can only prepare for the consequences.

If we are causing global warming by burning fossil fuels, we still cannot stop it, because the level of carbon dioxide already in the air is enough to cause a further temperature rise. All we can do is cut back heavily on new emissions of carbon dioxide, to stop warming getting out of control.

- Many people are trying to cut back on using fossil fuel, for example by using public transport or bikes, or walking, rather than going by car.
- Many countries have set targets for switching to clean ways to get electricity, such as windpower and solar power.
- Scientists are looking at ways to reduce the amount of carbon dioxide entering the atmosphere. For example by capturing it from power station chimneys, and burying it deep underground.

Some countries are starting to prepare for climate change. The poorest countries are likely to suffer most, since they do not have enough money to cope well with floods, drought, and other disasters.

▲ Melting ice in the Arctic means polar bears are under threat.

---

**Q**

1. **a** We need greenhouse gases. Why?
   **b** So why are they becoming a problem?
2. The two main greenhouse gases we are adding to the atmosphere are …?
3. Global warming could lead to the extinction of some species of living things. Explain why.

4. The more carbon dioxide in the air, the more will dissolve in the ocean. (A balance is reached.)
   **a** Which type of oxide is carbon dioxide: acidic or basic?
   **b i** How might the pH of the ocean be affected by our burning of fossil fuels? Explain.
      **ii** Do you think this could cause problems? Explain.

# 16.9 Limestone

## Limestone: from sea creatures

Most of the creatures that live in the sea have shells or skeletons made of calcium carbonate. When they die, their remains fall to the sea floor. Slowly, over millions of years, the layers of shells and bones become limestone rock. (The soft parts of sea organisms become oil and gas.)

Over millions of years, powerful forces raised some sea beds upwards, draining them to form land. That explains why plenty of limestone is found inland, miles from the sea!

▲ Limestone – made from the shells and skeletons of sea creatures.

## Making use of limestone

Around 5 billion tonnes of limestone are quarried from the Earth's crust every year. This is what it is used for:

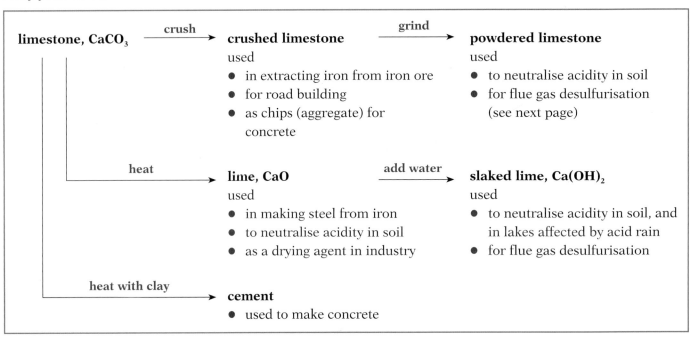

## Lime

When limestone is heated, it breaks down to **lime** (or **quicklime**):

$$CaCO_3\,(s) \rightleftharpoons CaO\,(s) + CO_2\,(g)$$

calcium carbonate     calcium oxide     carbon dioxide
(limestone)               (lime)

This is **thermal decomposition**.

The drawing shows a lime kiln. The kiln is heated. Limestone is fed in at one end. Lime comes out the other.

The reaction is reversible. So the calcium oxide and carbon dioxide *could* combine again. But air is blown through the kiln to carry the carbon dioxide away before it has a chance to react.

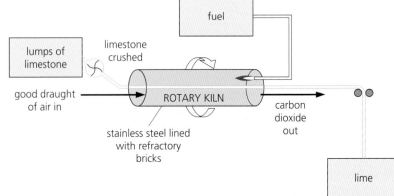

▲ A rotary kiln for making lime.

## Slaked lime

**Slaked lime** forms when water is added to lime. The reaction is exothermic, so the mixture hisses and steams. Conditions are controlled so that the slaked lime forms as a fine powder:

$$CaO\ (s)\ +\ H_2O\ (l)\ \longrightarrow\ Ca(OH)_2\ (s)$$
calcium oxide　　　　　　　　　　calcium hydroxide
(lime)　　　　　　　　　　　　　(slaked lime)

Slaked lime is used to neutralise acidity in soil, and in lakes. In the lab, we use it to test for carbon dioxide. **Limewater** is a weak solution of calcium hydroxide, which is sparingly soluble in water. (See the test on page 285.)

## Cement

**Cement** is made by mixing limestone with clay, heating the mixture strongly in a kiln, adding gypsum (hydrated calcium sulfate), and grinding up the final solid to give a powder.

▲ Cement: limestone heated with clay – and gypsum is added to slow down the 'setting' process.

## Flue gas desulfurisation

**Flue gas desulfurisation** means the removal of sulfur dioxide from the waste gases at power stations, before they go out the flue (chimney).

It is usually carried out using a runny mixture of powdered limestone, or slaked lime, and water. The mixture is sprayed through the waste gases, or the gases are bubbled through it.

When slaked lime is used, the reaction that removes the sulfur dioxide is:

$$Ca(OH)_2\ (s)\ +\ SO_2\ (g)\ \longrightarrow\ CaSO_3\ (s)\ +\ H_2O\ (l)$$
calcium hydroxide　　sulfur dioxide　　　calcium sulfite　　water

Then the calcium sulfite can be turned into hydrated calcium sulfate:

$$2CaSO_3\ (s)\ +\ O_2\ (g)\ +\ 4H_2O\ (l)\ \longrightarrow\ 2\ CaSO_4.2H_2O\ (s)$$
calcium sulfite　　oxygen　　　water　　　　　　hydrated
　　　　　　　　　　　　　　　　　　　　　calcium sulfate

Hydrated calcium sulfate is known as **gypsum**. It is used in making cement, plaster board, plaster for broken limbs, and other products. So the company that owns the power station can sell it, to earn some money.

▲ Flue gas desulfurisation removes sulfur dioxide from waste gases.

---

**Q**

**1** How was limestone formed?
**2 a** How is lime made? Write the equation.
　**b** Why is it important to remove the carbon dioxide?
　**c** How is the carbon dioxide removed, in a lime kiln?
**3** How is slaked lime made? Write the equation.
**4** Give two uses each for lime and slaked lime.
**5** Limewater is a solution of slaked lime. It is used for …?

**6** Slaked lime is more soluble in water than limestone is. Which of the two might be a better choice, for controlling soil acidity in a rainy area? Explain your choice.
**7 a** Explain the term *flue gas desulfurisation*.
　**b** Name a material used for this process.
　**c** Calcium sulfite from the process is often turned into gypsum. What is gypsum, and why do they make it?

SOME NON-METALS AND THEIR COMPOUNDS

# Checkup on Chapter 16

## Revision checklist

### Core curriculum
Make sure you can ...
- [ ] say how these can be prepared in the lab:
  *hydrogen*     *ammonia*
  and give two reactions for each of them
- [ ] give the equation for the reversible reaction between nitrogen and hydrogen
- [ ] name the three main elements plants need from the soil, and say why they need them
- [ ] – explain what fertilisers are
  – say why they are needed
  – and give examples of salts that act as fertilisers
- [ ] describe two problems associated with fertilisers
- [ ] give equations for three different types of reaction that produce carbon dioxide (including respiration)
- [ ] give three sources of methane
- [ ] explain what a *greenhouse gas* is, and how it works
- [ ] name two greenhouse gases
- [ ] explain these terms:
  *global warming*     *climate change*
- [ ] say that many scientists (but not all) believe that carbon dioxide plays a key role in climate change
- [ ] describe how limestone is converted to lime and slaked lime, and give equations
- [ ] give at least two uses each for limestone, lime and slaked lime
- [ ] explain what *flue gas desulfurisation* means, and describe how it is carried out
- [ ] say what *gypsum* is, and give some uses for it

### Extended curriculum
Make sure you can also ...
- [ ] explain why ammonia is an important chemical
- [ ] say how the raw materials (nitrogen and hydrogen) are obtained, for making ammonia
- [ ] state the conditions used in the manufacture of ammonia, and explain the choice of conditions
- [ ] name three sources of sulfur
- [ ] state three uses of sulfur dioxide
- [ ] describe the Contact process for making sulfuric acid, starting with sulfur, and state the conditions
- [ ] give the typical acid properties of dilute sulfuric acid
- [ ] sketch the carbon cycle, and give equations for these three reactions linked to the carbon cycle: respiration, combustion of a fuel such as methane, and photosynthesis

## Questions

### Core curriculum

1 An NPK fertiliser contains the three main elements that plants need, for healthy growth.
   a Name the three elements.
   b Describe *how* each element helps plants.
   c Which of the three elements are provided by the following fertilisers?
      i   ammonium phosphate
      ii  potassium nitrate
      iii ammonium sulfate
   d Write a formula for each fertiliser in c.

2 a Copy the diagram below. Then fill in:
      i  the common names of the substances
      ii their chemical formulae

   b Beside each arrow say how the change is carried out. One example is shown.
   c Give three reasons why limestone is an important raw material.

3 Limestone is calcium carbonate, $CaCO_3$. It is quarried on a huge scale.
   a Which elements does it contain?
   b Much of the quarried limestone is turned into lime (CaO) for the steel industry.
      i  What is the chemical name for lime?
      ii Describe how it is made from limestone.
   c Powdered limestone is used to improve the water quality in acidified lakes.
      i   How might the lakes have become acidified?
      ii  Why is limestone added?
      iii The limestone is used in powdered form, not lumps. Why? Try for more than one reason.
   d List other important uses of limestone.

4 Powdered limestone is used to treat the waste gases from power stations that burn coal and petroleum. The equation for the reaction that takes place is:

$CaCO_3\ (s) + SO_2\ (g) \longrightarrow CaSO_3\ (s) + CO_2\ (g)$

   a i Name the gas that is removed by this reaction.
     ii Why is it important to remove this gas?
   b Why are large lumps of limestone *not* used?
   c The process is called *flue gas desulfurisation*. Explain clearly what this means.
   d The calcium sulfite is usually turned into gypsum, which has the formula $CaSO_4.2H_2O$.
     i What is the *full* chemical name for gypsum?
     ii Which type of chemical reaction occurs when $CaSO_3$ is converted into $CaSO_4$?
     iii Give two uses for gypsum.
   e Name two chemicals that could be used to make calcium sulfate by precipitation.

**Extended curriculum**

5 This is about the manufacture of ammonia.
   a Which two gases react to give ammonia?
   b Why are the two gases scrubbed?
   c Why is the mixture passed over iron?
   d What happens to the *unreacted* nitrogen and hydrogen?
   e In manufacturing ammonia, is the chosen pressure high, low, or moderate? Explain why.

6 Nitrogen and hydrogen are converted to ammonia in the Haber process:

$N_2\ (g) + 3H_2\ (g) \rightleftharpoons 2NH_3\ (g)$

Below is the energy level diagram for the reaction.

   a What does this diagram tell you?
   b Explain why high temperatures are *not* used in the manufacture of ammonia.
   c The reaction is *reversible*, and *reaches equilibrium*. Explain very clearly what the two terms in italics mean.
   d i What effect does a catalyst have on an equilibrium reaction?
     ii Which catalyst is used in the Haber process?
     iii What effect does this catalyst have on the % yield of ammonia?

7 Sulfuric acid is made by the Contact process. The first stage is to make sulfur trioxide, like this:

$2SO_2\ (g) + O_2\ (g) \rightleftharpoons 2SO_3\ (g)$

The energy change in the reaction is $-97$ kJ/mol.
   a Name the catalyst used in this reaction.
   b Is the reaction exothermic, or endothermic?
   c What are the reaction conditions for making sulfur trioxide?
   d Will the yield of sulfur trioxide increase, decrease, or stay the same, if the temperature is raised? Explain your answer.
   e Describe how sulfur trioxide is changed into concentrated sulfuric acid.

8 Below is a flow chart for the Contact process:

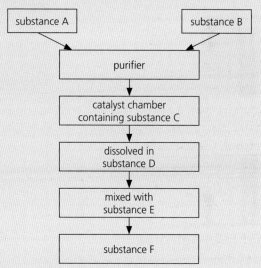

   a Name the substances **A**, **B**, **C**, **D**, **E**, and **F**.
   b Why is a catalyst used?
   c Write a chemical equation for the reaction that takes place on the catalyst.
   d The production of substance **F** is very important. Why? Give three reasons.
   e Copy out the flow chart, and write in the full names of the different substances.

9 Dilute sulfuric acid has typical acid properties. An excess of it is added to test-tubes **W**, **X**, **Y** and **Z**, which contain these powdered substances:
   **W** copper(II) oxide      **X** magnesium
   **Y** calcium hydroxide     **Z** sodium carbonate
   a In which test-tubes will you observe fizzing?
   b In which test-tube will a coloured solution form?
   c In which of the test-tubes does neutralisation take place?
   d Name the four salts obtained, after reaction.
   e Write balanced equations for the four reactions.

ORGANIC CHEMISTRY

## 17.1 Petroleum: a fossil fuel

### The fossil fuels

The **fossil fuels** are **petroleum** (or crude oil), **coal**, and **natural gas**. They are called fossil fuels because they are the remains of plants and animals that lived millions of years ago.

**Petroleum** formed from the remains of dead organisms that fell to the ocean floor, and were buried under thick sediment. High pressures slowly converted them to petroleum, over millions of years.

**Natural gas** is mainly **methane**. It is often found with petroleum. It is formed in the same way. But high temperatures and high pressures caused the compounds to break down to gas.

**Coal** is the remains of lush vegetation that grew in ancient swamps. The dead vegetation was buried under thick sediment. Pressure and heat slowly converted it to coal, over millions of years.

### What is in petroleum?

Petroleum is a smelly mixture of hundreds of different compounds. They are **organic compounds**, which means they contain carbon, and usually hydrogen.

In fact most are **hydrocarbons** – they contain *only* carbon and hydrogen. These drawings show molecules of three different hydrocarbons:

This is a molecule of pentane, $C_5H_{12}$. It has a straight chain of 5 carbon atoms.

This is a molecule of cyclohexane, $C_6H_{12}$. Here a chain of 6 carbon atoms form a ring.

This is a molecule of 3-methyl pentane, $C_6H_{14}$. Here 6 carbon atoms form a **branched** chain.

A formula drawn out in this way is called a **structural formula**.

Notice how the carbon atoms are bonded to each other, to make the spine of each molecule. The hydrogen atoms are bonded to the carbon atoms. In petroleum you will find hydrocarbon molecules of different shapes and sizes, with different numbers of carbon atoms, from 1 to over 70.

ORGANIC CHEMISTRY

## How we use petroleum
Over 13 billion litres of petroleum are used around the world every day.

Around half the petroleum pumped from oil wells is used for transport. It provides the fuel for cars, trucks, planes, and ships. You won't get far without it!

Most of the rest is burned for heat, in factories, homes, and power stations, as above. In a power station, the heat is used to turn water to steam, to drive turbines.

A small % is used as the starting chemicals to make many other things: plastics, shampoo, paint, thread, fabric, detergents, makeup, medical drugs, and more.

Many of the things you use every day were probably made from petroleum. Toothbrush, comb, and shampoo just for a start!

## A non-renewable resource
Petroleum is still forming, very slowly, under the oceans. But we are using it up much faster than it can form, which means it will run out one day.

So petroleum is called a **non-renewable resource**.

It is hard to tell when it will run out. At the present rate of use, some experts say the world's reserves will last about 40 more years. What will we do then?

▲ A platform for pumping petroleum from under the ocean.

**Q**
1. The other name for petroleum is … ?
2. Why is petroleum called a *fossil fuel*?
3. What is a *hydrocarbon*?
4. What is petroleum made of?
5. Explain why petroleum is such a valuable resource.
6. Petroleum is called a *non-renewable* resource. Why?
7. What do you think we will use for fuel, when petroleum runs out?

ORGANIC CHEMISTRY

## 17.2 Refining petroleum

### What does *refining* mean?
Petroleum contains *hundreds* of different hydrocarbons. But a big mixture like this is not very useful.

So the first step is to separate the compounds into groups with molecules of a similar size. This is called **refining** the petroleum. It is carried out by **fractional distillation**.

### Refining petroleum in the lab
The apparatus on the right can be used to refine petroleum in the lab.

1 As you heat the petroleum, the compounds start to evaporate. The ones with smaller lighter molecules go first, since it takes less energy to free these from the liquid.
2 As the hot vapours rise, so does the thermometer reading. The vapours condense in the cool test-tube.
3 When the thermometer reading reaches 100 °C, replace the first test-tube with an empty one. The liquid in the first test-tube is your first **fraction** from the distillation.
4 Collect three further fractions in the same way, replacing the test-tube at 150 °C, 200 °C, and 300 °C.

### Comparing the fractions
Now compare the fractions – how runny they are, how easily they burn, and so on. You can burn samples on a watch glass, like this:

fraction 1

fraction 2

fraction 3

fraction 4

It catches fire easily. The flame burns high, which shows that the liquid is **volatile** – it evaporates easily.

This catches fire quite easily. The flame burns less high – so this fraction is less volatile than fraction 1.

This seems less volatile than fraction 2. It does not catch fire so readily or burn so easily – it is not so **flammable**.

This one does not ignite easily. You need to use a wick to keep it burning. It is the least flammable of the four.

This table summarizes the results:

| Fraction | Boiling point range | How easily does it flow? | How volatile is it? | How easily does it burn? | Size of molecules |
|---|---|---|---|---|---|
| 1 | up to 100 °C | very runny | volatile | very easily | small |
| 2 | 100 – 150 °C | runny | less volatile | easily | |
| 3 | 150 – 200 °C | not very runny | even less volatile | not easily | |
| 4 | 200 – 300 °C | viscous (thick and sticky) | least volatile | only with a wick | large |

# ORGANIC CHEMISTRY

## The trends the fractions show
Those results show that, the larger the molecules in a hydrocarbon:

- the higher its boiling point will be
- the less volatile it will be
- the less easily it will flow (or the more viscous it will be)
- the less easily it will burn.

These trends help to dictate what the different fractions will be used for, as you will see below.

## In the petroleum refinery
In a refinery, the fractional distillation is carried out in a tower that is kept very hot at the base, and cooler towards the top. Look at the drawing.

Petroleum is pumped in at the base. The compounds start to boil off. Those with the smallest molecules boil off first, and rise to the top of the tower. Others rise only part of the way, depending on their boiling points, and then condense.

The table shows the fractions that are collected.

▲ The new road surface is bitumen mixed with fine gravel.

| Name of fraction | Number of carbon atoms | What fraction is used for |
|---|---|---|
| refinery gas | $C_1$ to $C_4$ | bottled gases for cooking and heating |
| gasoline (petrol) | $C_5$ to $C_6$ | fuel for cars |
| naphtha | $C_6$ to $C_{10}$ | starting point or **feedstock** for many chemicals and plastics |
| paraffin (kerosene) | $C_{10}$ to $C_{15}$ | fuel for aircraft, oil stoves, and lamps |
| diesel oil (gas oil) | $C_{15}$ to $C_{20}$ | fuel for diesel engines |
| fuel oil | $C_{20}$ to $C_{30}$ | fuel for power stations, ships, and for home heating systems |
| lubricating fraction | $C_{30}$ to $C_{50}$ | oil for car engines and machinery; waxes and polishes |
| bitumen | $C_{50}$ upwards | for road surfaces and roofs |

boiling points and viscosity increase ↓

As the molecules get larger, the fractions get less runny, or more viscous: from gas at the top of the tower to solid at the bottom. They also get less flammable. So the last two fractions in the table are not used as fuels.

## Q

1. Which two opposite processes take place, during fractional distillation?
2. A group of compounds collected during fractional distillation is called a ..........?
3. What does it mean?  **a** volatile  **b** viscous
4. List four ways in which the properties of different fractions differ.
5. Name the petroleum fraction that:  **a** is used for petrol  **b** has the smallest molecules  **c** is the most viscous  **d** has molecules with 20 to 30 carbon atoms

ORGANIC CHEMISTRY

## 17.3 Cracking hydrocarbons

### After fractional distillation ...
Petroleum is separated into fractions by fractional distillation. But that is not the end of the story. The fractions all need further treatment before they can be used.
1 They contain impurities – mainly sulfur compounds. If left in the fuels, these will burn to form harmful sulfur dioxide gas.
2 Some fractions are separated further into single compounds, or smaller groups of compounds. For example the gas fraction is separated into methane, ethane, propane, and butane. (We buy butane in canisters.)
3 Part of a fraction may be **cracked**.
   **Cracking breaks molecules down into smaller ones.**

▲ This sulfur was obtained from sulfur compounds removed from natural gas.

### Cracking a hydrocarbon in the lab
This experiment is carried out using a hydrocarbon oil from petroleum. The product is a gas, collected over water in the inverted test-tube:

The moment heating is stopped, the delivery tube must be lifted out of the water. Otherwise water will get sucked up into the hot test-tube. Now compare the reactant and product:

|  | The reactant | The product |
|---|---|---|
| Appearance | thick colourless liquid | colourless gas |
| Smell | no smell | pungent smell |
| Flammability | difficult to burn | burns readily |
| Reactions | few chemical reactions | many chemical reactions |

So the product is quite different from the reactant. Heating has caused the hydrocarbon to break down. A **thermal decomposition** has taken place. Note that:
- the reactant had a high boiling point and was not flammable – which means it had large molecules, with long chains of carbon atoms.
- the product has a low boiling point and is very volatile – so it must have small molecules, with short carbon chains.
- the product must also be a hydrocarbon, since nothing new was added.

So the molecules of the starting hydrocarbon have been cracked. And since the product is reactive, it could be a useful chemical.

▲ Some of the naphtha fraction from refining will be piped to the cracking plant.

# ORGANIC CHEMISTRY

## Cracking in the refinery
In the refinery, cracking is carried out in a similar way.
- The long-chain hydrocarbon is heated to vaporize it.
- The vapour is usually passed over a hot catalyst.
- Thermal decomposition takes place.

## Why cracking is important
- Cracking helps you make the best use of petroleum. Suppose you have too much of the naphtha fraction, and too little of the gasoline fraction. You can crack some naphtha to get molecules the right size for petrol.
- Cracking *always* produces short-chain compounds with a carbon–carbon double bond. This bond makes the compounds **reactive**. So they can be used to make plastics and other substances.

## Examples of cracking

▲ Plastic furniture: it all began with cracking.

1. **Cracking the naphtha fraction**  Compounds in the naphtha fraction are often cracked, since this fraction is used as the feedstock for making many useful chemicals. This is the kind of reaction that occurs:

decane, $C_{10}H_{22}$ from naphtha fraction

$\downarrow$ 540 °C, catalyst

pentane, $C_5H_{12}$ suitable for petrol     propene, $C_3H_6$     ethene, $C_2H_4$

So decane has been broken down into three smaller molecules. The propene and ethene molecules have carbon–carbon double bonds. These two compounds belong to the **alkene** family, and they are very reactive.

2. **Cracking ethane**  Ethane has very short molecules – but even it can be cracked, to give ethene and hydrogen:

ethane $\xrightarrow[>800\ °C]{\text{steam}}$ ethene + $H_2$

▲ Decane is one of the hydrocarbons in **white spirit**, a solvent used to thin oil-based paint, and clean paintbrushes.

The hydrogen can be used to make ammonia – see page 226.

**Q**

1. What happens during cracking?
2. Cracking is a *thermal decomposition*. Explain why.
3. Describe the usual conditions needed for cracking a hydrocarbon in the petroleum refinery.
4. What is *always* produced in a cracking reaction?
5. Explain why cracking is so important.
6. a  A straight-chain hydrocarbon has the formula $C_5H_{12}$. Draw the structural formula for its molecules.
   b  Now show what might happen when the compound is cracked.

ORGANIC CHEMISTRY

## 17.4 Families of organic compounds

### What their names tell you

There are *millions* of organic compounds. That could make organic chemistry confusing – but to avoid this, the compounds are named in a very logical way.

The rest of this chapter is about some families of organic compounds. **For these families, the name of the organic compound tells you:**
- which family it belongs to
- how many carbon atoms are in it.

Look at these two tables:

| If the name ends in ... | ... the compound belongs to this family ... | Example |
|---|---|---|
| -ane | the alkanes | ethane, $C_2H_6$ |
| -ene | the alkenes | ethene, $C_2H_4$ |
| -ol | the alcohols | ethanol, $C_2H_5OH$ |
| -oic acid | the carboxylic acids | ethanoic acid, $CH_3COOH$ |

| This in the name ... | ... means this many carbon atoms ... | Example from the alkane family |
|---|---|---|
| meth- | 1 | methane, $CH_4$ |
| eth- | 2 | ethane, $C_2H_6$ |
| prop- | 3 | propane, $C_3H_8$ |
| but- | 4 | butane, $C_4H_{10}$ |
| pent- | 5 | pentane, $C_5H_{12}$ |
| hex- | 6 | hexane, $C_6H_{14}$ |

▲ Natural gas burning at a cooker hob: it is mainly methane, the simplest alkane.

### The alkanes: the simplest family

Here again are the first four members of the alkane family. Note that methane is the simplest member. What patterns do you notice?

| Compound | methane | ethane | propane | butane |
|---|---|---|---|---|
| Formula | $CH_4$ | $C_2H_6$ | $C_3H_8$ | $C_4H_{10}$ |
| Structural formula | H–C(H)(H)–H | H–C(H)(H)–C(H)(H)–H | H–C(H)(H)–C(H)(H)–C(H)(H)–H | H–C(H)(H)–C(H)(H)–C(H)(H)–C(H)(H)–H |
| Number of carbon atoms in the chain | 1 | 2 | 3 | 4 |
| Boiling point/°C | −164 | −187 | −42 | −0.5 |

boiling point increases with chain length →

250

ORGANIC CHEMISTRY

## Comparing families

This table shows one member from each of the four families. Compare them.

| Family | A member | Structural formula | Comments |
|---|---|---|---|
| alkanes | ethane, $C_2H_6$ | H-C(H)(H)-C(H)(H)-H | • The alkanes contain only carbon and hydrogen, so they are **hydrocarbons**.<br>• The bonds between their carbon atoms are all **single bonds**. |
| alkenes | ethene, $C_2H_4$ | H₂C=CH₂ | • The alkenes are **hydrocarbons**.<br>• All alkenes contain **carbon – carbon double bonds**.<br>• The C=C bond is called their **functional group**. |
| alcohols | ethanol, $C_2H_5OH$ | H-C(H)(H)-C(H)(H)-O-H | • The alcohols are not hydrocarbons.<br>• They are like the alkanes, but with an **OH group**.<br>• The OH group is their functional group. |
| carboxylic acids | ethanoic acid, $CH_3COOH$ | H-C(H)(H)-C(=O)OH | • The carboxylic acids are not hydrocarbons.<br>• All carboxylic acids contain the **COOH group**.<br>• The COOH group is their functional group. |

## Functional groups

**A functional group is the part of a molecule that largely dictates how the molecule will react.**

For example, all the alkenes have similar reactions because they all have the same functional group, the C=C bond.

## Homologous series

Look back at the alkanes in the table at the bottom of page 250. They form a **homologous series**. In a homologous series:

- All the compounds fit the same general formula.
  For the alkanes the general formula is $C_nH_{2n+2}$, where $n$ is a number.
  For methane $n$ is 1, giving the formula $CH_4$.
  For ethane $n$ is 2, giving $C_2H_6$.
  For propane $n$ is 3, giving $C_3H_8$.
- The chain length increases by 1 each time.
- As the chain gets longer, the compounds show a gradual change in properties. For example, their boiling points rise, and they burn less easily.

As you will see later, all four families in this unit form homologous series.

> **In a homologous series ...**
> As the chain gets longer:
> - melting and boiling points rise
> - viscosity increases – the compounds flow less easily
> - flammability decreases – the compounds burn less easily.

**1** *Propanol* is an organic compound.
  **a** How many carbon atoms does it contain?
  **b** Which family does it belong to?
  **c** See if you can draw a structural formula for it.

**2** Draw a structural formula for the alkane called *hexane*.

**3** An alkane has 32 carbon atoms in each molecule. Give its formula.

**4** Try to draw the structural formula for *propanoic acid*.

ORGANIC CHEMISTRY

## 17.5 The alkanes

### Alkanes: a reminder
This is what you have learned about the alkanes so far:

- The alkanes are the simplest family of organic compounds.
- They are hydrocarbons: they contain only carbon and hydrogen.
- Their carbon – carbon bonds are all single bonds.
- They form a homologous series, with the general formula $C_nH_{2n+2}$.

This table shows the first four members of the alkane family. What patterns do you notice?

▲ The bonding in ethane.

| Compound | methane | ethane | propane | butane |
|---|---|---|---|---|
| Formula | $CH_4$ | $C_2H_6$ | $C_3H_8$ | $C_4H_{10}$ |
| Structural formula | H–C(H)(H)–H | H–C(H)(H)–C(H)(H)–H | H–C(H)(H)–C(H)(H)–C(H)(H)–H | H–C(H)(H)–C(H)(H)–C(H)(H)–C(H)(H)–H |
| Number of carbon atoms in the chain | 1 | 2 | 3 | 4 |
| Boiling point/°C | −164 | −87 | −42 | −0.5 |

boiling point increases with chain length →

### Key points about the alkanes

1. They are found in petroleum and natural gas. Petroleum contains alkanes with up to 70 carbon atoms. Natural gas is mainly methane, with small amounts of ethane, propane, butane, and other compounds.

2. The first four alkanes are gases at room temperature. The next twelve are liquids. The rest are solids. Boiling points increase with chain length because attraction between the molecules increases – so it takes more energy to separate them.

3. Since all their carbon – carbon bonds are single bonds, the alkanes are called **saturated**. Look at the bonding in ethane on the right above.

4. Generally, the alkanes are quite unreactive.

5. But alkanes do burn well in a good supply of oxygen, forming carbon dioxide and water vapour, and giving out plenty of heat. So they are used as **fuels**. Methane burns the most easily. Like this:

   $CH_4 (g) + 2O_2 (g) \longrightarrow CO_2 (g) + 2H_2O (l) +$ heat energy

6. If there is not enough oxygen, the alkanes undergo **incomplete combustion**, giving poisonous carbon monoxide. For example:

   $2CH_4 (g) + 3O_2 (g) \longrightarrow 2CO (g) + 4H_2O (l) +$ less heat energy

▲ Butane is used as fuel for cooking and heating, in many homes.

**7** Alkanes also react with chlorine in sunlight. For example:

$$CH_4 + Cl_2 \xrightarrow{light} CH_3Cl + HCl$$

methane → chloromethane

This is called a **substitution** reaction, because a chlorine atom takes the place of a hydrogen atom. If there is enough chlorine, all four hydrogen atoms will be replaced, one by one. Look at the panel on the right.

The reaction can be explosive in sunlight. But it will not take place in the dark, because it is also a **photochemical reaction**: light energy is needed to break the bonds in the chlorine molecules, to start the reaction off.

## Isomers

butane
boiling point 0 °C

2-methylpropane
boiling point −10 °C

Compare these alkane molecules. Both have the same formula, $C_4H_{10}$. But they have different structures. The first has a **straight** or unbranched chain. In the second, the chain is **branched**.

The two compounds are **isomers**.
**Isomers are compounds with the same formula, but different structures**.

The more carbon atoms in a compound, the more isomers it has. There are 75 isomers with the formula $C_{10}H_{22}$, for example.

Since isomers have different structures, they also have slightly different properties. For example branched isomers have lower boiling points, because the branches make it harder for the molecules to get close. So the attraction between them is less strong, and less heat is needed to overcome it.

> **Chlorine and methane**
> The hydrogen atoms can be replaced one by one:
>
> chloromethane $CH_3Cl$
> dichloromethane $CH_2Cl_2$
> trichloromethane $CHCl_3$
> tetrachloromethane $CCl_4$
>
> All four are used as solvents. But they can cause health problems, so are being used less and less.

▲ Branched-chain hydrocarbons burn less easily than their straight-chain isomers. So they are used in petrol to control combustion and stop engine 'knock' – and especially for racing cars.

1 Describe the bonding in ethane. A drawing will help!
2 Why are alkanes such as methane and butane used as fuels? See if you can give *at least* two reasons.
3 Butane burns in a similar way to methane. See if you write a balanced equation for its combustion.
4 a The reaction of chlorine with methane is called a *substitution* reaction. Why?
   b What special condition is needed, for this reaction?
5 Ethane reacts with chlorine, in a substitution reaction.
   a Draw the structural formula for each compound that can form, as the reaction proceeds. (Isomers too!)
   b Write the formula for each compound in **a**.
6 The compound $C_5H_{12}$ has *three* isomers.
   a Draw the structures of these three isomers.
   b Their boiling points are 9.5, 28, and 36 °C. Match these to your drawings, and explain your choice.

ORGANIC CHEMISTRY

## 17.6 The alkenes

### The alkene family
- The alkenes are hydrocarbons.
- They form a homologous series, with the general formula $C_nH_{2n}$.
- They all contain the C=C double bond. This is their functional group, and largely dictates their reactions. Look at the bonding in ethene.
- Because they contain C=C double bonds, they are called **unsaturated**. (Alkanes have only single carbon – carbon bonds, so are **saturated**.)

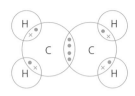

▲ The bonding in ethene.

Here are the first three members of the family. Note how ethene is drawn:

| Compound | ethene | propene | but-1-ene |
|---|---|---|---|
| Formula | $C_2H_4$ | $C_3H_6$ | $C_4H_8$ |
| Structural formula | H₂C=CH₂ | H₂C=CH–CH₃ | H₂C=CH–CH₂–CH₃ |
| Number of carbon atoms | 2 | 3 | 4 |
| Boiling point/°C | −102 | −47 | −6.5 |

Do you agree that these compounds fit the general formula $C_nH_{2n}$?

### Key points about the alkenes

1 The alkenes are made from alkanes by cracking. For example ethene is formed by cracking ethane. Hydrogen is also produced:

ethane $\xrightarrow{\text{steam, >800 °C}}$ ethene + $H_2$

2 Alkanes are much more reactive than alkanes, because the double bond can break, to add on other atoms. For example, ethene can add on hydrogen again, to form ethane:

ethene $(g) + H_2(g) \xrightarrow{\text{heat, pressure, a catalyst}}$ ethane $(g)$

It also adds on water (as steam) to form ethanol, an alcohol:

ethene $(g) + H_2O(g) \xrightleftharpoons{\text{heat, pressure, a catalyst}}$ ethanol $(l)$

These reactions are called **addition reactions**. Can you see why?
**An addition reaction turns an unsaturated alkene into a saturated compound.**

▲ Ethanol is a good solvent. It also kills many germs, so it is used in disinfectant gels for wiping your hands, in hospital.

## Polymerisation

Alkene molecules undergo a very useful **addition reaction,** where they add on *to each other* to form compounds with very long carbon chains.

The alkene molecules are called **monomers**. The long-chain compounds that form are called **polymers**. The reaction is called **polymerisation**. For example ethene polymerises like this:

**The short way to show it ...**
You can show polymerisation in a short way, like this. $n$ stands for a very large number.

The product is **poly(ethene)** or **polythene**. The chain can be many thousands of carbon atoms long!

## A test for unsaturation

You can use **bromine water** to test whether a hydrocarbon is unsaturated. It is an orange solution of bromine in water. If a C=C bond is present, an addition reaction takes place and the colour disappears. For example:

ethene (colourless) + $Br_2(aq)$ (orange) → 1,2–dibromoethane (colourless)

## Isomers in the alkene family

In alkenes, the chains can branch in different ways, *and* the double bonds can be in different positions.

Compare the three compounds below. All three have the formula $C_4H_8$, but they have different structures. So they are isomers.

but-1-ene    but-2-ene    2-methylpropene

▲ Polythene is used to make plastic bottles, plastic bags, plastic sheeting ...

Now look at the numbers in their three names. What do these tell you?

---

**1 a** Name the two simplest alkenes.
 **b** Now draw their structural formulae.
**2** What makes alkenes react so differently from alkanes?
**3** Ethene can *polymerise*. What does that mean?
**4 a** Propene is *unsaturated*. What does that mean?
 **b** Write an equation for its reaction with bromine.
**5** How would you turn propene into:
 **a** propane?   **b** propanol?

ORGANIC CHEMISTRY

## 17.7 The alcohols

### What are alcohols?
The **alcohols** are the family of organic compounds that contain the **OH** group. This table shows the first four members:

| Alcohol | methanol | ethanol | propan-1-ol | butan-1-ol |
|---|---|---|---|---|
| Formula | $CH_3OH$ | $C_2H_5OH$ | $C_3H_7OH$ | $C_4H_9OH$ |
| Structural formula | H–C(H)(H)–O–H | H–C(H)(H)–C(H)(H)–O–H | H–C(H)(H)–C(H)(H)–C(H)(H)–O–H | H–C(H)(H)–C(H)(H)–C(H)(H)–C(H)(H)–O–H |
| Number of carbon atoms | 1 | 2 | 3 | 4 |
| Boiling point/°C | 65 | 78 | 87 | 117 |

Note that:
- they form a homologous series, with the general formula $C_nH_{2n+1}OH$.
- their OH functional group means they will all react in a similar way.
- two of the names above have -1- in. This tells you that the OH group is attached to a carbon atom at one end of the chain.

### Ethanol, an important alcohol
- Ethanol is the alcohol in alcoholic drinks.
- It is a good solvent. It dissolves many substances that do not dissolve in water.
- It evaporates easily – it is **volatile**. That makes it a suitable solvent to use in glues, printing inks, perfumes, and aftershave

▲ Ethanol is used as a solvent for perfume. Why?

### Two ways to make ethanol
Ethanol is made in two ways, one biological and one chemical.

1 **By fermentation – the biological way**
   Ethanol is made from glucose using yeast, in the absence of air:

   $C_6H_{12}O_6$ (aq) $\xrightarrow{\text{enzymes in yeast}}$ $2C_2H_5OH$ (aq) + $2CO_2$ (g) + energy
   glucose                                           ethanol              carbon dioxide

   - Yeast is a mass of living cells. The enzymes in it catalyse the reaction. (See page 142.)
   - The process is called **fermentation**, and it is exothermic.
   - Ethanol can be made in this way from any substance that contains sugar, starch, or cellulose. (These break down to glucose.) For example it can be made from sugarcane, maize, potatoes, and wood.
   - The yeast stops working when the % of ethanol reaches a certain level, or if the mixture gets too warm.
   - The ethanol is separated from the final mixture by fractional distillation.

▲ Corn (maize) is widely grown in the USA to make ethanol, for car fuel. Fuel made from plant material, using yeast or bacteria, is called **biofuel**.

## 2 By the hydration of ethene – the chemical way

*Hydration* means water is added on. This is an **addition reaction**.

$$CH_2=CH_2\ (g) + H_2O\ (g) \rightleftharpoons CH_3CH_2OH\ (l)$$

at 570 °C, 60–70 atm, a catalyst (phosphoric acid); ethene → ethanol

- The reaction is reversible, and exothermic.
- High pressure and a low temperature would give the best yield. But in practice the reaction is carried out at 570 °C, to give a decent rate of reaction.
- A catalyst is also used, to speed up the reaction.

## Ethanol as a fuel

Ethanol burns well in oxygen, giving out plenty of heat:

$$C_2H_5OH\ (l) + 3O_2\ (g) \longrightarrow 2CO_2\ (g) + 3H_2O\ (l) + \text{heat}$$

It is increasingly used as a fuel for car engines because:
- it can be made quite cheaply from waste plant material
- many countries have no petroleum of their own, and have to buy it from other countries; it costs a lot, so ethanol is an attractive option
- ethanol has less impact on carbon dioxide levels than fossil fuels do.

▲ Traffic in Rio de Janeiro in Brazil: running mainly on ethanol made from sugar cane.

## Ethanol and global warming

Like the fossil fuels, ethanol does produce carbon dioxide when it burns. This is a greenhouse gas, linked to global warming. But ethanol has less impact on carbon dioxide levels in the atmosphere, because …

… although carbon dioxide is given out when ethanol burns …

… it is taken in by plants being grown to make more ethanol.

By contrast, the carbon dioxide given out when fossil fuels burn was taken in from the atmosphere many millions of years ago.

### The drawbacks
More and more crops are being grown to make ethanol, for cars.
- But that takes up a lot of land.
- It means less land to grow crops for food.
- A shortage of food crops means a rise in food prices.
- When food prices rise, it affects poor people the most.

Many people are against growing crops to make ethanol. They say: *Feed people*, not cars!

## Q

1. All alcohols react in a similar way. Why?
2. Draw the structural formula for ethanol.
3. Give three uses of ethanol.
4. In Brazil, sugarcane is used to make ethanol. Name the process used, and say what the catalyst is.
5. **a** Write a word equation for the combustion (burning) of:
   **i** ethanol  **ii** methane
   **b** Compare the equations. What do you notice?
6. There is another isomer with the same formula as propan-1-ol. Draw its structure, and suggest a name.

ORGANIC CHEMISTRY

# 17.8 The carboxylic acids

## The carboxylic acid family
Now we look at the family of organic acids: the carboxylic acids. Here are the first four members of the family:

| Name of acid | methanoic | ethanoic | propanoic | butanoic |
|---|---|---|---|---|
| Formula | HCOOH | $CH_3COOH$ | $C_2H_5COOH$ | $C_3H_7COOH$ |
| Structural formula | H—C(=O)—OH | H—C(H)(H)—C(=O)—OH | H—C(H)(H)—C(H)(H)—C(=O)—OH | H—C(H)(H)—C(H)(H)—C(H)(H)—C(=O)—OH |
| Number of carbon atoms | 1 | 2 | 3 | 4 |
| Boiling point/°C | 101 °C | 118 °C | 141 °C | 164 °C |

- The family forms a homologous series with the general formula $C_nH_{2n}O_2$. Check that this fits with the formulae in the table above.
- The functional group **COOH** is also called the **carboxyl group**.

We focus on ethanoic acid in the rest of this unit. But remember that other carboxylic acids behave in a similar way, because they all contain the carboxyl group.

## Two ways to make ethanoic acid
Ethanoic acid is made by oxidising ethanol:

H—C(H)(H)—C(H)(H)—OH  —[O]→  H—C(H)(H)—C(=O)—O—H

The oxidation can be carried out in two ways.

1. **By fermentation – the biological way**
   When ethanol is left standing in air, bacteria bring about its oxidation to ethanoic acid. This method is called **acid fermentation**.
   
   Acid fermentation is used to make vinegar (a dilute solution of ethanoic acid). The vinegar starts as foods such as apples, rice, and honey, which are first fermented to give ethanol.

2. **Using oxidising agents – the chemical way**
   Ethanol is oxidised much faster by warming it with the powerful oxidising agent potassium manganate(VII), in the presence of acid. The manganate(VII) ions are themselves reduced to $Mn^{2+}$ ions, with a colour change. The acid provides the $H^+$ ions for the reaction:
   
   $MnO_4^- + 8H^+ + 5e^- \longrightarrow Mn^{2+} + 4H_2O$
   purple                        colourless
   
   Potassium dichromate(VI) could also be used as the oxidising agent. As you saw on page 99, this gives a colour change from orange to green.

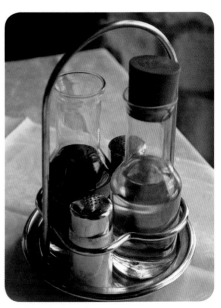

▲ Organic chemistry at the dinner table. Vinegar (on the left) is mainly a solution of ethanoic acid in water. Olive oil, on the right, is made of **esters**. (See page 275 for more about oils.)

# Ethanoic acid: typical acid reactions

Ethanoic acid shows typical acid reactions.
1. A solution of ethanoic acid turns litmus red.
2. A solution of ethanoic acid contains $H^+$ ions, because some of the ethanoic acid molecules dissociate in water, like this:

$$CH_3COOH\ (aq) \xrightarrow{\text{some molecules}} CH_3COO^-\ (aq) + H^+\ (aq)$$
ethanoic acid → ethanoate ions + hydrogen ions

Since only *some* molecules dissociate, ethanoic acid is a **weak acid**.

3. Ethanoic acid reacts with metals, bases, and carbonates, to form **salts**. It reacts with sodium hydroxide like this:

ethanoic acid + sodium hydroxide → sodium ethanoate (a salt) + water

or $CH_3COOH\ (aq) + NaOH\ (aq) \longrightarrow CH_3COONa\ (aq) + H_2O\ (l)$

Like all salts, sodium ethanoate is an ionic compound.

**The ethanoate ion**
This is its structural formula:

# Esters

Ethanoic acid also reacts with alcohols, to give compounds called **esters**. The alcohol molecule is reversed below, to help you see what is happening:

ethanoic acid + propanol ⇌ (conc $H_2SO_4$) propyl ethanoate (an ester) + $H_2O$ water

this group is called an **ester** linkage

or $CH_3COOH\ (l) + C_3H_7OH\ (l) \xrightleftharpoons{\text{conc. } H_2SO_4} CH_3COOC_3H_7\ (l) + H_2O\ (l)$

Note these points:
- Two molecules have joined to make a larger molecule, with the loss of a small molecule, water. So this is called a **condensation reaction**.
- The reaction is reversible, and sulfuric acid acts a catalyst.
- The alcohol part comes *first* in the name – but second in the formula.
- Propyl ethanoate smells of pears. In fact many esters have attractive smells and tastes. So they are added to shampoos and soaps for their smells, and to ice cream and other foods as flavourings.

▲ The smell and taste of the apple come from natural esters. Synthetic esters are used in the shampoo.

# Q

1. What is the functional group of the carboxylic acids?
2. Copy and complete. (Page 152 may help!)
   carboxylic acid + metal → _____ + _____
   carboxylic acid + alkali → _____ + _____
   carboxylic acid + alcohol ⇌ _____ + _____
3. Carboxylic acids are *weak* acids. Explain why.
4. Draw structural formulae to show the reaction between ethanol and ethanoic acid, and name the products.
5. What is a *condensation reaction*?
6. Esters are important compounds in industry. Why?

# ORGANIC CHEMISTRY

# Checkup on Chapter 17

## Revision checklist

### Core curriculum
Make sure you can …
- [ ] name the fossil fuels and say how they were formed
- [ ] explain what a *hydrocarbon* is
- [ ] explain why petroleum has to be refined, and:
  - describe the refining process
  - name the different fractions
  - say what these fractions are used for
- [ ] explain what *cracking* is, and why it is so useful, and give the equation for the cracking of ethane
- [ ] say what a *structural formula* is, and draw structural formulae for
  *methane   ethane   ethene   ethanol   ethanoic acid*
- [ ] say what family a compound belongs to, from its structural formula or its name
- [ ] give the functional groups for the alkenes, alcohols, and carboxylic acids
- [ ] describe alkanes as unreactive except for burning, and give an equation for the combustion of methane
- [ ] explain what *unsaturated* means and describe a test to identify an unsaturated hydrocarbon
- [ ] describe how ethene monomers add on to each other to form the polymer poly(ethene)
- [ ] describe the two ways of making ethanol
- [ ] give at least two uses for ethanol
- [ ] give the reaction for the combustion of ethanol
- [ ] give at least two advantages of ethanol as a fuel

### Extended curriculum
Make sure you can also …
- [ ] name, and draw the structural formulae for, the four simplest members of each family: alkanes, alkenes, alcohols, and carboxylic acids
- [ ] give the general properties of a homologous series
- [ ] explain what *isomers* are, and draw examples
- [ ] describe the substitution reactions of alkanes with chlorine
- [ ] describe the addition reactions of alkenes with hydrogen and steam
- [ ] describe the two ways to make ethanoic acid
- [ ] explain why ethanoic acid is a weak acid
- [ ] give examples of reactions to show that ethanoic acid is a typical acid
- [ ] describe the reaction of ethanoic acid with an alcohol, and name the products that form

## Questions

### Core curriculum
1 Petroleum is separated into fractions, like this:

  a i  What is this process called?
    ii It uses the fact that different compounds have different ……… ………. What is missing?
  b i  Is naphtha just one compound, or a group of compounds? Explain.
    ii Using the terms *evaporation* and *condensation*, explain how naphtha is produced.
  c Give one use for each fraction obtained.
  d A hydrocarbon has a boiling point of 200 °C.
    i Are its carbon chains shorter, or longer, than those found in naphtha?
    ii Is it more viscous, or less viscous, than the compounds found in naphtha?

2

A hydrocarbon can be cracked in the lab using the apparatus above.
  a What is cracking?
  b Which two things are needed, to crack the hydrocarbon?
  c The first tube of collected gas is discarded. Why? (What else is in the heated tube?)
  d At the end of the experiment, the delivery tube must be removed from the water immediately. Why is this?
  e Ethane, $C_2H_6$, can be cracked to give ethene, $C_2H_4$, and hydrogen. Write an equation for this.

## ORGANIC CHEMISTRY

**3** Answer these questions about the alkanes.
  **a** Which two elements do alkanes contain?
  **b** Which alkane is the main compound in natural gas?
  **c** After butane, the next two alkanes in the series are *pentane* and *hexane*. How many carbon atoms are there in a molecule of:
   **i** pentane?   **ii** hexane?
  **d** Will pentane react with bromine water? Explain.
  **e** Alkanes burn in a good supply of oxygen. Name the gases formed when they burn.
  **f** Write the word equation for the complete combustion of pentane in oxygen.
  **g** Name a harmful substance formed during *incomplete* combustion of pentane in air.

**4** When ethanol vapour is passed over heated aluminium oxide, a dehydration reaction occurs, and the gas ethene is produced.
  **a** Draw a diagram of suitable apparatus for carrying out this reaction in the lab.
  **b** What is meant by a *dehydration reaction*?
  **c** Write an equation for this reaction, using the structural formulae.
  **d i** What will you see if the gas that forms is bubbled through bromine water?
   **ii** You will not see this if ethanol vapour is passed through bromine water. Why not?

**5 a** Which of these could be used as monomers for addition polymers? Explain your choice.
   **i** ethene, $CH_2=CH_2$
   **ii** ethanol, $C_2H_5OH$
   **iii** propane, $C_3H_8$
   **iv** styrene, $C_6H_5CH=CH_2$
   **v** chloropropene, $CH_3CH=CHCl$
  **b** Suggest a name for each polymer obtained.

### Extended curriculum

**6** The saturated hydrocarbons form a homologous series with the general formula $C_nH_{2n+2}$.
  **a** What is a *homologous series*?
  **b** Explain what the term *saturated* means.
  **c** Name the series described above.
  **d i** Give the formula and name for a member of this series with two carbon atoms.
   **ii** Draw its structural formula.
  **e i** Name a homologous series of *unsaturated* hydrocarbons, and give its general formula.
   **ii** Give the formula and name for the member of this series with two carbon atoms.
   **iii** Draw the structural formula for the compound.

**7** Ethanol is a member of a homologous series.
  **a** Give two general characteristics of a homologous series.
  **b i** Which homologous series is ethanol part of?
   **ii** What is the general formula for the series?
   **iii** What does *functional group* mean?
   **iv** What is the functional group in ethanol's homologous series?
  **c** Write down the formula of ethanol.
  **d i** Draw the structural formula for the fifth member of the series, pentan-1-ol.
   **ii** Draw the structural formula for an isomer of pentan-1-ol.
   **iii** Describe how pent-1-ene could be made from pentan-1-ol.
   **iv** Name the organic product formed when pentan-1-ol is oxidised using acidified potassium manganate(VII).

**8** Ethanoic acid is a member of the homologous series with the general formula $C_nH_{2n}O_2$.
  **a** Name this series.
  **b** What is the functional group of the series?
  **c** Ethanoic acid is a *weak* acid. Explain what this means, using an equation to help you.
  **d** Ethanoic acid reacts with carbonates.
   **i** What would you *see* during this reaction?
   **ii** Write a balanced equation for the reaction with sodium carbonate.
  **e i** Name the member of the series for which $n = 3$, and draw its structural formula.
   **ii** Give the equation for the reaction between this compound and sodium hydroxide.

**9** Ethanoic acid reacts with ethanol in the presence of concentrated sulfuric acid.
  **a** Name the organic product formed.
  **b** Which type of compound is it?
  **c** How could you tell quickly that it had formed?
  **d** What is the function of the sulfuric acid?
  **e** The reaction is *reversible*. What does this mean?
  **f** Write an equation for the reaction.

**10** Hex-1-ene is an unsaturated hydrocarbon. It melts at $-140\,°C$ and boils at $63\,°C$. Its empirical formula is $CH_2$. Its relative molecular mass is 84.
  **a i** To which family does hex-1-ene belong?
   **ii** What is its molecular formula?
  **b i** Hex-1-ene reacts with bromine water. Write an equation to show this reaction.
   **ii** What is this type of reaction called?
   **iii** What would you *see* during the reaction?

# POLYMERS

## 18.1 Introducing polymers

### What is a polymer?

**A polymer** is any substance containing very large molecules, formed when lots of small molecules join together.

For example, look what happens when ethene molecules join:

This test tube contains ethene gas. When ethene is heated to 50 °C, at a few atmospheres pressure, and over a special catalyst …

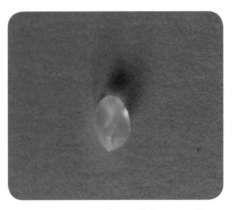

… it turns into a liquid that cools to a waxy white solid. This is found to contain very long molecules, made by the ethene molecules joining.

And it is really useful. It can be used to make toys, dustbins, tables and chairs, water pipes, buckets, crates, washing-up bowls and so on.

The reaction that took place is:

ethene molecules (monomers)

↓ polymerisation

part of a polythene molecule (a polymer)

The drawing shows six ethene molecules adding together. In fact many thousands add together, giving molecules with very long chains. These very large molecules are called **macromolecules**.

**A polymer is a substance made of macromolecules.**

The polymer made from ethene is called **poly(ethene)** or **polythene**. Poly- means *many*. The reaction is called a **polymerisation**.

**In a polymerisation reaction, thousands of small molecules join to give macromolecules. The small molecules are called monomers. The product is called a polymer.**

### Synthetic polymers

Polythene is a **synthetic polymer**. *Synthetic* means it is made in a factory. Other synthetic polymers include nylon, Terylene, lycra, chewing gum, and plastics such as polystyrene and perspex. Hair gels and shower gels contain water-soluble polymers.

▲ Hair gel: a water-soluble polymer. When you put it on, the water in it evaporates so the gel gets stiff.

## Natural polymers

Polythene was first made in 1935. But for billions of years, nature has been busy making natural polymers. Look at these examples:

**Starch** is a polymer made by plants. The starch molecules are built from molecules of **glucose**, a sugar. We eat plenty of starch in rice, bread, and potatoes.

Plants also use glucose to make another polymer called **cellulose**. Cotton T-shirts and denim jeans are almost pure cellulose, made by the cotton plant.

Your skin, hair, nails, bones and muscles are mostly polymers, made of macromolecules called **proteins**. Your body builds these up from **amino acids**.

The wood in trees is about 50% cellulose. Paper is made from wood pulp, so this book is mainly cellulose. The polymer in your hair and nails, and in wool and silk, and animal horns and claws, is called **keratin**. The polymer in your skin and bones is called **collagen**.

So – you contain polymers, you eat polymers, you wear polymers, and you use polymers. Polymers play a big part in your life!

## The reactions that produce polymers

All polymers, natural and synthetic, consist of macromolecules, formed by small molecules joined together.

But these macromolecules are not all made in the same way. There are two types of reaction: **addition polymerisation** and **condensation polymerisation**. You can find out more about these in the next two units.

▲ Wood: over 75% cellulose. This wood may end up as paper.

---

**Q**

**1** What is:
  a  a macromolecule?   b  a polymer?
  c  a natural polymer? d  a synthetic polymer?
  e  polymerisation?

**2** Name the natural polymer found in:
  a  your hair   b  this book

**3** Name at least three items you own, that are made of polymers.

## POLYMERS

# 18.2 Addition polymerisation

## Another look at the polymerisation of ethene

Here again is the reaction that produces polythene:

[diagram of ethene molecules (monomers) with C=C double bonds, undergoing polymerisation to form part of a polythene molecule (a polymer) with single C-C bonds]

The reaction can be shown in a short form like this:

$$n \begin{pmatrix} H & H \\ \phantom{}\backslash & / \\ C=C \\ / & \backslash \\ H & H \end{pmatrix} \xrightarrow{\text{heat, pressure, a catalyst}} \begin{pmatrix} H & H \\ | & | \\ -C-C- \\ | & | \\ H & H \end{pmatrix}_n$$

where $n$ stands for a large number. It could be many thousands. The catalyst for the reaction is usually a mixture of titanium and aluminium compounds.

### It's an addition reaction

The reaction above takes place because the double bonds in ethene break, allowing the molecules to add on to each other. So this is called **addition polymerisation**.

**In addition polymerisation, double bonds in molecules break and the molecules add on to each other.**

### The monomer

The small starting molecules in a polymerisation are called **monomers**. In the reaction above, ethene is the monomer.

**For addition polymerisation to take place, the monomers must have C=C double bonds.**

### The chain lengths in polythene

In polythene, all the macromolecules are long chains of carbon atoms, with hydrogen atoms attached. So they are all similar. But they are not all identical. The chains are not all the same length. That is why we can't write an exact formula for polythene.

By changing the reaction conditions, the *average* chain length can be changed. But the chains will never be all the same length.

The relative atomic mass ($M_r$) of an ethene molecule is 28.
The *average* $M_r$ of the macromolecules in a sample of polyethene can be 500 000 or more. In other words, when making polythene, at least 17 000 ethene molecules join, on average!

▲ Polythene for packaging is made and sold as pellets like these. Later they will be heated to soften or melt them, and turned into plastic bottles and bags.

▲ To make a bottle, polythene pellets are melted. A little molten polymer is fed into a mould. A jet of air forces it into the shape of the mould. Then the mould is opened – and out comes a bottle!

## Making other polymers by addition

Look at the polymers in this table. You have probably heard of them all. They are all made by addition polymerisation. Compare them:

| The monomer | Part of the polymer molecule | The equation for the reaction |
|---|---|---|
| chloroethene (vinyl chloride) — $CH_2=CHCl$ | poly(chloroethene) or polyvinyl chloride (PVC) | $n(CH_2=CHCl) \longrightarrow (-CH_2-CHCl-)_n$ ; $n$ stands for a large number! |
| tetrafluoroethene — $CF_2=CF_2$ | poly(tetrafluoroethene) or Teflon | $n(CF_2=CF_2) \longrightarrow (-CF_2-CF_2-)_n$ |
| phenylethene (styrene) — $C_6H_5CH=CH_2$ | poly(phenylethene) or poly(styrene) | $n(C_6H_5CH=CH_2) \longrightarrow (-CH(C_6H_5)-CH_2-)_n$ |

## Identifying the monomer

If you know the structure of the addition polymer, you can work out what the monomer was. Like this:

- Identify the repeating unit. (It has two carbon atoms side by side, in the main chain.) You could draw brackets around it.
- Then draw the unit, but put a double bond between the two carbon atoms. That is the monomer.

For example:

This shows part of a molecule of **poly(propene)**. The unit within brackets is the repeating unit.

So this is the monomer that was used. It is the alkene **propene**. Note the C=C double bond.

▲ PVC is light and flexible so is widely used for hoses and water pipes, and as an insulating cover for electrical wiring.

### Q

1. Why was *addition polymerisation* given that name?
2. **a** What is a *monomer*?
   **b** Could methane ($CH_4$) be used as a monomer for addition polymerisation? Explain your answer.
3. It is not possible to give an exact formula for the macromolecules in polythene. Why not?
4. Draw a diagram to show the polymerisation of:
   **a** ethene   **b** chloroethene   **c** phenylethene
5. A polymer has the general formula shown on the right. Draw the monomer that was used to make it.

# POLYMERS

## 18.3 Condensation polymerisation

### Condensation polymerisation
In addition polymerisation, there is only one monomer.
Double bonds break, allowing the monomer molecules to join together.
But in condensation polymerisation, no double bonds break. Instead:
- two *different* monomers join.
- each has *two functional groups* that take part in the reaction.
- the monomers join at their functional groups, by getting rid of or **eliminating** small molecules.

Let's look at two examples.

### 1 Making nylon
Below are the two monomers used in making nylon. We will call them **A** and **B**, for convenience:

**A** 1,6-diaminohexane

**B** hexan-1,6-dioyl chloride

**A** has an $NH_2$ group *at each end*. **B** has a COCl group *at each end*.
*Only* these functional groups take part in the reaction.
So we can show the rest of the molecules as blocks, for simplicity.

### The reaction
This shows the reaction between the two monomer molecules:

HCl is lost

Then another **B** reacts here ... ... and another **A** reacts here ... and so on.

So the nitrogen atom at one end of **A** has joined to the carbon atom at one end of **B**, by eliminating a molecule of hydrogen chloride.

The reaction continues at the other ends of A and B. In this way, thousands of molecules join, giving a macromolecule of nylon. Here is part of it:

this linking group of atoms is called an **amide linkage**

The group where the monomers joined is called the **amide** linkage. So nylon is called a **polyamide**. (Proteins have this link too, as you will see.)

Nylon can be drawn into tough strong fibres that do not rot away. So it is used for thread, ropes, fishing nets, car seat belts, and carpets.

▲ Making nylon in the school lab. **A** is dissolved in water. **B** is dissolved in an organic solvent that does not mix with water. Nylon forms where the solutions meet.

▲ Nylon thread: tough, strong, great for flying kites.

# POLYMERS

## 2 Making Terylene

Like nylon, Terylene is made by condensation polymerisation, using two different monomers. This time we call them **C** and **D**:

C   benzene-1,4-dicarboxylic acid     D   ethane-1,2-diol

**C** has two COOH (carboxyl) groups, and **D** has two OH (alcohol) groups. Only these functional groups take part in the reaction.
So once again we can show the rest of the molecules as blocks.

### The reaction
This shows the reaction between the two monomer molecules:

▲ Fibres of nylon and Terylene are made by pumping the melted polymer through a spinneret (like a shower head). As it comes out through the holes, the polymer hardens into long threads.

So a carbon atom at one end of **C** has joined to an oxygen atom at one end of **D**, by eliminating a water molecule.

The reaction continues at the other ends of **C** and **D**. In this way thousands of molecules join, giving a macromolecule of Terylene. Here is part of it:

*this group is called an **ester linkage***

In fact the reaction is the same as the reaction between the acid and alcohol on page 259, giving an ester. (See the last section on that page.)

So the group where the monomers have joined is called an **ester linkage**. Terylene is called a **polyester**.

Terylene is used for shirts and other clothing, and for bedlinen. It is usually woven with cotton. The resulting fabric is more hard-wearing than cotton, and does not crease so easily. Terylene is also sold as polyester thread.

▲ Shirts made from Terylene woven with cotton.

> **Q**
> 1. How many products are there, in condensation polymerisation?
> 2. In condensation polymerisation, each monomer molecule must have *two* functional groups. Explain why.
> 3. List the differences between condensation and addition polymerisation.
> 4. a Draw a diagram to show the reaction that produces nylon. (You can show the carbon chains as blocks.)
>    b Circle the *amide linkage* in your drawing.
>    c Nylon is called a *polyamide*. Why?
> 5. Draw part of a Terylene macromolecule in a simple way, using blocks as above. Circle the *ester linkage*.

# POLYMERS

## 18.4 Making use of synthetic polymers

### Plastics are synthetic polymers

Synthetic polymers are usually called **plastics**. (*Plastic* means *can be moulded into shape without breaking*, and this is true of all synthetic polymers while they are being made.) But when they are used in fabrics, and for thread, we still call them synthetic polymers.

Most plastics are made from chemicals found in the naphtha fraction of petroleum (pages 247 and 249). They are usually quite cheap to make.

### The properties of plastics

Most plastics have these properties:
1 They do not usually conduct electricity or heat.
2 They are unreactive. Most are not affected by air, water, acids, or other chemicals. This means they are usually safe for storing things in, including food.
3 They are usually light to carry – much lighter than wood, or stone, or glass, or most metals.
4 They don't break when you drop them. You have to hammer most rigid plastics quite hard, to get them to break.
5 They are strong. This is because their long molecules are attracted to each other. Most plastics are hard to tear or pull apart.
6 They do not catch fire easily. But when you heat them, some soften and melt, and some char (go black as if burned).

▲ A synthetic polymer for sewing. (It is polyester.)

as molecules get longer, the force of attraction between them increases

### Changing the properties

By choosing monomers and reaction conditions carefully, you can make plastics with exactly the properties you want. For example, look at how you can change the properties of polythene:

At about 50 °C, 3 or 4 atmospheres pressure, and using a catalyst, you get long chains like these. They are packed close together so the polythene is quite **dense**.

At about 200 °C, 2000 atmospheres pressure, and with a little oxygen present, the chains will branch. Now they can't pack closely, so the polythene is far less dense.

So by choosing the right conditions, you can change the density of the polythene, and make it 'heavy' or 'light' to suit your needs.

The **high-density** polythene is hard and strong, which is why it is used for things like bowls and dustbins. The **low-density** polythene is ideal for things like plastic bags, and 'cling film' for wrapping food.

# POLYMERS

## Uses for synthetic polymers

Given all those great properties, it is not surprising that plastics have thousands of uses. Here are some examples.

| Polymer | Examples of uses |
|---|---|
| polythene | plastic bags and gloves, clingfilm (*low density*) mugs, bowls, chairs, dustbins (*high density*) |
| polychloroethene (PVC) | water pipes, wellingtons, hoses, covering for electricity cables |
| polypropene | crates, ropes |
| polystyrene | used as expanded polystyrene in fast-food cartons, packaging, and insulation for roofs and walls (to keep homes warm) |
| Teflon | coated on frying pans to make them non-stick, fabric protector, windscreen wipers, flooring |
| nylon | ropes, fishing nets and lines, tents, curtains |
| Terylene | clothing (especially mixed with cotton), thread |

▲ Another use of nylon: for parasails like this one, and parachutes.

▲ Polystyrene is an insulator: it helps to prevent heat loss. So it is used under floors, and in fast-food cartons.

▲ Teflon – a slippery polymer. It is used to coat irons to help them glide, and on frying pans to stop food sticking.

**Q**

1  Look at the properties of plastics, on page 268. Which *three* properties do you think are the most important for:
   a  plastic bags?        b  kitchen bowls and utensils?
   c  water pipes?         d  fishing nets?
   e  hair dryers?         f  polystyrene fast-food containers?
2  What is *low-density* polythene, and how is it made?
3  Teflon is used to coat frying pans, to make them non-stick. So what properties do you think Teflon has? List them.
4  a  What is *expanded polystyrene*?
   b  Give three uses of this material.

5  a  Now make a table with these headings:

| Item | Properties of the plastic in it | Disadvantages of this plastic | Name of this plastic |
|---|---|---|---|

   b  i   Fill in the first column of your table, giving three or four plastic items you own or use.
      ii  In the second column, give the properties you observe, for that plastic. (You are a scientist!) For example is the plastic rigid? Or flexible?
      iii In the third column give any disadvantages you notice, for this plastic.
      iv  Then see if you can name it. If you can, well done!

POLYMERS

## 18.5 Plastics: here to stay?

### Plastics: the problem
There were only a few plastics around before the 1950s. Since then, dozens of new ones have been developed, and more are on the way.

Now it is hard to imagine life without them. They are used everywhere.

One big reason for their success is their unreactivity. But this is also a problem. They do not break down or rot away. Most of the plastics thrown out in the last 50 years are still around – and may still be here 50 years from now. A mountain of waste plastic is growing.

### Polythene: the biggest problem
Polythene is the biggest problem. It is the most-used plastic in the world, thanks to its use in plastic bags and food packaging. Around 5 trillion polythene bags are made every year. (That's 5 million million.) Most are used only once or twice, then thrown away.

In many places, rubbish is collected and brought to **landfill sites**. The plastic bags fill up these sites. In other places, rubbish is not collected. So the plastic bags lie around and cause many problems. For example:

- they choke birds, fish and other animals that try to eat them. Or they fill up the animals' stomachs so that they cannot eat proper food, and starve to death. (Animals cannot digest plastics.)
- they clog up drains, and sewers, and cause flooding.
- they collect in rivers, and get in the way of fish. Some river beds now contain a thick layer of plastic.
- they blow into trees and onto beaches. So the place looks a mess. Tourists are put off – and many places depend on tourists.

Because of these problems, plastic bags have been banned in many places. For example in Bangladesh, Rwanda and several states in India.

▲ Plastic bags – here today, still here tomorrow …

▲ A stomach full of plastic means the bird will starve to death.

▲ Nice for visitors …

## Recycling plastics

Some waste plastics do get reused. For example:

- some are melted down and made into new plastic bags, and things like soles for shoes, and fleeces.
- some are melted and their long chains cracked, to make small molecules that can be polymerised into new plastics.
- some are burned, and the heat is used to produce electricity.

But only a small % of waste plastic is reused in these ways. One problem is the many different types of plastic. These must be separated before reusing them, but that is not easy to do. Burning also poses problems, since some plastics give off poisonous gases.

▲ A degradeable plastic bag: it will break down along with the vegetable peelings and scrap paper inside.

## Degradeable plastics

Degradeable polythene is already here. Some is **biodegradeable**: it contains additives such as starch that bacteria can feed on. Some is **photodegradeable**: it contains additives that break down in sunlight. In both cases, the result is that the polythene breaks down into tiny flakes.

The amount of additive can be varied for different purposes – for example to make rubbish sacks that will break down within weeks.

## Bio-polymers: the future?

In future, the plastics you use could be **bio-polymers** – grown inside plants, or made in tanks by bacteria.

For example, one strain of bacteria can feed on sugar from crops such as maize, to produce polyesters.

Plants that can make plastics in their cells have already been developed. When the plants are harvested, the plastic is extracted using a solvent. Then the solvent is evaporated.

Work on bio-polymers is still at an early stage. But when oil runs out, we will be glad of bio-polymers. And they have two advantages for the environment: they are a renewable resource, and biodegradeable.

▲ This little cress plant has been genetically modified to produce a plastic in its cells.

**Q**

1. Describe some negative effects of plastics on the environment.
2. Polythene is responsible for most of the environmental problems caused by plastics. Explain why.
3. Explain what these are, in your own words:
   a photodegradable polythene   b bio-polymers
4. See if you can come up with some ideas, to help prevent pollution by plastic bags.

POLYMERS

## 18.6 The macromolecules in food (part I)

### What's in your food?
No matter what kind of food you eat, its main ingredients are the same: **carbohydrates**, **proteins** and **fats**. All three are made of macromolecules. And plants can produce them all.

### Plants: the polymer factories
1. Plants take in carbon dioxide from the air, and water from the soil.
2. Using energy from sunlight, and **chlorophyll** as a catalyst, they turn them into glucose and oxygen, in a process called **photosynthesis**:

$$6CO_2\,(g) + 6H_2O\,(l) \longrightarrow C_6H_{12}O_6\,(s) + 6O_2\,(g)$$

carbon dioxide — water — glucose (a sugar) — oxygen

3. Then they turn the glucose molecules into macromolecules of starch and cellulose, by polymerisation. These natural polymers are called **carbohydrates**. Plants use cellulose to build stems and other structures. They use starch as an energy store.
4. Using glucose, and minerals from the soil, they also produce macromolecules of **proteins** and **fats**.

Enzymes in plant cells act as catalysts, for the reactions in **3** and **4**.

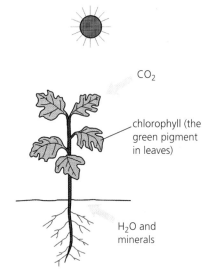

▲ A plant: a natural chemical factory.

### From plants to you
This is how the macromolecules from plants reach you:

1. Animals eat plants, and seeds of plants. They digest them, and build their own carbohydrates, proteins, and fats from them.

2. You eat animal carbohydrates, proteins, and fats, in animal produce such as eggs, milk, and cheese.

3. You eat them in meat and fish too.

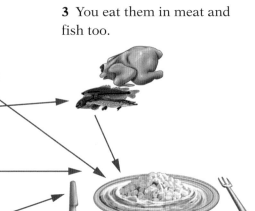

4. You also eat parts of plants. For example maize, rice, potatoes and other vegetables, and fruit, …

5. … and things like bread and pasta, made from plant products.

6. During digestion, you break the macromolecules back down to their building blocks. You use some for energy, and the rest to build up new carbohydrates, proteins, and fats.

We will now look more closely at the carbohydrates, proteins, and fats that plants produce, in this unit and the next one.

# Carbohydrates

**Carbohydrates** contain just carbon, hydrogen and oxygen. Glucose is called a **simple carbohydrate**. It is also called a **monosaccharide**, which means *a single sugar unit*.

The structure of a glucose molecule is shown on the right. Now let's see how glucose molecules join:

◀ A molecule of glucose, $C_6H_{12}O_6$.

**1** We can draw a glucose molecule like this, showing the two groups that react:

HO—☐—OH

**2** Two glucose molecules can join like this, giving maltose, a **disaccharide**:

HO—☐—OH  HO—☐—
↓ water molecule eliminated
HO—☐—O—☐—OH

**3** Hundreds or thousands can join in the same way, giving **starch**, a **complex carbohydrate**. It is also called a **polysaccharide**:

HO—☐—OH  HO—☐—OH  HO—☐—OH  HO—☐—OH
↓ water molecules eliminated
—O—☐—O—☐—O—☐—O—☐—

In reaction 2, two molecules join, eliminating a small molecule (water). So it is a condensation reaction. Reaction 3 is a **condensation polymerisation**, so starch is a **polymer**.

# Cellulose

Cellulose is also a polysaccharide. Its molecules are built from at least 1000 glucose units. But they are joined differently than those in starch, so cellulose has quite different properties.

The cell walls in plants are made of cellulose. So we eat cellulose every time we eat cereals, vegetables, and fruit. We can't digest it, but it helps to clean out our digestive systems. We call it **fibre**.

▲ It can do something we can't do: digest cellulose. Grass is mainly cellulose.

# The importance of carbohydrates

Your body *can* digest starch. It breaks it back down to glucose. It uses some of this for respiration, which provides you with energy (page 235).

It builds the rest into a complex carbohydrate called **glycogen**, which acts as an energy store.

So carbohydrates are an important part of your diet. Rice, wheat, pasta, potatoes, and bananas are all rich in starch. Honey and fruit juices are rich in glucose.

▶ Before races, marathon runners eat plenty of carbohydrate to build up their glycogen levels.

**Q**

1. All life depends on photosynthesis. Explain why.
2. Explain what it is, and name one example:
   a. a carbohydrate   b. a monosaccharide
   c. a disaccharide   d. a polysaccharide
3. In what ways is cellulose:
   a. like starch?   b. different from starch?
4. The cellulose in vegetables is good for us. Why?
5. Name three foods you eat, that are rich in starch.

# POLYMERS

## 18.7 The macromolecules in food (part II)

### Proteins – built from amino acids

Proteins are **polymers**, built up from molecules of **amino acids**. Amino acids contain carbon, hydrogen, oxygen, and nitrogen, and some contain sulfur. The general structure of an amino acid molecule is shown on the right. Note the COOH and NH₂ functional groups.

▲ An amino acid is a carboxylic acid with an amino (NH₂) group. R stands for the rest of the molecule.

There are twenty common amino acids. Here are three of them – with the COOH bonds drawn vertically, to help you see how the amino acids join:

glycine   alanine   cysteine

### How amino acids join up to make proteins

This shows four different amino acids combining:

water molecules eliminated → the **amide linkage** as in nylon

▲ Proteins are large and complex. The chains are often coiled. The genes in the cells of plants and animals control which amino acids join up, and in what order.

From 60 to 6000 amino acid units can join to make a macromolecule of protein. They can be different amino acids, joined in different orders – so there are a huge number of proteins!

The reaction is a **condensation polymerisation**, with loss of water molecules. Note the **amide linkage**, as in nylon (page 266).

### The importance of proteins in your food

When your body digests food, it breaks the proteins back down to amino acids. These then join up again to make proteins your body needs. For example all these substances in your body are proteins:

- the enzymes that act as catalysts for reactions in your body cells
- the collagen in your skin, bones, and teeth
- the keratin that forms your hair
- haemoglobin, the red substance in blood, that carries oxygen
- hormones, the chemicals that dictate how you grow and develop.

Your body needs all 20 amino acids to make these proteins. It *can* make 11 by itself. But there are 9 **essential amino acids** that it cannot make. To be healthy, you must eat foods that can provide these.

**Rich in proteins**
chicken and other meats
fish
cheese
yoghurt
milk
eggs
soya beans
lentils
beans and peas
spinach
nuts
seeds (such as sunflower seeds)

▲ Proteins from animals usually have all 20 amino acids. So do those from soya beans. But in other plant proteins, some essential amino acids are often missing.

## Fats

Foods also contain natural **fats** and **oils** (liquid fats).

Complex carbohydrates, and proteins, are polymers. But fats are not made by polymerisation, so they are not polymers. They are esters: compounds formed from an **alcohol** and an **acid**.

- The alcohol is always **glycerol**, a natural alcohol with three OH groups. (Its chemical name is propan-1,2,3-triol.)
- The acids are natural carboxylic acids, usually with long carbon chains. They are called **fatty acids**. For example palmitic acid, $C_{15}H_{31}COOH$.

### How fats are formed

This shows the reaction between glycerol and a fatty acid. R stands for the long chain of carbon atoms with hydrogen atoms attached, in the acid:

3 molecules of a fatty acid + 1 molecule of glycerol → 1 macromolecule of a fat

This is a **condensation reaction**, with the elimination of water. Each OH group in a glycerol molecule can react with a *different* fatty acid, so you can get many different esters. Note the ester linkage, as in Terylene (page 267).

### The importance of fats in your food

In your body, fats and oils in food are broken down to fatty acids and glycerol. Some of these are used for energy. Some are combined into new fats, to make the membranes in your body cells. Some cells also store fat droplets. These cells form a layer under your skin, which keeps you warm.

So you need some fats in your diet. But runny **unsaturated fats** (containing carbon–carbon double bonds) are better for you than the hard, saturated, fats found in meat and cheese. Saturated fats have been linked to heart disease.

◀ A molecule of glycerol.

▲ Making palm oil. She will crush the boiled palm fruit to release the oil, and use it for cooking. Palm oil is a mixture of esters, mainly from palmitic acid.

**Rich in fats**
meat
oily fish
butter, cheese, cream
avocados
nuts and seeds
vegetable oils (such as palm oil, olive oil, sunflower oil)
margarine and other spreads

▲ Fish oil and vegetable oils contain unsaturated fats. These are better for you than saturated fats.

**Q**
1. What is: **a** an amino acid? **b** a protein?
2. Describe in your own words a protein macromolecule.
3. Give three examples of the important roles proteins play in your body.
4. Name six foods that are rich in protein.
5. Show how palmitic acid reacts with propan-1,2,3-triol to give an ester found in palm oil.
6. What happens to fats when you eat them?
7. Compare the reactions that produce carbohydrates, proteins, and fats. What do they have in common?

# POLYMERS

## 18.8 Breaking down the macromolecules

### What happens during digestion?
You saw earlier how the natural macromolecules in food were built up by condensation reactions, with the loss of water molecules.

The opposite happens when you eat them. In your mouth, stomach and small intestine, the macromolecules are broken down again, by reacting with water. This is called **hydrolysis**.

**Hydrolysis is a reaction in which molecules are broken down by reaction with water.**

### Hydrolysis in the digestive system
This is what happens in your body, during digestion:

- **Starch** and any disaccharides get broken down to glucose. Your cells then use the glucose to provide energy, in a process called respiration. It is the reverse of photosynthesis:

  $C_6H_{12}O_6$ (aq) + $6O_2$ (g) $\rightarrow$ $6CO_2$ (g) + $6H_2O$ (l) + energy
  glucose + oxygen $\rightarrow$ carbon dioxide + water + energy

- **Proteins** get broken down to **amino acids** which your body then uses to build up the proteins it needs.

- **Fats** and **oils** (which are esters) get broken down into **glycerol** and **fatty acids**. These are used for energy, or to make new fats for cell membranes, or to be stored.

**All the 'breaking down' reactions during digestion are hydrolyses.**

### Example: hydrolysis of an ester during digestion
This shows the hydrolysis of an ester in a vegetable oil, in your digestive system. R represents long chains of carbon atoms:

[diagram: ester in a vegetable oil + water → (enzyme as catalyst) → 3R—C(=O)OH (fatty acid) + glycerol (HO—CH₂, HO—CH, HO—CH₂)]

an ester in a vegetable oil    water    fatty acid    glycerol

Compare it with the reaction shown on page 275. What do you notice?

### Enzymes as catalysts
Enzymes act as catalysts, in building up the macromolecules in food. In digestion, other enzymes act as catalysts to break them down again. (Look at the hydrolysis above.) Enzymes called **amylases** act on starch, **lipases** act on fats and oils, and **proteinases** act on proteins.

▲ The hydrolysis of starch starts in your mouth, where the enzyme *amylase* in saliva start breaking it down.

▲ All the macromolecules will be broken down, with help from enzymes.

▲ Getting ready for hydrolysis.

POLYMERS

## Hydrolysis in the lab

You can also carry out hydrolysis of starch, proteins and fats in the lab.
This table shows the conditions, and the results for complete hydrolysis.

| Macromolecule | Conditions for the hydrolysis | Complete hydrolysis gives ... |
|---|---|---|
| starch | heat with dilute hydrochloric acid | glucose |
| proteins | boil with 6M hydrochloric acid for 24 hours | amino acids |
| fats | boil with dilute sodium hydroxide | glycerol plus the sodium salts of the fatty acids (R—COO$^-$ Na$^+$) |

Note that:
- the products are the same as for digestion, except for fats, where you obtain sodium salts of the fatty acids.
- the hydrolyses in your digestive system take place in much milder conditions, at much lower temperatures, thanks to enzymes.
- if the hydrolysis of starch and proteins is *not* complete, you will obtain a mixture of molecules of different sizes. Partial hydrolysis of starch can give glucose, maltose (made of two glucose units), maltotriose (three glucose units), and dextrins (many glucose units).

You can use paper chromatography to identify the products of the hydrolyses, as shown in Unit 2.5. They are colourless, so you need to use locating agents.

## Making soap from fats and oils

The sodium salts of fatty acids are used as **soap**. So soap is made in factories by boiling fats and oils with sodium hydroxide, as above. For example:

```
R—COO—CH₂                    HO—CH₂
    |                            |
R—COO—CH  (l) + 3NaOH(aq)  →  HO—CH  (l) + 3R—COONa(aq)
    |                            |
R—COO—CH₂                    HO—CH₂
an ester in a                glycerol        soap – the sodium
vegetable oil                                salt of a fatty acid
```

▲ Sodium salts that keep you clean!

The soap you buy may be made from vegetable oil – like palm oil or coconut oil – or even from fish oil or animal fat. Chemicals are added to make it smell nice. These are usually artificial esters. (As you saw on page 259, many esters have attractive smells.)

1 a What does *hydrolysis* mean?
  b See if you can draw a diagram to show that the complete hydrolysis of starch to glucose, in the lab, is the opposite of a condensation polymerisation.
  c If you carry out an incomplete hydrolysis of starch in the lab, you get a mixture of products. Explain.

2 Hydrolysis of a protein in the lab will give a mixture of products. Explain why, and how to identify them.

3 Oils are broken down in your digestive system. And oils are used to make soaps, in industry.
  a What do these two processes have in common?
  b In what way are they different?

# POLYMERS

## Checkup on Chapter 18

### Revision checklist

**Extended curriculum**

Make sure you can …

☐ explain these terms:
  *monomer    polymer    polymerisation*
  *macromolecule    natural polymer    synthetic polymer*

☐ describe addition polymerisation, and
  – say what the key feature of the monomer is
  – draw part of a polymer molecule, formed from a given monomer
  – identify the monomer, for a given polymer

☐ name at least three polymers formed by addition polymerisation, and give uses for them

☐ describe condensation polymerisation, and
  – say what the key features of the monomers are
  – state the differences between condensation and addition polymerisation

☐ draw simple diagrams to show the monomers, and part of the macromolecule, for:
  *nylon    Terylene*
  using blocks to represent carbon chains

☐ explain what the *amide* and *ester* linkages are, and identify them on a drawing

☐ give uses for nylon and Terylene

☐ give at least five general properties of plastics

☐ describe some of the environmental problems caused by plastics

☐ name and describe the three main groups of macromolecules in food

☐ explain what these are:
  *amino acids    fatty acids    glycerol    esters*

☐ draw simple diagrams to show how these are formed by condensation polymerisation:
  *starch    proteins*

☐ draw a simple diagram to show how fats are formed by a condensation reaction

☐ explain what *hydrolysis* is

☐ describe the products, when starch, proteins, and fats are broken down in your digestive system

☐ describe how the hydrolysis of starch, proteins, and fats is carried out in the lab, and name the products of complete hydrolysis

☐ describe how to carry out paper chromatography to identify products of hydrolysis

☐ explain how fats and oils are used to make soaps

### Questions

**Extended curriculum**

1 This diagram represents two units of an addition polymer called polyacrylamide:

$$-\underset{H}{\overset{H}{C}}-\underset{CONH_2}{\overset{H}{C}}-\underset{H}{\overset{H}{C}}-\underset{CONH_2}{\overset{H}{C}}-$$

  a Draw the structure of the monomer.
  b Suggest a name for the monomer.
  c Is the monomer saturated, or unsaturated?

2 The polymer 'Teflon' is obtained from the monomer tetrafluoroethene, which has this structure:

$$\underset{F}{\overset{F}{>}}C=C\underset{F}{\overset{F}{<}}$$

  a Which feature of the monomer makes polymerisation possible?
  b Which type of polymerisation occurs?
  c Draw three units in the structure of the macromolecule that forms.
  d Give the chemical name for this polymer.

3 The polymer poly(dichloroethene) has been used to make 'cling film', for covering food to keep it fresh. This shows the structure of the polymer:

$$\left(\underset{H}{\overset{H}{-}}\underset{|}{\overset{|}{C}}-\underset{Cl}{\overset{Cl}{\underset{|}{C}}}-\right)_n$$

  a What does *n* represent?
  b Name the monomer, and draw its structural formula.
  c Which type of polymerisation takes place?
  d One property of poly(dichlorothene) is its *low permeability* to moisture and gases.
    i See if you can explain what the term in italics means.
    ii That property is important in keeping food fresh. Why?
    iii Give three other *physical* properties a polymer would need, to be suitable for use as 'cling film'.
  e Poly(dichloroethene) is *non-biodegradable*.
    i Explain the term in italics.
    ii Describe two environmental problems caused by the disposal of such plastics.

**4 Polyamides** are polymers made by condensation polymerisation. One polyamide was developed for use in puncture-resistant bicycle tyres. The two monomers for it are:

The hexagon with the circle in the middle stands for a ring of 6 carbon atoms, with 3 double bonds.
a What is *condensation polymerisation*?
b Show in detail how the monomers join.
c Name the other product of the reaction.
d i In what way is this polymer similar to nylon? (See page 266.)
  ii But its properties are different from those of nylon. Why?

A similar polymer has been developed as a fabric for fireproof clothing. Its structure is:

e Draw the structures of the two monomers that could be used to make this polymer.

**5** Many synthetic polymers contain the amide linkage.
a Draw the structure of the amide linkage.
b Which important natural macromolecules also contain the amide linkage?

The substances in **b** will undergo hydrolysis in the laboratory, in the presence of acid.
c i What does *hydrolysis* mean?
  ii What are the products of the hydrolysis?
  iii How can the products be separated?

**6** One very strong polymer has this structure:

a Which type of polymerisation produced it?
b Which type of linkage joins the monomers?
c Draw the structures of the two monomers from which this polymer could be made.
d Compare the structure above with that for Terylene (page 267). What may be responsible for the greater strength of this polymer?
e i Which *natural* macromolecules have the same linkage as this polymer?
  ii Hydrolysis of these macromolecules, using an alkali, gives a useful product. Name it.

**7** Starch is a carbohydrate. It is a natural polymer. This shows part of a starch macromolecule:

a What is a *macromolecule*?
b What is a *carbohydrate*?
c Which type of polymerisation gives starch?
d What do the blocks represent, above?
e i Draw a diagram showing the structure of the monomer for starch. (Use a block.)
  ii Name this monomer.
f Starch is also called a *polysaccharide*. Why?
g Starch can be broken down by hydrolysis.
  i Describe two ways in which the hydrolysis is carried out. (One occurs in your body.)
  ii One takes place at a far lower temperature than the other. What makes this possible?

**8** In the lab, *partial* hydrolysis of starch gives a mixture of colourless products. They can be identified using chromatography. A locating agent is needed.
a Draw diagrams showing at least two of the products. (Use blocks like those in question **7**.)
b What is a *locating agent* and why is it needed?
c Outline the steps in carrying out the chromatography. (Page 25 may help.)

**9** Soaps are salts of fatty acids.
a Name one fatty acid.
b In which way is a fatty acid different from ethanoic acid? In which way is it similar?
c Below is one example of a compound found in vegetable oil, and used to make soap.

$$H_2C - OOC(C_{17}H_{35})$$
$$HC - OOC(C_{15}H_{31})$$
$$H_2C - OOC(C_{14}H_{29})$$

i This compound is an *ester*. Explain that term.
ii To make soap, the oil is usually reacted with a sodium compound. Which one?
iii Which type of reaction takes place?
d i The reaction in c will give *four* different products. Write down their formulae.
  ii Which ones can be used as soap?
  iii One product is an alcohol. Name it.
  iv In which way is this product similar to ethanol? In which way is it different?
e Name three vegetable oils used to make soap.

IN THE LAB

## 19.1 Chemistry: a practical subject

### The lab: the home of chemistry
All the information in this book has one thing in common. It is all based on real experiments, carried out in labs around the world, over the years – and even over the centuries. The lab is the home of chemistry!

### How do chemists work?
Like all scientists, chemists follow the **scientific method**. This flowchart shows the steps. The handwritten notes are from a student.

1. You observe something that makes you ask yourself a question.
   *Kitchen cleaner X is better at removing grease than kitchen cleaner Y. Why?*

2. Come up with a **hypothesis** – a reasonable statement that you can test. You might need to do some research in books or on the internet, to help you.
   *Sodium hydroxide is used in kitchen cleaners to help remove grease.*
   *The labels on X and Y say they contain sodium hydroxide.*
   *My hypothesis: X may contain more sodium hydroxide than Y does.*

3. Plan an experiment to test the hypothesis.
   *I plan to do a titration to test my hypothesis. See the details in the next unit.*

4. Carry out the experiment, and record the results.
   *See the results in the next unit.*

5. Analyse the results.
   *You can help me do this, in the next unit.*

6. Did the results support your hypothesis?
   *See the next unit.*

7. Share your conclusions with other people.
   *The teacher wants to see them!*

▲ Step into the lab, and try the scientific method ....

### Planning an experiment: the variables
Suppose you want to investigate how the rate of a reaction changes with temperature.

- The temperature is under your control. So it is called the **independent variable**. It is the *only* thing *you* change as you do the experiment.
- If the rate changes as you change the temperature, the rate is a **dependent variable**. It **depends on** the temperature.

In many experiments you do, there will be an independent variable. You control it – and keep everything else unchanged.

> **That golden rule …**
> When you investigate something in the lab, **change only one thing at a time**, and see what effect it has.

# IN THE LAB

## The skills you use

When you plan and carry out an experiment, you use many different skills:

**Thinking** Use your brain before, during, and after the experiment. That is what brains are for. (They really like being used.)

**Observing** This is a very important skill. Chemists have made some amazing discoveries just by watching very carefully.

**Using apparatus and techniques** Weigh things, measure out volumes of liquids, measure temperature, do titrations, prepare crystals ....

**Working accurately** Sloppy work will ruin an experiment. Follow the instructions. Measure things carefully. Think about safety too.

**Doing some maths** You often have to do some calculations using your results. And drawing a graph can help you see what is going on.

**Writing up** You may have to write a report on your experiment, and give conclusions. And say how the experiment could be improved?

## The experiments you do

Often, you will not get a chance to plan an experiment for yourself. Instead, the teacher will tell you what to do. So you might miss out steps **1–3** in the flowchart on page 280.

But even if you pick up at step **4**, you are still using the scientific method, and gaining practice in it. And you are following in the footsteps of many famous scientists, who have changed our lives by their careful work in the lab.

One day, you may become a scientist yourself. Even a famous one!

---

**Q**

1. Do you think this counts as a *hypothesis*?
   a  I am late for class again.
   b  If I add more yeast, the fermentation may go faster.
   c  December follows November.
   d  The rate of photosynthesis may change with temperature.

2. Explain in your own words what an *independent variable* is.
3. Which would be the independent variable, in an experiment to test the statement in **1b**?
4. Do you think the scientific method would be useful to:
   a  a doctor?    b  a detective?
   Explain your answer.

IN THE LAB

# 19.2 Example of an experiment

## Comparing those kitchen cleaners

In step **2** of the scientific method in the last unit, a student put forward a hypothesis. Here you can read how the student tested the hypothesis. But the report is not quite finished. That is *your* task.

---

An experiment to compare the amount of sodium hydroxide in two kitchen cleaners

### Introduction

I noticed that kitchen cleaner X is better at removing grease than kitchen cleaner Y is. The labels show that both kitchen cleaners contain sodium hydroxide. This chemical is used in many cleaners because it reacts with grease to form soluble sodium salts, which go into solution in the washing-up water.

### My hypothesis

Kitchen cleaner X may contain more sodium hydroxide than kitchen cleaner Y does.

### Planning my experiment

I plan to titrate a sample of each cleaner against dilute hydrochloric acid, using methyl orange as indicator. This is a suitable method because the sodium hydroxide in the cleaner will neutralise the acid. The indicator will change colour when neutralisation is complete.

To make sure it is a fair test, I will use exactly the same volume of cleaner, and the same concentration of acid, and the same number of drops of indicator each time, and swirl the flask in the same way. The only thing I will change is the type of cleaner.

I will wear safety goggles, since sodium hydroxide and hydrochloric acid are corrosive.

### The experiment

25 cm$^3$ of cleaner X were measured into a conical flask, using a pipette. 5 drops of methyl orange were added, and the solution turned yellow.

A burette was filled to the upper mark with hydrochloric acid of concentration 1 mol/dm$^3$. The initial level of the acid was noted.

The acid was allowed to run into the conical flask. The flask was continually and carefully swirled. As the acid dripped in, the solution showed flashes of pink. When the end point was near the acid was added drop by drop. When the solution changed from yellow to pink, the titration was stopped. The final level of the acid was recorded.

The experiment was repeated with cleaner Y.

▲ The colour of the solution has changed: the titration is complete.

## IN THE LAB

### The results

For X:
Initial level of acid in the burette   0.0 cm³
Final level                           22.2 cm³
Volume of acid used                   22.2 cm³

For Y:
Initial level of acid in the burette  22.2 cm³
Final level                           37.5 cm³
Volume of acid used                   15.3 cm³

### Analysis of the results

The same volume of each cleaner was used. The sodium hydroxide in X neutralised 22.2 cm³ of acid. The sodium hydroxide in Y neutralised 15.3 cm³ of acid. This means that solution …

### My conclusion

These results …

### To improve the reliability of the results

I would …

▲ One is better at removing grease. Might it have a higher concentration of sodium hydroxide?

In the question section below, you will have the chance to complete the student's analysis and conclusions, and come up with suggestions for ensuring that the results were reliable.

**Q**

1 In this experiment, was there:
   a an independent variable? If so, what was it?
   b a dependent variable? If so, what was it?
2 a Look at the apparatus below.
     Which pieces did the student use in the experiment? Give their letters and names.
   b When measuring out solutions for titration, a pipette is used instead of a measuring cylinder. Why is this?
   c Why is a conical flask used rather than a beaker, for the titration?
   d Why are burettes used for titrations?
   e Which is *more* accurate for measuring liquids?
     i  a burette    ii  a pipette
     Explain clearly why you think so.

3 Why is an indicator needed, for titrations?
4 a Suggest another indicator the student could have used, in place of methyl orange. (Hint: page 149.)
   b What colour change would be observed at the end-point, for the indicator you suggested?
5 Now complete the student's *Analysis of the results*.
6 Complete the *Conclusion*, by saying whether or not the results supported the hypothesis.
7 How would you improve the reliability of the results?
8 How would you modify the experiment, to compare liquid scale-removers for kettles? (They contain acid.)
9 Next week the student will do an experiment to see whether neutralisation is exothermic or endothermic. Which item below will the student *definitely* use?

A     B     C     D     E   F

IN THE LAB

## 19.3 Working with gases in the lab

### Preparing gases in the lab

You might have to prepare a gas in the lab, one day. The usual way to make a gas is to displace it from a solid or solution, using apparatus like this. The table below gives some examples.

| To make ... | Place in flask .... | Add .... | Reaction |
|---|---|---|---|
| carbon dioxide | calcium carbonate (marble chips) | dilute hydrochloric acid | $CaCO_3 (s) + 2HCl (aq) \longrightarrow CaCl_2 (aq) + H_2O (l) + CO_2 (g)$ |
| hydrogen | pieces of zinc | dilute hydrochloric acid | $Zn (s) + 2HCl (aq) \longrightarrow ZnCl_2 (aq) + H_2 (g)$ |
| oxygen | manganese(IV) oxide (as a catalyst) | hydrogen peroxide | $2H_2O_2 (aq) \longrightarrow 2H_2O (l) + O_2 (g)$ |

But to make ammonia, you can heat any ammonium compound with a base such as sodium hydroxide or calcium hydroxide – using both reactants in solid form.

### Collecting the gases you have prepared

The table below shows four ways of collecting a gas you have prepared. The method depends on whether the gas is heavier or lighter than air, whether you need it dry, and what you want to do with it.

> **Using a measuring cylinder**
> - You can use a gas jar to collect a gas over water.
> - But if you want to measure the volume of the gas, roughly, use a measuring cylinder instead.
> - If you want to measure its volume accurately, use a gas syringe.

| Method | upward displacement of air | downward displacement of air | over water | gas syringe |
|---|---|---|---|---|
| Use when ... | the gas is heavier than air | the gas is lighter than air | the gas is sparingly soluble in water | you want to measure the volume accurately |
| Apparatus | gas jar | gas jar | gas jar / water |  gas syringe |
| Examples | carbon dioxide, $CO_2$ sulfur dioxide, $SO_2$ hydrogen chloride, HCl | ammonia, $NH_3$ hydrogen, $H_2$ | carbon dioxide, $CO_2$ hydrogen, $H_2$ oxygen, $O_2$ | any gas |

IN THE LAB

# Tests for gases

You have a sample of gas. You think you know what it is, but you're not sure. So you need to do a test. Below are some tests for common gases. Each is based on particular properties of the gas, including its appearance, and sometimes its smell.

| Gas | Description and test details |
|---|---|
| **Ammonia, $NH_3$** | |
| Properties | Ammonia is a colourless alkaline gas with a strong sharp smell. |
| Test | Hold damp indicator paper in it. |
| Result | The indicator paper turns blue. (You may also notice the sharp smell.) |
| **Carbon dioxide, $CO_2$** | |
| Properties | Carbon dioxide is a colourless, weakly acidic gas. It reacts with **limewater** (a solution of calcium hydroxide in water) to give a white precipitate of calcium carbonate: $$CO_2\,(g) + Ca(OH)_2\,(aq) \longrightarrow CaCO_3\,(s) + H_2O\,(l)$$ |
| Test | Bubble the gas through limewater. |
| Result | Limewater turns cloudy or milky. |
| **Chlorine, $Cl_2$** | |
| Properties | Chlorine is a green poisonous gas which bleaches dyes. |
| Test | Hold damp indicator paper in the gas, *in a fume cupboard*. |
| Result | Indicator paper turns white. |
| **Hydrogen, $H_2$** | |
| Properties | Hydrogen is a colourless gas which combines violently with oxygen when lit. |
| Test | Collect the gas in a tube and hold a lighted splint to it. |
| Result | The gas burns with a squeaky pop. |
| **Oxygen, $O_2$** | |
| Properties | Oxygen is a colourless gas. Fuels burn much more readily in it than in air. |
| Test | Collect the gas in a test-tube and hold a glowing splint to it. |
| Result | The splint immediately bursts into flame. |

**Q**

**1 a** Sketch the *complete* apparatus you will use to prepare and collect carbon dioxide. Label all the parts.
  **b** How will you then test the gas to confirm that it is carbon dioxide?
  **c** Write the equation for a positive test reaction.
**2 a** Hydrogen cannot be collected by upward displacement of air. Why not?
  **b** Hydrogen burns with a squeaky pop. Write a balanced equation for the reaction that takes place.
**3 a** Name two substances you could use to make ammonia.
  **b** Ammonia cannot be collected over water. Why not?
  **c** The test for ammonia is …… ?
**4** It is not a good idea to rely on smell, to identify a gas. Suggest at least two reasons why.
**5** To measure the rate of the reaction between magnesium and hydrochloric acid, you will collect the hydrogen that forms. Which is better to use for this: a measuring cylinder over water, or a gas syringe? Give more than one reason.

IN THE LAB

## 19.4 Testing for ions in the lab

### Time for detective work!
You have an unknown salt, and you want to find out what it is.
This unit gives some tests you can do. But first, note these points:

- Positive ions are also called **cations**. Negative ions are called **anions**.
- In each test, *either* a precipitate forms *or* a gas you can test.

### Tests for cations
This table shows tests for the ammonium ion, and several metal ions.

- To test for the ammonium ion you can use the unknown salt as a solid, or in aqueous solution. But for metal ions, use their aqueous solutions.
- To test for metal cations, you can use dilute sodium hydroxide or ammonia solution, since both provide hydroxide ions. But the results are not always the same, as you will see below.

> **Remember CAP!**
> **C**ations **A**re **P**ositive.
> They would go to the cathode (−).

> **Complex ions**
> - In complex ions, a metal ion is surrounded by several negative ions, or molecules.
> - Many transition elements form complex ions.
> - The copper ion $Cu(NH_3)_4^{2+}$ is an example. (See below.)

| Cation | Test | If the cation is present | Ionic equation for the reaction |
|---|---|---|---|
| ammonium $NH_4^+$ | Add a little dilute sodium hydroxide solution. Heat gently. | Ammonia gas is given off. (It turns litmus red.) | $NH_4^+ (aq) + OH^- (aq) \longrightarrow NH_3 (g) + H_2O (l)$ |
| copper(II) $Cu^{2+}$ | Add dilute sodium hydroxide or ammonia solution. | A pale blue precipitate forms. But it dissolves on adding more ammonia, giving a deep blue solution. | $Cu^{2+} (aq) + 2OH^- (aq) \longrightarrow Cu(OH)_2 (s)$ The precipitate dissolves again in ammonia solution because a soluble complex ion forms: $[Cu(NH_3)_4]^{2+} (aq)$. |
| iron(II) $Fe^{2+}$ | Add dilute sodium hydroxide or ammonia solution. | A pale green precipitate forms. | $Fe^{2+} (aq) + 2OH^- (aq) \longrightarrow Fe(OH)_2 (s)$ |
| iron(III) $Fe^{3+}$ | Add dilute sodium hydroxide or ammonia solution. | A red-brown precipitate forms. | $Fe^{3+} (aq) + 3OH^- (aq) \longrightarrow Fe(OH)_3 (s)$ |
| aluminium $Al^{3+}$ | Add dilute sodium hydroxide or ammonia solution. | A white precipitate forms. It dissolves again on adding excess sodium hydroxide, giving a colourless solution. But it *will not* dissolve if more ammonia is added instead. | $Al^{3+} (aq) + 3OH^- (aq) \longrightarrow Al(OH)_3 (s)$ The precipitate dissolves in excess sodium hydroxide because aluminium hydroxide is amphoteric. The soluble aluminate ion forms: $(Al(OH)_4^-)$. |
| zinc $Zn^{2+}$ | Add dilute sodium hydroxide or ammonia solution. | A white precipitate forms. It dissolves again on adding more sodium hydroxide *or* ammonia, giving a colourless solution. | $Zn^{2+} (aq) + 2OH^- (aq) \longrightarrow Zn(OH)_2 (s)$ The precipitate dissolves again in sodium hydroxide because zinc hydroxide is amphoteric. The soluble zincate ion forms: $(Zn(OH)_4^{2-})$ It dissolves again in ammonia solution because a soluble complex ion forms: $[Zn(NH_3)_4]^{2+} (aq)$. |
| calcium $Ca^{2+}$ | Add dilute sodium hydroxide solution. | A white precipitate forms. It will not dissolve on adding excess sodium hydroxide. | $Ca^{2+} (aq) + 2OH^- (aq) \longrightarrow Ca(OH)_2 (s)$ |
| | Add dilute ammonia solution. | No precipitate, or very slight white precipitate. | |

# IN THE LAB

## Tests for anions

### Halide ions (Cl⁻, Br⁻, I⁻)
- To a small amount of the solution, add an equal volume of dilute nitric acid. Then add silver nitrate solution.
- Silver halides are insoluble. So if halide ions are present a precipitate will form. The colour tells you which one. Look at this table:

| Precipitate | Indicates presence of … | Ionic equation for the reaction |
|---|---|---|
| white | chloride ions, Cl⁻ | $Ag^+ (aq) + Cl^- (aq) \longrightarrow AgCl (s)$ |
| cream | bromide ions, Br⁻ | $Ag^+ (aq) + Br^- (aq) \longrightarrow AgBr (s)$ |
| yellow | iodide ions, I⁻ | $Ag^+ (aq) + I^- (aq) \longrightarrow AgI (s)$ |

▲ Halides are present. From left to right: chloride, bromide, iodide.

### Sulfate ions ($SO_4^{2-}$)
- To a small amount of the solution add an equal volume of dilute hydrochloric acid. Then add barium nitrate solution.
- Barium sulfate is insoluble. So if sulfate ions are present a **white** precipitate will form. The ionic equation for the reaction is:

$Ba^{2+}(aq) + SO_4^{2-} (aq) \longrightarrow BaSO_4 (s)$

### Nitrate ions ($NO_3^-$)
- To a small amount of the unknown solid or solution, add a little sodium hydroxide solution. Then add some small pieces of aluminium foil, and heat gently.
- If ammonia gas is given off, the unknown substance contained nitrate ions. The ionic equation for the reaction is:

$8Al (s) + 3NO_3^- (aq) + 5OH^- (aq) + 2H_2O (l) \longrightarrow 3NH_3 (g) + 8AlO_2^- (aq)$

▲ You need only a small amount of the unknown substance, in testing for cations and anions.

### Carbonate ions ($CO_3^{2-}$)
- To a small amount of the unknown solid or solution, add a little dilute hydrochloric acid.
- If the mixture bubbles and gives off a gas that turns limewater milky, the unknown substance contained carbonate ions. The gas is carbon dioxide. The ionic equation for the reaction is:

$2H^+ (aq) + CO_3^{2-} (aq) \longrightarrow CO_2 (g) + H_2O (l)$

▶ The carbonate test: that is limewater on the right, and it is turning milky.

**Q**

1. The other name for a positive ion is … ?
2. Which two cations on page 286 cannot be identified using only sodium hydroxide? Which further test could be done?
3. Sodium hydroxide and ammonia solutions cannot be used to identify Na⁺ or K⁺ ions. Why not?
4. Silver nitrate is used in the test for halides. Why?
5. Nitrates are not tested by forming a precipitate. Why not?
6. Where do the OH⁻ ions come from, in the test for nitrate ions?
7. a Why is acid used, in testing for carbonates?
   b Limewater is also used in the test. What is limewater?

IN THE LAB

# Checkup on Chapter 19

## Revision checklist

### For all students
Make sure you can ...
- [ ] identify these common pieces of laboratory apparatus, and say what they are used for:
  beaker    test-tube    conical flask
  pipette    burette    measuring cylinder
  gas jar    gas syringe    condenser
  thermometer    filter funnel    water trough
- [ ] arrange these pieces of apparatus in order of accuracy (as here) for measuring out a volume of liquid:
  beaker    measuring cylinder    burette    pipette
- [ ] describe how to carry out these procedures:
  – filtration
  – simple distillation
  – fractional distillation
  – crystallisation
  – paper chromatography
  – titration
- [ ] describe the scientific method
- [ ] explain what these are:
  independent variable    dependent variable
- [ ] explain why measurements are often repeated, in experimental work
- [ ] describe how to prepare these gases in the lab:
  hydrogen    oxygen    carbon dioxide    ammonia
  – and name suitable reactants to use
  – give the equations for the reactions
  – draw the apparatus
- [ ] give the test for these gases:
  hydrogen    oxygen    carbon dioxide
  ammonia    chlorine
- [ ] give another term for:    cation    anion
- [ ] explain that in the tests for anions and cations, either a precipitate is formed, or a gas is given off
- [ ] describe tests to identify these cations:
  $Cu^{2+}$    $Fe^{2+}$    $Fe^{3+}$    $Al^{3+}$    $Zn^{2+}$    $Ca^{2+}$    $NH_4^+$
- [ ] describe tests to identify these anions:
  halide ions ($Cl^-$, $Br^-$, $I^-$)    sulfate ion, $SO_4^{2-}$
  nitrate ion, $NO_3^-$    carbonate ion, $CO_3^{2-}$

Make sure you can also ...
- [ ] describe a test for water (page 124)
- [ ] explain that melting and boiling points can be used to test whether a substance is pure (page 19)

## Questions

### For all students

**1** A sample of soil from a vegetable garden was thoroughly crushed, and water added as shown:

**a** Using a conical flask, filter funnel, filter paper, universal indicator, and dropping pipette, show how you would measure the pH of the soil.
**b** How would you check that the results for this sample were valid for the whole garden?

**2** This apparatus is used to collect gases in the lab.
**a** Make a drawing of the apparatus, labelling the water, trough, measuring cylinder, delivery tube, flask, and dropping funnel.

**b** This apparatus can be used for preparing the gases hydrogen and carbon dioxide, but *not* sulfur dioxide. Explain why.

**3** This apparatus is used to measure rate of a reaction.

**a** Suggest a suitable reagent to use as Y.
**b** Which other piece of apparatus is needed?
**c** Outline the procedure for this experiment.
**d** You must be careful not to use too much of the reagents. Why?

**4** A sample of a potassium salt was contaminated with potassium chloride. These tests were carried out on the contaminated sample.
   **TEST A**
   Dilute nitric acid is added to the solid. The mixture bubbles. The gas given off turns limewater milky.
   **a i** Name the gas given off.
     **ii** Which anion is present in the potassium salt?
   **TEST B**
   An equal volume of barium nitrate solution is added to a solution of the solid. A precipitate forms.
   **b i** What colour will the precipitate be?
     **ii** Name the precipitate, and explain why it forms.
     **iii** The precipitate will disappear if dilute nitric acid is added. Why?
   **TEST C**
   An equal volume of silver nitrate solution is added to a solution of the solid. A precipitate forms.
   **c i** What colour will this precipitate be?
     **ii** This precipitate confirms the presence of the impurity. Explain why.
   **d** Give the formulae for both the potassium salt and the impurity.

**5** Two solutions W and X are tested with universal indicator paper.
   Solution W: the indicator paper turns red
   Solution Y: the indicator paper turns orange
   **a i** Which solution could have a pH of 1, and which could have a pH of 5?
     **ii** Which type of solution is Y?

   Further tests are carried out in test-tubes.
   **TEST A**
   A piece of magnesium is added to solution W.
   **b i** What will you observe in the test-tube?
     **ii** What is formed as a result of the reaction
     **iii** How will solution Y compare, in this reaction?
   **TEST B**
   A solid, which is a sodium compound, is added to solution W. A gas is given off. It turns limewater milky.
   **c i** What colour will the solid be?
     **ii** Name the gas released.
     **iii** Suggest a name for the solid.
   **TEST C**
   A few drops of barium nitrate solution are added to a solution of W. A white precipitate forms.
   **d i** Name the white precipitate.
     **ii** Identify solution W.

**6** Ammonium nitrate ($NH_4NO_3$) is an important fertiliser. The ions in it can be identified by tests.
   **a** Name the cation present, and give its formula.
   **b** Which of these tests will confirm its presence?
     **A** When aqueous sodium hydroxide is added to a solution of the compound, a white precipitate forms. This does not dissolve in excess sodium hydroxide.
     **B** On heating the solid with solid sodium hydroxide, a gas is given off. It turns damp red litmus paper blue.
     **C** On heating the solid with dilute hydrochloric acid, a gas is given off. It turns damp blue litmus paper red.
   **c** Name the anion present, and give its formula.
   **d** Which of these tests will confirm its presence?
     **A** When dilute hydrochloric acid is added the solid fizzes, and releases a gas which relights a glowing splint.
     **B** When a solution of barium ions is added to a solution of the compound, a white precipitate forms.
     **C** When sodium hydroxide solution and aluminium foil are added to the solid, ammonia is given off after gentle heating.

**7** A sample of mineral water contained these ions:

| Name of ion | Concentration (milligrams/dm$^3$) |
|---|---|
| calcium | 55 |
| chloride | 37 |
| hydrogen carbonate | 248 |
| magnesium | 19 |
| nitrate | 0.05 |
| potassium | 1 |
| sodium | 24 |
| sulfate | 13 |

   **a** Make two lists, one for the anions and the other for the cations present in this mineral water.
   **b i** Which metal ion is present in the highest concentration?
     **ii** What mass of that metal would be present in a small bottle of water, volume 50 cm$^3$?
   **c** Which of the ions will react with barium nitrate solution to give a white precipitate?
   **d** Of the metal ions, only calcium can be identified by a precipitation test. Why is this?
   **e** A sample of the water is heated with sodium hydroxide and aluminium foil. Ammonia gas could not be identified, even though the nitrate ion is present. Suggest a reason.

# Answers to the numerical questions in this book

**page 33**  **6 C** 6p 6e 6n; **O** 8p 8e 8n; **Mg** 12p 12e 12n; **Al** 13p 13e 14n; **Cu** 29p 29e 35n

**page 44**  **3 b i** 60 **ii** 34 **iii** 0 **iv** 10 **v** 146

**page 45**  **9 a i** 38 **ii** 40

**page 71**  **2** 127 **4** 32 **b** 254 **c** 16 **d** 71 **e** 58 **f** 46 – all relative molecular mass **g** 132, relative formula mass

**page 73**  **1 a** 95 g **b** 35.5 g **c** 47.5 g **2** 75% carbon 25% hydrogen **3 a** 90% **b** 1.5 g

**page 75**  **10 a** 18 **b** 17 **c** 46 **d** 80 **e** 98 **f** 36.5 **g** 142
**11 a** 40 **b** 239 **c** 78 **d** 58.5 **e** 170 **f** 132 **g** 138 **h** 278
**12 a** 27.2 g **ii** 2.72 g **b** 50 g **c** 80%
**13 a** 17.5% **b i** 2185 kg **ii** 375 kg **c** 91.7%

**page 77**  **4 a** 1 g **b** 127 g **c** 35.5 g **d** 71 g **5 a** 32 g **b** 64 g **6** 138 g
**7 a** 9 moles **b** 3 moles **8 a** $6.02 \times 10^{23}$ **b** 35.5 g

**page 79**  **1 b** 2 **c i** 32 g **ii** 8 g **2 b** $CuCO_3$, 124 g; CuO, 80 g; $CO_2$, 44 g; **c i** 11 g **ii** 20 g

**page 81**  **3** 24 $dm^3$ **4 a** 168 $dm^3$ **b** 12 $dm^3$ **c** 0.024 $dm^3$ (24 $cm^3$)
**5 a** 12 $dm^3$ **b** 2.4 $dm^3$ **6 a** 12 $dm^3$ **b** 12 $dm^3$ **7 a** 12 $dm^3$ **b** 6 $dm^3$

**page 83**  **1 a** 1 mole **b** 1 mole **2 a** 2 mol/$dm^3$ **b** 1.5 mol/$dm^3$
**3 a** 0.5 $dm^3$ (500 $cm^3$) **b** 0.005 $dm^3$ (5 $cm^3$) **4 a** 20 g **b** 0.5 g
**5 a** 0.5 moles per litre **b** 0.25 moles per litre

**page 85**  **1 a** 4 **b** 4 **g** 3 **FeS** 4 $SO_3$

**page 87**  **4** CH **5** $C_2H_4$ **6 a** $C_7H_{16}$ **b** $C_7H_{16}$ **7** $P_4O_6$

**page 89**  **2** 76.7% **3** 63% **4** 172 g **5** 88%

**page 90**  **1 b** 160 g **c** 2000 moles **d** 2 moles **e** 4000 moles **f** 224 kg
**2b** 0.5 moles **c i** 11.2 g **ii** 8.8 g **iii** 4.8 $dm^3$ or 4800 $cm^3$ **3 a i** 4 moles
**ii** 19 moles **b** 4.75 moles **c** 114 $dm^3$ **d** 227 g **e** 502.2 $dm^3$

**page 91**  **4 a** 0.5 moles **b** 25 $cm^3$ **c** 75 $cm^3$ **d** 50 $cm^3$
**5 a** 1.4 g **b** 0.025 moles **c** 0.025 moles **d** $Fe^{2+}$ **e** 0.6 $dm^3$
**6 a** 106.5 g **b** 3 moles **c** 1 mole **d** $AlCl_3$ **e** 0.1 mol/$dm^3$
**7 a** 45.5 $cm^3$ **b** 41.7 $cm^3$ **c** 62.5 $cm^3$
**8 a** $P_2O_3$ **b** 41.3 g **c** $P_4O_{10}$ **d** $P_4O_6$ (or $P_2O_3$)
**9 a** $Zn_3P_2$ **b** 24.1% **10 a** 64 g **b** 4 moles **c** 2 moles **d** $MnO_2$
**e** 632.2g **11 a** $N_2H_4$ **b** $C_2N_2$ **c** $N_2O_4$ **d** $C_6H_{12}O_6$ **12 b** $CH_2$ **c** A is $C_3H_6$ B is $C_6H_{12}$ **13 a** 217 g **b** 20.1 g of mercury, 1.6 g of oxygen
**c** 94.5 % **14 a** 0.0521 moles **b** 4.375 g **c** 87.5%

**page 117**  **4** −486 kJ/mol

**page 128**  **2 b** drop of 4 °C for ammonium nitrate, rise of 20 °C for calcium chloride, **d i** 17 °C for $NH_4NO_3$, 65 °C for $CaCl_2$, **ii** 23 °C for $NH_4NO_3$, 35 °C for $CaCl_2$, **iii** 21 °C for $NH_4NO_3$, 45 °C for $CaCl_2$

**page 129**  **4 d** 55.6 kJ **6 c i** 2220 kJ/mol **ii** 2801 kJ/mol
**d** −581 kJ/mol

**page 133**  **3 a i** 29 $cm^3$ **ii** 39 $cm^3$ **b** 1.5 minutes **c i** 5 $cm^3$ of hydrogen per minute **ii** 0 $cm^3$ of hydrogen per minute

**page 135**  **1 a i** 60 $cm^3$ **ii** 60 $cm^3$

**page 137**  **1 a** experiment 1, 0.55 g; experiment 2, 0.95 g
**b** experiment 1, 0.33 g per minute; experiment 2, 0.5 g per minute

**page 146**  **2 c i** 14 $cm^3$ of hydrogen/minute **ii** 9 $cm^3$ of hydrogen/minute **iii** 8 $cm^3$ of hydrogen/minute **e** 40 $cm^3$ **f** 5 minutes
**g** 8 $cm^3$ of hydrogen/minute

**page 147**  **5 i** 0.5 g

**page 163**  **2** 50 $cm^3$ **3** 1.6 mol/$dm^3$

**page 165**  **8 c** 0.014 moles **d** 0.007 moles **e** 0.742g **i** 1.258g
**f** 0.07 moles **g** 10 moles

**page 180**  **2b** Accept rough values around melting point 40 °C, boiling point 725 °C **c i** 5 **ii** 37 **iii** 1

**page 193**  **8 c i** 2.45 volts **ii** 0.65 volts

**page 209**  **3 c i** 990 kg **ii** 0.25%

**page 222**  **3 d** 20.8%

**page 223**  **4 a i** 36% **ii** It is greater. **b** 78%

**page 227**  **3 c** 28%

**page 229**  **4 a** 21.2%

**page 289**  **7 b ii** 2.75 milligrams

**Questions from past exam papers**

**page 296**  **13 c** 80

**page 298**  **2 b i** 7.7% **ii** 1:1 **iii** empirical formula is CH, molecular formula is $C_6H_6$

**page 299**  **5 a i**

|  | copper | iron | sulfur |
|---|---|---|---|
| composition by mass/g | (4.80) | (4.20) | 4.80 |
| number of moles of atoms | 0.075 | 0.075 | 0.15 |
| simplest mole ratio of atoms | 1 | 1 | 2 |

**ii** $CuFeS_2$

**page 299**  **7 b i** 100 **ii** 56 **iii** 12.5 kg

**page 301**  **13 a** energy to break bonds = 436 + 158 = +594 kJ
energy released in forming bonds = 2 × 562 = −1124 kJ
energy in - energy out = −530 kJ, so the reaction is exothermic

**page 302**  **15 d i** 0.033 moles **ii** 0.033 moles **iii** 5.67 g **iv** 70.6%

**page 305**  **3 a**

| expt | initial temperature /°C | final temperature /°C | difference /°C |
|---|---|---|---|
| 1 | 24 | 27 | +3 |
| 2 | 26 | 22 | −4 |
| 3 | 21 | 11 | −10 |
| 4 | 29 | 23 | −6 |

**3 e i** −8 °C **ii** −3 °C

**page 305**  **4 a**

| time/s | volume of oxygen/$cm^3$ | |
|---|---|---|
|  | catalyst W | catalyst X |
| 0 | 0 | 0 |
| 20 | 16 | 29 |
| 40 | 32 | 34 |
| 60 | 36 | 36 |
| 80 | 37 | 37 |
| 100 | 37 | 37 |

**page 306**  **5 b** 47 ± 1 g / 100 g of water

**page 307**  **9 b** and **c**

| | Burette readings/$cm^3$ | |
|---|---|---|
| | Experiment 1 | Experiment 2 |
| final reading | 26.0 | 29.0 |
| initial reading | 0.0 | 16.0 |
| difference | 26.0 | 13.0 |

# About the Cambridge IGCSE Chemistry exam

## The exam papers
For your Cambridge IGCSE Chemistry exam, you must take three papers out of six. Look at this list.

> **Everyone must take this:**
> **Paper 1,** multiple choice questions (45 minutes, 30% of the total marks)
>
> **Then *one* from these two:**
> **Paper 2,** for the Core syllabus (1 hour 15 minutes, 50% of the total marks)
> **Paper 3,** for the Extended syllabus (1 hour 15 minutes, 50% of the total marks)
>
> **And *one* from these three:**
> **Paper 4**   Coursework (20% of total marks)
> **Paper 5**   Practical test (1 hour 15 minutes, 20% of the total marks)
> **Paper 6**   Alternative to Practical written paper (1 hour, 20% of the total marks)

Notice that Paper 1 carries 30% of the total marks for the exam. Your choice of Papers 2 or 3 carries 50%. Your remaining paper carries 20%.

## Getting ready for the exam
- First, do you know which papers you will take? If you are not sure, ask your teacher.
- For Paper 3 (Extended) you will need to revise everything in this book, except the extra material on the yellow pages.
- For Paper 2 (Core), you can ignore all the material with a red line beside it, and the extra material.
- The CD with this book has lots of material to help you revise, including tests, revision advice, and sample exam papers with real questions from past papers. Check it out!

## Doing past exam questions
Doing questions is a great way to revise. And before your exam, it is really important to work through questions from past papers.

The next eighteen pages of this book have real questions from past exams, for Papers 2, 3 and 6.

- If you are taking Paper 2 in the exam, you can ignore the Paper 3 questions.
- If you are taking Paper 3, you should concentrate on the Paper 3 questions – but you may also find the Paper 2 questions helpful.
- The Paper 6 questions are for *everyone*. Even if you are not taking Paper 6 in the exam, you will find that these questions help with the other papers.

Note that answers for all the *numerical* questions in this book, including exam questions, are on page 290. Your teacher can provide answers for all the other questions, from both the book and CD, so that you can check your progress.

Good luck with your revision!

# CAMBRIDGE IGCSE EXAM QUESTIONS from Paper 2

**1 a** The table gives some information about five elements, A, B, C, D and E.
Copy and complete the table by writing either *metal* or *non-metal* in the last column.

| element | properties | metal or non-metal |
|---|---|---|
| A | shiny solid which conducts electricity | |
| B | reddish brown liquid with a low boiling point | |
| C | a form of carbon which is black in colour and conducts electricity | |
| D | white solid which is an insulator and has a high melting point | |
| E | dull yellow solid which does not conduct heat | |

[5]

**b** Describe how metallic character changes across a Period. [1]

**c** Sodium is in Group I of the Periodic Table.
   **i** Draw a diagram to show the full electronic structure of sodium. [1]
   **ii** Complete the equation to show what happens when a sodium atom forms a sodium ion.
   $Na \longrightarrow Na^+ + \ldots\ldots$ [1]

**d** Complete these sentences about properties of the Group I elements using words from the list.
*acidic     basic     decrease     hard
increase     lithium     potassium     soft*
The Group I elements are relatively ......... metals which ........ in reactivity going down the Group. Sodium reacts more violently with water than ......... The Group I metals all form ......... oxides. [4]

*Cambridge IGCSE Chemistry 0620 Paper 2 Q1 November 2008*

**2** Lavandulol is found in lavender plants.
The formula of lavandulol is shown on the right.

**a** Which is the alcohol functional group in this formula? [1]

**b** Is lavandulol a saturated or unsaturated compound? Give a reason for your answer. [1]

**c** State the names of the two products formed when lavandulol is burnt in excess oxygen. [2]

**d** Lavandulol can be extracted from lavender flowers by distillation using the apparatus shown below. The lavandulol is carried off in small droplets with the steam.

   **i** State the name of the piece of apparatus labelled A. [1]
   **ii** What is the temperature of the water at point X in the diagram? [1]
   **iii** The lavandulol and water are collected in the beaker. What information in the diagram shows that lavandulol is less dense than water? [1]

**e** Lavender flowers contain a variety of different pigments (colourings). A student separated these pigments using paper chromatography. The results are shown in the diagram below.

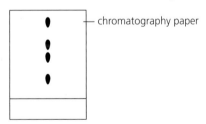

   **i** Copy the diagram and put an X to show where the mixture of pigments was placed at the start of the experiment. [1]
   **ii** How many pigments have been separated? [1]
   **iii** Draw a diagram to show how the chromatography apparatus was set up. Label
   • the solvent     • the origin line [1]

**iv** During chromatography, the solvent evaporates and then diffuses through the chromatography jar. What do you understand by *diffusion*? [1]

**v** Ethanol can be used as a solvent in chromatography. Draw the formula for ethanol showing all atoms and bonds. [1]

**vi** Which two of the following statements about ethanol are true?

*It is a carboxylic acid.*
*It is a product of the fermentation of glucose.*
*It is an unsaturated compound.*
*It is formed by the catalytic addition of steam to ethene.* [1]

*Cambridge IGCSE Chemistry 0620 Paper 2 Q3 November 2006*

**3** The diagram shows the structures of some substances containing carbon.

**a** Answer these questions using the letters A–F.
  **i** Which one of these structures is ionic? [1]
  **ii** Which one of them represents ethanol? [1]
  **iii** Which one of these structures represents a gas which turns limewater milky? [1]
  **iv** Which one of these structures is an unsaturated hydrocarbon? [1]
**b** Describe a chemical test for an unsaturated hydrocarbon. Give the result of the test. [2]
**c** State the chemical name of structure B. [1]
**d** Structure F has several uses. Which one of the following is a correct use of structure F?
*for cutting metals     as a lubricant*
*for filling balloons     as an insulator* [1]
**e** The structures A to E are compounds. What do you understand by the term compound? [1]
**f** State the type of bonding in structure A. [1]

*Cambridge IGCSE Chemistry 0620 Paper 2 Q1 June 2008*

**4** The table shows observations about the reactivity of various metals with dilute hydrochloric acid.

| metal | observations |
|---|---|
| calcium | many bubbles produced rapidly with much spitting |
| copper | no bubbles formed |
| iron | a few bubbles produced very slowly |
| magnesium | many bubbles produced rapidly with no spitting |

**a** Put the metals in order of increasing reactivity. [1]
**b** Zinc is between iron and magnesium in reactivity. Suggest what observations will be made when zinc reacts with dilute hydrochloric acid. [1]
**c** Magnesium is extracted by the electrolysis of molten magnesium chloride.

  **i** What information in the diagram suggests that magnesium is less dense than molten magnesium chloride? [1]
  **ii** Magnesium is extracted by electrolysis rather than by heating its oxide with carbon. Why? [1]
  **iii** Suggest why a stream of inert gas is blown over the surface of the molten magnesium. [1]
  **iv** Name a gaseous element which is inert. [1]
**d** In some old magnesium manufacturing plants, coal gas is blown over the surface of the magnesium. The list shows the main substances in coal gas.
*carbon monoxide     ethene     hydrogen*
*hydrogen sulfide     methane*
  **i** Draw the structure of ethene showing all atoms and bonds. [1]
  **ii** Suggest two hazards of using coal gas by referring to two specific substances in the list. [2]
**e** Carbon monoxide can be removed from coal gas by mixing it with steam and passing the mixture over a catalyst of iron(III) oxide at 400 °C.
$CO + H_2O \rightleftharpoons CO_2 + H_2$
  **i** Write a word equation for this reaction. [1]
  **ii** What does the symbol $\rightleftharpoons$ mean? [1]
  **iii** Iron(III) oxide reacts with acids to form a solution containing iron(III) ions. Describe a test for aqueous iron(III) ions. Give the result. [2]

*Cambridge IGCSE Chemistry 0620 Paper 2 Q2 June 2009*

**5** Hydrogen chloride can be made by burning hydrogen in chlorine.
  **a** Complete the equation for this reaction.
  $H_2 +$ ............... $\longrightarrow$ .... HCl   [2]
  **b** Draw a dot and cross diagram for a molecule of hydrogen chloride. Show all the electrons.
  *Use o for an electron from a hydrogen atom.*
  *Use x for an electron from a chlorine atom.* [2]
  **c** Hydrochloric acid is formed when hydrogen chloride gas dissolves in water. Suggest the pH of hydrochloric acid. Choose from this list.
  pH1   pH7   pH9   pH13   [1]
  **d** Complete the equation for the reaction of hydrochloric acid with zinc.
  zinc + hydrochloric acid $\longrightarrow$ zinc chloride + ...............   [1]
  **e** Describe how dry crystals of zinc chloride can be obtained from a solution of zinc chloride. [2]
  **f** A student electrolysed molten zinc chloride. State the name of the product formed at
  **i** the anode,   **ii** the cathode.   [2]
  *Cambridge IGCSE Chemistry 0620 Paper 2 Q7 June 2009*

**6** When Group I elements react with water, hydrogen gas is given off. The diagram shows the reaction of lithium, potassium and sodium with water.

A     B     C

  **a** Which one of these three elements A, B or C is lithium?   [1]
  **b i** Balance the equation for the reaction of sodium with water by completing the lefthand side.
  .......Na + .......$H_2O \longrightarrow 2NaOH + H_2$   [1]
  **ii** Apart from fizzing, describe two things that you would see when sodium reacts with water.   [2]
  **iii** After the sodium had reacted with the water, the solution was tested with red litmus paper. What colour did the litmus paper turn? Give a reason for your answer.   [2]
  **iv** Which two of the following statements about sodium are true?
  *It is made by reducing sodium oxide with carbon.*
  *It reacts with chlorine to form sodium chloride.*
  *It reacts readily with oxygen.*
  *It only conducts electricity when molten.*   [2]

  **c** Rubidium also reacts with water. How does the speed of reaction of rubidium with water compare with that of potassium with water?   [1]
  **d** Sodium has only one stable isotope whereas potassium has several isotopes.
  **i** What do you understand by the term isotopes?   [1]
  **ii** How many protons does sodium have in its nucleus? Use the Periodic Table to help you.   [1]
  **iii** How many electrons are there in an atom of potassium?   [1]
  **iv** Uranium has many isotopes. One of these is uranium-235 ($^{235}U$). What is the main use of this isotope of uranium?   [1]
  *Cambridge IGCSE Chemistry 0620 Paper 2 Q1 November 2006*

**7** The table shows the concentration of some ions present in seawater.

| name of ion | formula of ion | concentration of ion in g/dm³ |
|---|---|---|
| bromide | $Br^-$ | 0.07 |
| calcium | $Ca^{2+}$ | 0.04 |
| chloride | $Cl^-$ | 19.1 |
| magnesium | $Mg^{2+}$ | 1.2 |
| potassium | $K^+$ | 0.3 |
| sodium | $Na^+$ | 10.6 |
|  | $SO_4^{2-}$ | 0.8 |

  **a** Which negative ion has the highest concentration in seawater?   [1]
  **b** Name the ion with the formula $SO_4^{2-}$   [1]
  **c** Which two ions in the table are formed from Group I elements?   [1]
  **d** When seawater is evaporated a number of different compounds are formed. State the name of the compound present in the greatest quantity.   [1]
  **e** Names the four ions in the table which move to the cathode when seawater is electrolysed.   [2]
  **f** When concentrated seawater is electrolysed, chlorine is formed at one of the electrodes.
  **i** To which Period in the Periodic Table does chlorine belong?   [1]
  **ii** Draw the electronic structure of a chlorine molecule. Show only the outer electrons.   [2]
  **g** Drinking water can be obtained by purifying seawater. Explain why distillation rather than filtration is used to purify seawater for drinking. [2]
  *Cambridge IGCSE Chemistry 0620 Paper 2 Q3 June 2008*

**8** Iron is extracted from its ore in a blast furnace.
  **a** State the name of the ore from which iron is extracted. [1]
  **b** The diagram shows a blast furnace.

   **i** Which one of the raw materials is added to the blast furnace to help remove the impurities from the iron ore? [1]
   **ii** The impurities are removed as a slag. Which letter on the diagram shows the slag? [1]
  **c** Carbon monoxide is formed in the blast furnace by reaction of coke with oxygen.
   **i** Complete the equation for this reaction.
   ...... C + ............... $\longrightarrow$ ...... CO [2]
   **ii** State the adverse affect of carbon monoxide on human health. [1]
  **d** In the hottest regions of the blast furnace the following reaction takes place.
   $Fe_2O_3 + 3C \longrightarrow 2Fe + 3CO$
   Which two of these five sentences correctly describe this reaction?
   *The iron oxide gets reduced.*
   *The reaction is a thermal decomposition.*
   *The carbon gets oxidised.*
   *The carbon gets reduced.*
   *Carbon neutralises the iron oxide.* [1]
  **e** Aluminium cannot be extracted from aluminium oxide in a blast furnace. Explain why aluminium cannot be extracted in this way. [2]
  **f i** State the name of the method used to extract aluminium from its oxide ore. [1]
   **ii** State one use of aluminium. [1]
  *Cambridge IGCSE Chemistry 0620 Paper 2 Q4 November 2008*

**9** Calcium carbonate, $CaCO_3$, is the raw material used in the manufacture of lime, CaO.
  **a i** Describe how lime is manufactured from calcium carbonate. [1]
   **ii** Write a symbol equation for this reaction. [1]
   **iii** State one large scale use of lime. [1]
  **b** A student investigated the speed of reaction of calcium carbonate with hydrochloric acid using the apparatus shown below.
   **i** Name the pieces labelled A to C. [3]

   **ii** The equation for the reaction is
   $CaCO_3 + 2HCl \longrightarrow CaCl_2 + CO_2 + H_2O$
   Write the word equation for this reaction. [2]
   **iii** The student carried out the reaction at 40 °C using large pieces of calcium carbonate. The results are shown below. At what time did the reaction stop? [1]

   **iv** The student repeated the experiment using the same mass of powdered calcium carbonate. All other conditions were kept the same. Copy the grid above and sketch the graph for the reaction with calcium carbonate powder. [2]
   **v** How does the speed of reaction change when
   – the concentration of hydrochloric acid is decreased,
   – the temperature is increased? [2]
  *Cambridge IGCSE Chemistry 0620 Paper 2 Q3 November 2008*

**10** The apparatus below can be used to measure the energy released when a liquid fuel is burnt. The amount of energy released is calculated from the increase in temperature of a known amount of water.

**a i** Explain how this experiment shows that the burning of ethanol is exothermic. [1]
 **ii** Complete the word equation for the complete combustion of ethanol.
   ethanol + oxygen ⟶ ............ + ............ [2]
**b** Ethanol is a fuel containing carbon. Name two other commonly used fuels containing carbon. [2]
**c** Give the formula of the functional group present in ethanol. [1]
**d** The can contains water. Describe a chemical test for water. Give the result of the test. [2]
**e** The iron can used in this experiment rusts easily.
 **i** Describe a method which can be used to prevent iron from rusting. [1]
 **ii** Rust contains hydrated iron(III) oxide. What do you understand by the term hydrated? [1]
 **iii** Iron is a transition metal. State two typical properties of transition metals. [2]

*Cambridge IGCSE Chemistry 0620 Paper 2 Q5 November 2008*

**11** Some coal dust was heated with copper(II) oxide using the apparatus shown below.

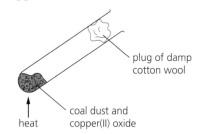

**a** Coal contains carbon and various hydrocarbons. Carbon reduces the copper(II) oxide when heated.
 **i** What do you understand by *reduction*? [1]
 **ii** At the end of the experiment a reddish-brown solid remained in the tube. State its name. [1]
 **iii** The reddish brown solid conducts electricity. How could you show that it does so? [2]

**b** During the experiment, water collected on the cooler parts of the test-tube.
 **i** Suggest where the hydrogen in the water comes from. [1]
 **ii** Water is a liquid. Describe the arrangement and motion of the particles in a liquid. [2]

*Cambridge IGCSE Chemistry 0620 Paper 2 Q5 November 2007*

**12** Some sunglasses are made from glass which darkens in bright sunlight. The glass contains tiny crystals of silver chloride and copper(I) chloride.
**a** In bright sunlight, in the presence of copper(I) chloride, the silver chloride breaks down to solid silver which darkens the glass.
   $Ag(s) + e^- \longrightarrow Ag(s)$
   Name the particle with the symbol $e^-$. [1]
**b** Silver is a metal. State two physical properties which are characteristic of all metals. [2]
**c** In bright sunlight, the copper(I) chloride in the sunglasses is converted to copper(II) chloride. What do the roman numerals (I) and (II) show in these copper compounds? Choose one from
   *the number of copper atoms in the compounds*
   *the number of neutrons in the compounds*
   *whether the copper is in the solid, liquid or gaseous state*
   *the oxidation state of the copper in the compounds* [1]
**d** Describe a test for aqueous copper(II) ions. Give the result of the test. [3]
**e** Give a common use of copper. [1]

*Cambridge IGCSE Chemistry 0620 Paper 2 Q5 June 2007*

**13** This question is about compounds.
**a** What do you understand by the term *compound*? [1]
**b** Copy and complete the table below to show the formulae and uses of some compounds. [6]

| compound | relative number of atoms present | formula | use |
|---|---|---|---|
| calcium oxide | Ca = 1<br>O = 1 | CaO | |
| sodium chloride | Na = 1<br>Cl = 1 | | table salt |
| calcium carbonate | Ca = 1<br>C = 1<br>O = 3 | | |
| | | $NH_4NO_3$ | in fertilisers |

**c** Calculate the relative formula mass of $NH_4NO_3$. [1]

*Cambridge IGCSE Chemistry 0620 Paper 2 Q4 November 2006*

**14** The list shows part of the reactivity series.

strontium — more reactive  
calcium  
magnesium  
iron  
copper — less reactive

**a** Calcium is manufactured by the electrolysis of molten calcium chloride. Suggest why calcium is extracted by electrolysis. [1]

**b** Equal sized pieces of magnesium, strontium and calcium are placed in water. Some observations about these reactions are shown in the table. Complete the box for strontium.

| metal | observations |
|---|---|
| magnesium | Gives off a few bubbles of gas with hot water. Dissolves very slowly. |
| calcium | Gives off bubbles steadily with cold water. Dissolves slowly. |
| strontium | |

[2]

**c** When water is added to calcium carbide, acetylene and calcium hydroxide are formed. State a use for acetylene. [1]

**d** A solution of calcium hydroxide is alkaline.
  **i** Complete and balance the equation for the reaction of calcium hydroxide with hydrochloric acid.
  
  $Ca(OH)_2 + 2HCl \longrightarrow CaCl_2 + $ ............ [1]
  
  **ii** What type of chemical reaction is this? [1]

**e** A student used the apparatus shown below to calculate the concentration of a solution of calcium hydroxide.

**i** Name the piece of apparatus labelled A. [1]
**ii** Describe how the pH of the solution in the flask changes as the hydrochloric acid is added. [2]

*Cambridge IGCSE Chemistry 0620 Paper 2 Q5 November 2006*

**15** Petroleum is a mixture of hydrocarbons. It is separated into fractions such as petrol, paraffin and diesel.

**a** Name the process used to separate the fractions. [1]
**b** Name two other fractions which are obtained from petroleum. [2]
**c** Give one use for the paraffin fraction. [1]
**d** Many of the compounds from petroleum are alkanes. Which two of these are alkanes? [1]

A     B     C     D

**e** Use words from the list below to complete the following sentence.

*ethane  ethene  hydrogen  nitrogen*
*oxygen  reactive  unreactive  water*

Alkanes such as **i**............... are generally **ii**............... but they can be burnt in **iii**............... to form carbon dioxide and ............... [4]

**f** Alkanes are saturated hydrocarbons. What do you understand by **i** saturated, **ii** hydrocarbon? [2]

*Cambridge IGCSE Chemistry 0620 Paper 2 Q3 June 2009*

**16 a** Choose from the list of compounds to answer questions **i** to **v**. Each compound can be used once, more than once, or not at all.

*calcium carbonate  carbon dioxide*
*hydrogen chloride  iron(III) oxide*
*lead(II) bromide  methane  sodium hydroxide*

Name the compound which
  **i** is a transition metal compound, [1]
  **ii** produces brown fumes at the anode when electrolysed, [1]
  **iii** is used to manufacture lime, [1]
  **iv** forms an alkaline solution in water, [1]
  **v** is the main constituent of natural gas. [1]

**b** At a high temperature iron(III) oxide is reduced by carbon: $Fe_2O_3 + 3C \longrightarrow 2Fe + 3CO$
  **i** Explain how the equation shows that iron(III) oxide is reduced by carbon. [1]
  **ii** Copy and complete these sentences about the extraction of iron using words from the list.
  
  *bauxite  blast  converter  hematite*
  *lime  limestone  sand  slag*
  
  Iron is extracted from ........ by mixing the ore with coke and ........ in a ........ furnace. The iron ore is reduced to iron. Impurities in the ore react with calcium oxide to form ........ [4]

*Cambridge IGCSE Chemistry 0620 Paper 2 Q1 June 2009*

# CAMBRIDGE IGCSE EXAM QUESTIONS from Paper 3

1. Use your copy of the periodic table to help you answer these questions.
   a. Predict the formula of each of the following compounds.
      i. barium oxide  ii. boron oxide  [2]
   b. Give the formula of the following ions.
      i. sulfide   ii. gallium  [2]
   c. Draw a diagram showing the arrangement of the valency electrons in one molecule of the covalent compound nitrogen trichloride.
      *Use x to represent an electron from a nitrogen atom. Use o to represent an electron from a chlorine atom.*  [3]
   d. Potassium and vanadium are elements in Period IV.
      i. State two differences in their physical properties.  [2]
      ii. Give two differences in their chemical properties.  [2]
   e. Fluorine and astatine are halogens. Use your knowledge of the other halogens to predict the following:
      i. the physical state of fluorine at r.t.p. the physical state of astatine at r.t.p.  [2]
      ii. two similarities in their chemical properties.  [2]
   
   *Cambridge IGCSE Chemistry 0620 Paper 3 Q4 June 2007*

2. Across the world, food safety agencies are investigating the presence of minute traces of the toxic hydrocarbon, benzene, in soft drinks.

   It is formed by the reduction of sodium benzoate by vitamin C.

   a. Sodium benzoate is a salt. It has the formula $C_6H_5COONa$. It can be made by neutralising benzoic acid using sodium hydroxide.
      i. Deduce the formula of benzoic acid.  [1]
      ii. Write a word equation for the reaction between benzoic acid and sodium hydroxide.  [1]
      iii. Name two other compounds that would react with benzoic acid to form sodium benzoate.  [2]
   b. Benzene contains 92.3% of carbon and its relative molecular mass is 78.
      i. What is the percentage of hydrogen in benzene?  [1]
      ii. Calculate the ratio of moles of C atoms: moles of H atoms in benzene.  [2]
      iii. Calculate its empirical formula and then its molecular formula.  [2]
   c. This shows the structural formula of Vitamin C.

      i. What is its molecular formula?  [1]
      ii. Name the two functional groups that are circled.  [2]
   
   *Cambridge IGCSE Chemistry 0620 Paper 3 Q4 November 2008*

3. The following is a list of the electron distributions of atoms of unknown elements.

   | element | electron distribution |
   |---------|----------------------|
   | A | 2,6 |
   | B | 2,8,4 |
   | C | 2,8,8,2 |
   | D | 2,8,18,8 |
   | E | 2,8,18,8,1 |
   | F | 2,8,18,18,7 |

   a. Choose an element from the list for each of the following descriptions.
      i. It is a noble gas.
      ii. It is a soft metal with a low density.
      iii. It can form a covalent compound with element A.
      iv. It has a giant covalent structure similar to diamond.
      v. It is a diatomic gas with molecules of the type $X_2$.  [5]
   b. Elements C and A can form an ionic compound.
      i. Draw a diagram that shows the formula of this compound, the charges on the ions, and the arrangement of the valency electrons around the negative ion.
      Use o to represent an electron from an atom of C. Use x to represent an electron from an atom of A.  [3]
      ii. Predict two properties of this compound.  [2]
   
   *Cambridge IGCSE Chemistry 0620 Paper 32 Q3 June 2009*

4 The results of experiments on electrolysis using inert electrodes are given in the table.

| electrolyte | change at negative electrode | change at positive electrode | change to electrolyte |
|---|---|---|---|
| molten lead(II) bromide | lead formed | bromine formed | used up |
| a ............... | potassium formed | iodine formed | used up |
| dilute aqueous sodium chloride | b ............ | c ............ | d ......... |
| aqueous copper(II) sulfate | e ............ | f ............ | g ............ |
| h ............ | hydrogen formed | bromine formed | potassium hydroxide formed |

Complete the table; the first line has been completed as an example. [8]

*Cambridge IGCSE Chemistry 0620 Paper 3 Q2 June 2009*

5 An ore of copper is the mineral, chalcopyrite. This is a mixed sulfide of iron and copper.
   a Analysis of a sample of this ore shows that 13.80 g of the ore contained 4.80 g of copper, 4.20 g of iron and the rest sulfur.
     i Copy and complete the table. [3]

|  | copper | iron | sulfur |
|---|---|---|---|
| composition by mass /g | 4.80 | 4.20 |  |
| number of moles of atoms |  |  |  |
| simplest mole ratio of atoms |  |  |  |

     ii Find the empirical formula of chalcopyrite. [1]
   b Impure copper is extracted from the ore. This copper is refined by electrolysis.
     i Name
        A: the material used for the positive electrode (anode),
        B: the material used for the negative electrode (cathode),
        C: a suitable electrolyte. [3]
     ii Write an ionic equation for the reaction at the negative electrode. [1]
     iii One use of this pure copper is electrical conductors, another is to make alloys. Name the metal that is alloyed with copper to make brass. [1]

   c Two of the elements in chalcopyrite are the metal, copper, and the non-metal, sulfur. These have different properties. Copper is an excellent conductor of electricity and is malleable. Sulfur is a poor conductor and is not malleable, it is brittle. Explain, in terms of their structures, why this is so.
     i difference in electrical conductivity [2]
     ii difference in malleability [2]

*Cambridge IGCSE Chemistry 0620 Paper 3 Q6 November 2006*

6 There are three types of giant structure – ionic, metallic and macromolecular.
   a Sodium nitride is an ionic compound. Draw a diagram that shows the formula of the compound, the charges on the ions, and the arrangement of the valency electrons around the negative ion.
   *Use x to represent an electron from a sodium atom.*
   *Use o to represent an electron from a nitrogen atom.* [3]
   b i Describe metallic bonding. [3]
     ii Use the above ideas to explain why metals are
        A: good conductors of electricity [1]
        B: metals are malleable. [2]
   c Silicon(IV) oxide has a macromolecular structure.
     i Describe the structure of silicon(IV) oxide (a diagram is not acceptable). [3]
     ii Diamond has a similar structure and consequently similar properties. Give two physical properties common to both diamond and silicon(IV) oxide. [2]

*Cambridge IGCSE Chemistry 0620 Paper 3 Q2 November 2008*

7 Calcium carbonate is an important raw material.
   a Name a rock made up of calcium carbonate. [1]
   b When calcium carbonate is heated strongly, it decomposes: $CaCO_3 \longrightarrow CaO + CO_2$
     i Calculate the relative formula mass of $CaCO_3$. [1]
     ii Calculate the relative formula mass of CaO. [1]
     iii 7.00 kg of calcium oxide was formed. What mass of calcium carbonate was heated? [2]
   c Calcium carbonate is used to control soil acidity.
     i Why is it important to control soil acidity? [1]
     ii Both calcium carbonate, insoluble in water, and calcium oxide, slightly soluble, are used to increase soil pH. Suggest two advantages of using calcium carbonate. [2]
     iii Give one use of calcium carbonate other than for making calcium oxide and controlling soil pH. [1]

*Cambridge IGCSE Chemistry 0620 Paper 3 Q3 November 2006*

**8** Iron is a transition element.
  **a** Which three of the following six statements about transition elements are correct?
  **i** The metals are highly coloured e.g. yellow, green, blue.
  **ii** The metals have low melting points.
  **iii** Their compounds are highly coloured.
  **iv** Their compounds are colourless.
  **v** The elements and their compounds are often used as catalysts.
  **vi** They have more than one oxidation state. [3]
  **b i** In which Period in the Periodic Table is iron? [1]
  **ii** Use the Periodic Table to work out the number of protons and the number of neutrons in one atom of iron. [1]
  **c** Iron is extracted in a blast furnace. The list below gives some of the substances used or formed in the extraction.
  *carbon monoxide  coke  iron ore  limestone  slag*
  **i** Which substance is a mineral containing largely calcium carbonate? [1]
  **ii** Which substance is formed when impurities in the ore react with calcium oxide? [1]
  **iii** Which substance is also called hematite? [1]
  **d** State two functions of the coke used in the blast furnace. [2]
  **e** Most of the iron is converted into mild steel or stainless steel. Give one use for each. [2]
  *Cambridge IGCSE Chemistry 0620 Paper 3 Q1 June 2006*

**9** The first three elements in Group IV are carbon, silicon, germanium.
  **a** The element germanium has a diamond-type structure. Describe the structure of germanium. A diagram is acceptable. [2]
  **b** Unlike diamond, graphite is soft and is a good conductor of electricity.
  **i** Explain why graphite has these properties. [3]
  **ii** Give a use of graphite that depends on one of these properties. [1]
  **c** Carbon dioxide and silicon(IV) oxide have similar formulae but different types of structure.
  **i** Give the formulae of these oxides. [1]
  **ii** How are their structures different? [2]
  **d** All these elements form compounds with hydrogen called hydrides. The saturated hydrides of carbon are the alkanes. Predict the formula of the hydride of germanium which contains two germanium atoms.
  *Cambridge IGCSE Chemistry 0620 Paper 3 Q4 June 2006*

**10** Zinc is extracted from zinc blende, ZnS.
  **a** Zinc blende is heated in air to give zinc oxide and sulfur dioxide.
  Most of the sulfur dioxide is used to make sulfur trioxide. This is used to manufacture sulfuric acid. Some of the acid is used in the plant, but most of it is used to make fertilisers.
  **i** Give another use of sulfur dioxide. [1]
  **ii** Describe how sulfur dioxide is converted into sulfur trioxide. [3]
  **iii** Name a fertiliser made from sulfuric acid. [1]
  **b** Some of the zinc oxide was mixed with an excess of carbon and heated to 1000 °C. Zinc distils out of the furnace.
  $$2ZnO + C \rightleftharpoons 2Zn + CO_2$$
  $$C + CO_2 \longrightarrow 2CO$$
  **i** Name the two changes of state involved in the process of distillation. [2]
  **ii** Why is it necessary to use an excess of carbon? [2]
  **c** The remaining zinc oxide reacts with sulfuric acid to give aqueous zinc sulfate. This is electrolysed with inert electrodes (the electrolysis is the same as that of copper(II) sulfate with inert electrodes).
  Ions present: $Zn^{2+}(aq)$   $SO_4^{2-}(aq)$   $H^+(aq)$   $OH^-(aq)$.
  **i** Zinc forms at the negative electrode (cathode). Write the equation for this reaction. [1]
  **ii** Write the equation for the reaction at the positive electrode (anode). [2]
  **iii** Complete this sentence:
  The electrolyte changes from aqueous zinc sulfate to ................ [1]
  **d** Give two uses of zinc. [2]
  *Cambridge IGCSE Chemistry 0620 Paper 3 Q4 November 2007*

**11** Aluminium is extracted by the electrolysis of a molten mixture that contains alumina, which is aluminium oxide, $Al_2O_3$.
  **a** The ore of aluminium is bauxite. This contains alumina, which is amphoteric, and iron(III) oxide, which is basic. The ore is heated with aqueous sodium hydroxide.
  Complete the following sentences.
  The **i**.......... dissolves to give a solution of **ii**..........
  The **iii**..........does not dissolve and can be removed by **iv**........... [4]

b Complete the labelling of the diagram.

c The ions involved in the electrolysis are $Al^{3+}$ and $O^{2-}$.
  i Write an equation for the reaction at the cathode. [2]
  ii Explain how carbon dioxide is formed at the anode. [2]
d Give an explanation for each of the following.
  i Aluminium is used extensively in the manufacture of aircraft. [1]
  ii Aluminium is used for food containers. [2]
  iii Aluminium electricity cables have a steel core. [1]

*Cambridge IGCSE Chemistry 0620 Paper 3 Q6 June 2007*

12 a Four bottles were known to contain aqueous ammonia, dilute hydrochloric acid, sodium hydroxide solution and vinegar, which is dilute ethanoic acid. The bottles had lost their labels. The pH values of the four solutions were 1, 4, 10 and 13. Complete the table. [2]

| solution | pH |
|---|---|
| aqueous ammonia |  |
| dilute hydrochloric acid |  |
| sodium hydroxide solution |  |
| vinegar |  |

b The following apparatus was set up to investigate the electrical conductivity of dilute acids.

Dilute sulfuric acid is a strong acid. If it was replaced by a weak acid, what two differences in the observations would you expect to make? [2]

c When nitric acid is added to water the following reaction occurs.

$$HNO_3 + H_2O \longrightarrow NO_3^- + H_3O^+$$

Give the name and the formula of the particle transferred from nitric acid to water. [2]

d This question is about the following oxides.

aluminium oxide $Al_2O_3$
calcium oxide $CaO$
carbon dioxide $CO_2$
carbon monoxide $CO$
magnesium oxide $MgO$
sulfur dioxide $SO_2$

  i Which will react with hydrochloric acid but not with aqueous sodium hydroxide? [1]
  ii Which will react with aqueous sodium hydroxide but not with hydrochloric acid? [1]
  iii Which will react both with hydrochloric acid and aqueous sodium hydroxide? [1]
  iv Which will react neither with hydrochloric acid nor with aqueous sodium hydroxide? [1]

*Cambridge IGCSE Chemistry 0620 Paper 3 Q3 June 2006*

13 Hydrogen reacts with the halogens to form hydrogen halides.
a Bond energy is the amount of energy, in kJ, that must be supplied (endothermic) to break one mole of a bond.

| bond | bond energy in kJ/mol |
|---|---|
| H–H | +436 |
| F–F | +158 |
| H–F | +562 |

Use the above data to show that the following reaction is exothermic.

$$H\!-\!H + F\!-\!F \longrightarrow 2H\!-\!F \quad [3]$$

b They react with water to form acidic solutions.

$$HCl + H_2O \rightleftharpoons H_3O^+ + Cl^-$$
$$HF + H_2O \rightleftharpoons H_3O^+ + F^-$$

  i Explain why water behaves as a base in both of these reactions. [2]
  ii At equilibrium, only 1% of the hydrogen chloride exists as molecules, the rest has formed ions. In the other equilibrium, 97% of the hydrogen fluoride exists as molecules, only 3% has formed ions. What does this tell you about the strength of each acid? [2]
  iii How would the pH of these two solutions differ? [1]

*Cambridge IGCSE Chemistry 0620 Paper 32 Q7 June 2009*

**14** The reactivity series lists metals in order of reactivity. To find out which is more reactive, zinc or tin, this experiment could be carried out.

**a** The experiment could be carried out with other metals and the results recorded in a table. Then the order of reactivity can be deduced.

  **i** The order was found to be:

  manganese   most reactive
  zinc          ↑
  tin
  silver       least reactive

  Copy and complete this table of results from which the order was determined. [3]

| aqueous solution | tin Sn | manganese Mn | silver Ag | zinc Zn |
|---|---|---|---|---|
| tin(II) nitrate | | R | NR | R |
| manganese(II) nitrate | | | | |
| silver(I) nitrate | | | | |
| zinc nitrate | | | | |

  **ii** Write the ionic equation for the reaction between tin atoms and silver(I) ions. [2]

  **iii** The following is a redox reaction.

  $Mn + Sn^{2+} \longrightarrow Mn^{2+} + Sn$

  Indicate on the equation the change which is oxidation. Give a reason for your choice. [2]

  **iv** Explain why experiments of this type cannot be used to find the position of aluminium in the reactivity series. [2]

**b** Potassium and calcium are very reactive metals at the top of the series. Because their ions have different charges, $K^+$ and $Ca^{2+}$, their compounds behave differently when heated.

  **i** Explain why the ions have different charges. [2]

  **ii** Their hydroxides are heated. If the compound decomposes, complete the word equation. If it does not decompose, write 'no reaction'.

  Potassium hydroxide ⟶ ..............
  Calcium hydroxide ⟶ .............. [2]

  **iii** Complete the equations for the decomposition of their nitrates.

  $2KNO_3 \longrightarrow$ .................. + ...............
  $2Ca(NO_3)_2 \longrightarrow$ .......... + .............. + ............. [4]

*Cambridge IGCSE Chemistry 0620 Paper 31 Q6 November 2008*

**15** The fractional distillation of crude oil usually produces large quantities of the heavier fractions. The market demand is for the lighter fractions and for the more reactive alkenes. The heavier fractions are cracked to form smaller alkanes and alkenes as in the following example.

$C_8H_{18} \longrightarrow C_4H_{10} + C_4H_8$
octane       butane    butenes

**a** **i** Write a different equation for the cracking of octane.

  $C_8H_{18} \longrightarrow$ ............. + ................... [1]

  **ii** The cracking of octane can produce isomers with the molecular formula $C_4H_8$. Draw the structural formulae of two of these isomers. [2]

**b** **i** Give the essential condition for the reaction between chlorine and butane. [1]

  **ii** What type of reaction is this? [1]

  **iii** This reaction produces a mixture of products. Give the names of two products that contain four carbon atoms per molecule. [2]

**c** Alkenes are more reactive than alkanes and are used to make a range of organic chemicals. Propene, $CH_3–CH=CH_2$, is made by cracking. Give the structural formula of the addition product when propene reacts with these.

  **i** water   **ii** bromine [2]

**d** Propene reacts with hydrogen iodide to form 2-iodopropane.

$CH_3–CH=CH_2 + HI \longrightarrow CH_3–CHI–CH_3$

1.4 g of propene produced 4.0 g of 2-iodopropane. Calculate the percentage yield.

  **i** moles of $CH_3–CH=CH_2$ reacted =
  **ii** maximum moles of $CH_3\text{-}CHI\text{-}CH_3$ that could be formed =
  *mass of one mole of $CH_3\text{-}CHI\text{-}CH_3$ is 170 g*
  **iii** maximum mass of 2-iodopropane that could be formed =
  **iv** percentage yield = [4]

*Cambridge IGCSE Chemistry 0620 Paper 3 Q7 June 2006*

**16** The alcohols form a homologous series. The first four members are methanol, ethanol, propan-1-ol and butan-1-ol.

**a** One characteristic of a homologous series is that the physical properties vary in a predictable way. The table below gives the heats of combustion of the first three alcohols.

| alcohol | formula | heat of combustion in kJ/mole |
|---|---|---|
| methanol | $CH_3OH$ | −730 |
| ethanol | $CH_3\text{-}CH_2\text{-}OH$ | −1370 |
| propan-1-ol | $CH_3\text{-}CH_2\text{-}CH_2\text{-}OH$ | −2020 |
| butan-1-ol | $CH_3\text{-}CH_2\text{-}CH_2\text{-}CH_2\text{-}OH$ | |

  **i** The minus sign indicates there is less chemical energy in the products than in the reactants. In what form is the energy given out? [1]
  **ii** Is the reaction exothermic or endothermic? [1]
  **iii** Complete the equation for the complete combustion of ethanol.

  $C_2H_5OH$ + ....$O_2$ ⟶ ............... + ............... [2]

  **iv** Determine the heat of combustion of butan-1-ol by plotting the heats of combustion of the first three alcohols against the number of carbon atoms per molecule. Label your graph as on the right.

  What is the heat of combustion of butan-1-ol in kJ/mol? [3]

  **v** Describe two other characteristics of homologous series. [2]

**b** Give the name and structural formula of an isomer of propan-1-ol. [2]

**c** Methanol is made from carbon monoxide.

$CO\ (g) + 2H_2\ (g) \rightleftharpoons CH_3OH\ (g)$

The forward reaction is exothermic.

**i** Describe how hydrogen is obtained from alkanes. [2]
**ii** Suggest a method of making carbon monoxide from methane. [2]
**iii** Which condition, high or low pressure, would give the maximum yield of methanol? Give a reason for your choice. [2]

**d** For each of the following predict the name of the organic product.
  **i** reaction between methanol and ethanoic acid [1]
  **ii** oxidation of propan-1-ol by potassium dichromate(VI) [1]
  **iii** removal of $H_2O$ from ethanol (dehydration) [1]

*Cambridge IGCSE Chemistry 0620 Paper 3 Q6 November 2007*

**17** Copy and complete the following table.

| gas | test for gas |
|---|---|
| ammonia | |
| | bleaches damp litmus paper |
| hydrogen | |
| | relights a glowing splint |
| | turns limewater milky |

[5]

*Cambridge IGCSE Chemistry 0620 Paper 3 Q1 November 2008*

## CAMBRIDGE IGCSE EXAM QUESTIONS from paper 6

**1** The colours present in some blackcurrant sweets can be separated by chromatography. The colours are water-soluble dyes. The diagrams show how the colours can be extracted from the sweets.

**a** Name the pieces of apparatus labelled A to C. [3]

The apparatus below was used to carry out the chromatography.

**b i** Name the solvent used. [1]
**ii** Label, with an arrow, the origin on the diagram. [1]
**c** On a larger copy of this rectangle, sketch the chromatogram you would expect if two different colours were present in the sweets. [1]

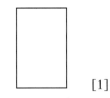

*Cambridge IGCSE Chemistry 0620 Paper 6 Q1 November 2008*

**2** Describe a chemical test to distinguish between each of the following pairs of substances.

**Example**: hydrogen and carbon dioxide
*test*: lighted splint
*result*: with hydrogen gives a pop
*result*: with carbon dioxide splint is extinguished

**a** zinc carbonate and zinc chloride [2]
**b** ammonia and chlorine [3]
**c** aqueous iron(II) sulfate and aqueous iron(III) sulfate [3]

*Cambridge IGCSE Chemistry 0620 Paper 6 Q3 June 2009*

**3** A student investigated the addition of four different solids, A, B, C and D, to water. Five experiments were carried out.

*Experiment 1*

By using a measuring cylinder, 30 cm$^3$ of distilled water was poured into a polystyrene cup and the initial temperature of the water was measured. 4 g of solid A was added to the cup and the mixture stirred with a thermometer. The temperature of the solution was measured after 2 minutes.

*Experiment 2*
Experiment 1 was repeated using 4 g of solid B.

*Experiment 3*
Experiment 1 was repeated using 4 g of solid C.

*Experiment 4*
Experiment 1 was repeated using 4 g of solid D.

## Experiment 5

A little of the solution from Experiment 4 was added to a little of the solution from Experiment 2 in a test-tube. The observations were recorded.

**observations**

a fast reaction
vigorous effervescence and bubbles produced

a Copy out the table and use the thermometer diagrams for Experiments 1–4 to record the initial and final temperatures. Calculate and record the temperature difference in the table.

| expt | initial temperature / °C | final temperature / °C | difference / °C |
|---|---|---|---|
| 1 | | | |
| 2 | | | |
| 3 | | | |
| 4 | | | |

[4]

b Draw a labelled bar chart of the results to Experiments 1, 2, 3 and 4 on graph paper. Label your bar chart as shown below. [4]

Use the results and observations from Experiments 1 – 5 to answer the following questions.

c  i Which solid dissolves in water to produce an exothermic reaction? [1]
   ii Give a reason why you chose this solid. [1]
d Which Experiment produced the largest temperature change? [1]
e Predict the temperature change that would happen if
   i 8 g of solid B was used in Experiment 2, [1]
   ii 60 cm³ of water was used in Experiment 4. [1]
   iii Explain your answer to **e ii**. [2]
f Suggest an explanation for the observations in Experiment 5. [2]

*Cambridge IGCSE Chemistry 0620 Paper 6 Q4 November 2008*

4 Hydrogen peroxide breaks down to form oxygen. The volume of oxygen given off can be measured using this apparatus.

Solids W and X both catalyse the breakdown of hydrogen peroxide. The syringe diagrams show the volume of oxygen formed every 20 seconds using these catalysts at 25 °C.

a Copy the table. Use the gas syringe diagrams to complete it.

| time / s | volume of oxygen / cm³ | |
|---|---|---|
| | catalyst W | catalyst X |
| 0 | | |
| 20 | | |
| 40 | | |
| 60 | | |
| 80 | | |
| 100 | | |

[3]

b Plot a graph to show each set of results. Clearly label the curves. [6]
c Which solid is the better catalyst in this reaction? Give a reason for your choice. [2]
d Why is the final volume of oxygen the same in each experiment? [1]
e Sketch a line on the grid to show the shape of the graph you would expect if the reaction with catalyst X was repeated at 40 °C. [2]

*Cambridge IGCSE Chemistry 0620 Paper 6 Q6 June 2007*

**5** An experiment was carried out to determine the solubility of potassium chlorate at different temperatures. The solubility is the mass of potassium chlorate that dissolves in 100 g of water. The results obtained are shown in the table below.

| temperature / °C | 0 | 10 | 20 | 30 | 40 | 50 | 60 |
|---|---|---|---|---|---|---|---|
| solubility in g / 100 g of water | 14 | 17 | 20 | 24 | 29 | 34 | 40 |

  **a** Draw a smooth line graph to show the solubility of potassium chlorate at different temperatures. Label your graph as shown below.

[4]

  **b** Use your graph to determine the solubility of potassium chlorate at 70 °C. Show clearly on the graph how you obtained your answer. [2]

  **c** What would be the effect of cooling a saturated solution of potassium chlorate from 60 °C to 20 °C? [2]

*Cambridge IGCSE Chemistry 0620 Paper 6 Q6 November 2008*

**6** The diagram shows the formation of a solution of magnesium hydroxide from magnesium.

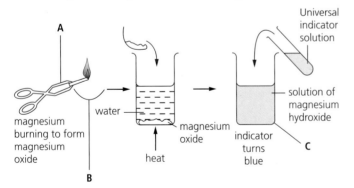

  **a** Name the pieces of apparatus labelled A – C. [3]

  **b** What type of chemical reaction is the burning of magnesium? [1]

  **c** Suggest a pH for the solution of magnesium hydroxide. [1]

*Cambridge IGCSE Chemistry 0620 Paper 6 Q1 November 2006*

**7** The diagram shows an experiment to pass electricity through lead bromide. Electricity has no effect on solid lead bromide.

  **a i** Copy the diagram and clearly label the electrodes. [1]

    **ii** Suggest a suitable material to make the electrodes. [1]

  **b** Give two observations expected when the lead bromide is heated to melting point. [2]

  **c** State two different safety precautions when carrying out this experiment. [2]

*Cambridge IGCSE Chemistry 0620 Paper 6 Q2 June 2008*

**8** Concentrated hydrochloric acid can be electrolysed using the apparatus shown.

  **a** Copy the diagram and label the position of the electrodes. [1]

  **b** Give two observations when the circuit is switched on. [2]

  **c i** Name the product at the positive electrode. [1]

    **ii** State a test for this product, and the result. [2]

*Cambridge IGCSE Chemistry 0620 Paper 6 Q2 June 2007*

**9** A student investigated the reaction between potassium manganate(VII) and a metallic salt solution. Two experiments were carried out.

*Experiment 1*

**a** About 1 cm³ of aqueous sodium hydroxide was added to a little of the salt solution **A** and the observation noted.

**observation**   *green precipitate formed*

**b** A burette was filled with potassium manganate(VII) solution up to the 0.0 cm³ mark.
By using a measuring cylinder, 25 cm³ of solution **A** of the salt was placed into a conical flask. The flask was shaken to mix the contents. The potassium manganate(VII) solution was added to the flask, and shaken to mix thoroughly. Addition of potassium manganate(VII) solution was continued until there was a pale pink colour in the contents of the flask.

Copy the table of results below.

| Burette readings / cm³ | | |
|---|---|---|
| | Experiment 1 | Experiment 2 |
| final reading | | |
| initial reading | | |
| difference | | |

Use this burette diagram to record the volume and complete the column for Experiment 1 in the table.

final reading

*Experiment 2*

**c** Experiment **1b** was repeated using a different solution **B** of the salt, instead of solution A. Use the burette diagrams to record the volumes in the table and complete the table.

initial reading      final reading

[4]

**d** About 1 cm³ of aqueous sodium hydroxide was added to a little of the solution in the flask and the observation noted.

**observation**   *red-brown precipitate*

**e i** In which experiment was the greatest volume of potassium manganate(VII) solution used? [1]
  **ii** Compare the volumes of potassium manganate(VII) solution used in Experiments 1 and 2. [2]
  **iii** Suggest an explanation for the difference in the volumes. [2]

**f** Predict the volume of potassium manganate(VII) solution which would be needed to react completely with 50 cm³ of solution **B**. [2]

**g** Explain one change that could be made to the experimental method to obtain more accurate results. [2]

**h** What conclusion can you draw about the salt solution from:
  **i** experiment **1a**, [1]
  **ii** experiment **2d**? [1]

*Cambridge IGCSE Chemistry 0620 Paper 6 Q4 June 2008*

**10** This label is from a container of 'Bite Relief' solution.

> **BITE RELIEF**
> **FOR FAST RELIEF FROM INSECT BITES AND STINGS**
>
> Active ingredient: ammonia
> Also contains water and alcohol
>
> DIRECTIONS FOR USE: Use cotton wool to dab the solution on the affected area of the skin.

**a** Give a chemical test to show the presence of ammonia in Bite Relief solution. Give the result of the test. [2]
**b** What practical method could be used to separate the mixture of alcohol (b.p. 78 °C) and water (b.p. 100 °C)? [2]
**c** Give a chemical test to show the presence of water. Give the result of the test. [2]
**d** What would be the effect of touching the alcohol with a lighted splint? [1]

*Cambridge IGCSE Chemistry 0620 Paper 6 Q7 June 2008*

# CAMBRIDGE IGCSE EXAM QUESTIONS

**11** Two solids, **S** and **V**, were analysed. **S** was copper(II) oxide. The tests on the solids, and some of the observations, are in the following table. Write down the observations that are missing from the table.

| test | observation |
|---|---|
| **tests on solid S** | |
| **a** Appearance of solid **S** | black solid |
| **b** Hydrogen peroxide was added to solid **S** in a test-tube. A glowing splint was inserted into the tube | slow effervescence<br>splint relit |
| **c** Dilute sulfuric acid was added to solid **S** in a test-tube. The mixture was heated to boiling point.<br>The solution was divided into three equal portions into test-tubes.<br>  **i** To the first portion of the solution excess sodium hydroxide was added.<br>  **ii** To the second portion of the solution, about 1 cm³ of aqueous ammonia solution was added.<br>    Excess ammonia solution was then added.<br>  **iii** To the third portion of the solution, dilute hydrochloric acid was added followed by barium chloride solution. | blue solution formed<br><br><br>.................................................. [1]<br><br>.................................................. [2]<br>.................................................. [2]<br><br>.................................................. [2] |
| **tests on solid V** | |
| **d** Appearance of solid **V** | black solid |
| **e** Hydrogen peroxide was added to solid **V** in a test-tube. A glowing splint was inserted into the tube. | rapid effervescence<br>splint relit |

  **f i** Compare the reactivity of solid **S** and solid **V** with hydrogen peroxide. [1]
    **ii** Identify the gas given off in test **e**. [1]
  **g** What conclusions can you draw about solid **V**? [2]

*Cambridge IGCSE Chemistry 0620 Paper 6 Q5 June 2009*

**12** A sample of solid **C** was analysed. **C** is a mixture of two salts, **D** and **E**. Solid **D** is insoluble lead carbonate and solid **E** is water-soluble.

The tests on **C**, and some of the observations are in the following tables. Complete the observations that would go in the tables.

| tests | observations |
|---|---|
| **a** Describe the appearance of **C**. | pale green solid |
| **b** Using a spatula, place a little of **C** in a hard glass test-tube. Inside the top of the tube suspend a piece of damp indicator paper. Heat **C** gently until gas comes out of the tube. | paper turns blue pH 8 to 11 |
| **c** Using a spatula, place a little of **C** in a test-tube. Add about 2 cm$^3$ of dilute nitric acid and test the gas. | **c** ................................................. [3] |

Solid **C** was added to a boiling tube containing distilled water. The tube was shaken to mix the contents. The contents of the boiling tube were filtered.

| tests on the residue in the filter paper | observations |
|---|---|
| **d** Place the funnel in a test-tube. Pour dilute nitric acid onto the residue contained in the funnel. Add 2 cm$^3$ of potassium iodide to the solution collected in the tube. | **d** ................................................. [2] |

| tests on the filtrate | observations |
|---|---|
| **e** Divide the filtrate into three test tubes. **i** To the first portion add dilute hydrochloric acid and about 1 cm$^3$ of aqueous barium nitrate. **ii** To the second portion of solution add excess aqueous ammonia. **iii** To the third portion of solution, add an equal volume of aqueous sodium hydroxide. Warm the mixture gently. Test the gas with indicator paper. | white precipitate green precipitate paper turns blue pH 8 to 11 |

**f** Name the gas given off in **c**. [1]
**g** Name the gas given off in **e iii**. [1]
**h** What conclusions can you draw about salt **E**? [4]

*Cambridge IGCSE Chemistry 0620 Paper 6 Q5 June 2007*

# REFERENCE

# Glossary

## A

**acetylene** a gas (formula $C_2H_2$) used as a fuel, for example in the oxy-acetylene torch

**acid rain** rain that is acidic because gases such as sulfur dioxide are dissolved in it (from burning fossil fuels)

**acidic solution** has a pH less than 7; an acidic solution contains $H^+$ ions

**acid fermentation** the process in which bacteria convert ethanol to ethanoic acid

**addition reaction** where a molecule adds onto an alkene, and the C = C double bond of the alkene changes to a single bond

**addition polymerisation** where small molecules join to form a very large molecule, by adding on at double bonds

**alcohols** a family of organic compounds, similar to the alkanes but with the OH functional group; ethanol is an example

**alkali** a soluble base; for example sodium hydroxide

**alkali metals** the Group I elements of the Periodic Table

**alkaline earth metals** the Group II elements of the Periodic Table

**alkaline solution** has a pH above 7; alkaline solutions contain $OH^-$ ions

**alkanes** a family of saturated hydrocarbons with the general formula $C_nH_{2n+2}$; 'saturated' means they have only single C–C bonds

**alkenes** a family of unsaturated hydrocarbons with the general formula $C_nH_{2n}$; their molecules contain a carbon = carbon double bond

**allotropes** different forms of an element; diamond and graphite are allotropes of carbon

**alloy** a mixture where at least one other substance is added to a metal, to improve its properties; the other substance is often a metal too (but not always)

**amphoteric** can be both acidic and basic in its reactions; for example aluminium oxide is an amphoteric oxide

**anion** another name for a negative ion

**anode** the positive electrode of a cell

**aquifer** underground rocks holding a large volume of water; it can be pumped out to give a water supply

**atmosphere** the layer of gases around the Earth; here at the Earth's surface, we call it air

**atoms** elements are made up of atoms, which contain protons, neutrons, and electrons

**Avogadro constant** the number of particles in one mole of an element or compound; it is $6.02 \times 10^{23}$

## B

**back reaction** the reaction in which the product breaks down again, in a reversible reaction

**bacteria** tiny organisms, some of which can cause disease; others break down dead plant and animal material

**balanced equation** a chemical equation in which the number of each type of atom is the same on both sides of the arrow

**base** a metal oxide or hydroxide; a base will neutralise an acid, to form a salt and water

**battery** a portable electrical cell; for example a torch battery

**biodegradable** will decay naturally in the soil, with the help of bacteria

**biopolymer** a polymer made by bacteria

**blast furnace** the chemical plant in which iron is extracted from its ore, iron(III) oxide

**boiling** the change from a liquid to a gas, which takes place at the boiling point

**boiling point** the temperature at which a substance boils

**bond energy** the energy needed to break a bond, or released when the bond is formed; it is given in kilojoules (kJ) per mole

**bonding** how the atoms are held together in an element or compound; there are three types of bonds: ionic, covalent, and metallic

**brittle** breaks up easily when struck

**brine** the industrial name for a concentrated solution of sodium chloride in water; it can be made by dissolving rock salt

**burette** a piece of laboratory equipment for delivering a measured volume of liquid

**burning** an exothermic chemical reaction in which the reactant combines with oxygen to form an oxide; also called combustion

## C

**carbon cycle** the way carbon moves non-stop between the atmosphere, living things, the land, and the ocean; it moves in the form of carbon dioxide

**carboxylic acids** a family of organic acids, which have the COOH functional group; ethanoic acid is an example

**cast iron** iron from the blast furnace that is run into molds to harden; it contains a high % of carbon, which makes it brittle

**catalyst** a substance that speeds up a chemical reaction, without itself being used up in the process

**catalytic converter** a device in a car exhaust, in which catalysts are used to convert harmful gases to harmless ones

**catalytic cracking** where large molecules of hydrocarbons are split up into smaller ones, with the help of a catalyst

**cathode** the negative electrode of an electrolysis cell

**cation** another name for a positive ion

**cell (biological)** the building blocks for animals and plants

**cell (electrical)** a device that converts chemical energy to electrical energy

**cement** a substance used in building, made from limestone and clay

**ceramic** a hard, ureactive material that can withstand high temperatures, made by baking clay in a kiln; ceramics are non-conductors

**chalk** a rock made of calcium carbonate

**change of state** a change in the physical state of a substance – for example from solid to liquid, or liquid to gas

**chemical change** a change in which a new chemical substance forms

**chemical equation** uses chemical symbols to describe a chemical reaction in a short way

**chemical reaction** a process in which chemical change takes place

**chromatogram** the paper showing the separated coloured substances, after paper chromatography has been carried out

**climate change** how climates around the Earth are changing, because of the rise in average air temperatures

**coagulant** a substance that will make small particles stick together; coagulants are used in cleaning up water, ready for piping to homes

**coke** a form of carbon made by heating coal

**combination** where two or more substances react to form a single substance

**combustible** can catch fire and burn very easily

**combustion** another name for burning

**compound fertiliser** it provides nitrogen, potassium, and phosphorus for plants

**compound ion** an ion containing more than one element; for example the nitrate ion $NO_3^-$

**compound** a substance in which two or more elements are chemically combined

**concentration** tells you how much of one substance is dissolved in another; usually given as grams or moles per $dm^3$

**condensation** the physical change in which a gas turns into a liquid on cooling

**condensation polymerisation** where molecules join to make very large molecules, by eliminating small molecules (such as water molecules)

**condenser** a piece of laboratory equipment used to cool a gas rapidly, and turn it into a liquid

**conductor** a substance that allows heat or electricity to pass through it easily

**Contact process** the industrial process for making sulfuric acid

**corrosion** where a substance is attacked by air or water, from the surface inwards; the corrosion of iron is called rusting

**covalent bond** the chemical bond formed when two atoms share electrons

**covalent compound** a compound made of atoms joined by covalent bonds

**cracking** reactions in which long-chain hydrocarbon molecules are broken down to shorter, more useful molecules

**cross-linking** the chemical bonds between the long-chain molecules in some polymers, that hold the chains together

**crude oil** the fossil fuel formed over millions of years from the remains of tiny sea plants and animals; it is also called petroleum

**crystallisation** the process in which crystals form, as a saturated solution cools

## D

**decomposition reaction** where a substance breaks down to give two or more products

**denature** to destroy the structure of an enzyme by heat, or a change in pH

**degradeable** will break down naturally (for example through the action of bacteria)

**density** tells you how 'heavy' something is; the density of a substance is its mass per unit volume; for water it is 1 g/cm$^3$

**diatomic** a substance is called diatomic if its molecules contain two atoms joined by a covalent bond

**diffusion** the process in which particles mix by colliding randomly with each other, and bouncing off in all directions

**displacement reaction** a reaction in which a more reactive element takes the place of a less reactive one, in a compound

**dissolving** the process in which a soluble substance forms a solution

**distillation** separating a liquid from a mixture by boiling it off, then condensing it

**double bond** a covalent bond in which two atoms share two pairs of electrons

**ductile** can be drawn out into a wire; for example copper is ductile

**dynamic equilibrium** where forward and back reactions take place at the same rate, so there is no *overall* change

## E

**electrodes** the conductors used to carry current into and out of an electrolyte; they could be graphite rods, for example

**electrolysis** the process of breaking down a compound by passing a current through it

**electrolyte** the liquid through which the current is passed, in electrolysis; the current is carried by ions in the electrolyte

**electron distribution** how the electrons in an atom are arranged in shells (2 + 8 + …)

**electron shells** the different energy levels which electrons occupy, around the nucleus

**electronic configuration** another term for electron distribution

**electrons** the particles with a charge of 1– and almost no mass, in an atom

**electroplating** coating one metal with another, using electrolysis

**element** a substance that cannot be split into anything simpler, in a chemical reaction

**empirical** found by experiment

**empirical formula** shows the simplest ratio in which the atoms in a compound are combined

**endothermic** takes in energy from the surroundings

**enzymes** proteins made by living cells, that act as biological catalysts

**equation** it uses symbols to describe a chemical reaction (but a *word equation* uses just words)

**equilibrium** the state where the forward and back reactions are taking place at the same rate, in a reversible reaction; so there is no *overall* change

**ester** a compound formed when an alcohol reacts with a carboxylic acid; esters often smell of fruit or flowers

**evaporation** the physical change where a liquid turns to a gas at a temperature below its boiling point

**exothermic** gives out energy

**extract** to remove a metal from its ore

## F

**fermentation** the process in which the enzymes in yeast break down sugars, to form ethanol and carbon dioxide

**fertilisers** substances added to soil to help crops grow well

**filtering** separating solids from liquids by pouring the mixture through filter paper

**filtrate** the liquid obtained from filtration (after the solid has been removed)

**flammable** burns easily

**flue gas desulfurisation** the removal of sulfur dioxide from the waste gases at power stations, to stop it getting into the atmosphere

**formula** uses symbols and numbers to tell you what elements are in a compound, and the ratio in which they are combined

**forward reaction** the reaction in which the product is made, in a reversible reaction

**fossil fuels** petroleum (crude oil), natural gas, and coal; they are called the fossil fuels because they were formed from the remains of living things, millions of years ago

**fractional distillation** a method used to separate two or more liquids that have different boiling points

**fractions** the different groups of compounds that a mixture is separated into, by fractional distillation; fractions are collected one by one

**freezing** the change from liquid to solid, that occurs at the freezing point (= melting point)

**fuel** a substance we use to provide energy; most fuels are burned to release their energy (but nuclear fuels are not)

**fuel cell** a cell in which a chemical reaction provides electricity (to light homes and so on)

**functional group** the part of the molecule of an organic compound, that largely dictates how it reacts; for example the OH group in molecules of the alcohol family

## G

**galvanising** coating iron with zinc, to prevent the iron from rusting

**giant structure** where a very large number of atoms or ions are held in a lattice by strong bonds; metals, diamond and ionic solids such as sodium chloride are all giant structures

**global warming** the rise in average temperatures taking place around the world; many scientists believe that carbon dioxide (from burning fossil fuels) is the main cause

**greenhouse gas** a gas in the atmosphere that traps heat, preventing its escape into space; carbon dioxide and methane are examples

**group** a column of the Periodic Table; elements in a group have similar properties

## H

**Haber process** the process for making ammonia from nitrogen and hydrogen, in industry

**half-equation** an equation that shows the reaction taking place at an electrode

**halogens** the Group VII elements of the Periodic Table

**heating curve** a graph showing how the temperature of a substance changes on heating, while it goes from solid to liquid to gas

**homologous series** a family of organic compounds, that share the same general formula and have similar properties

**hydrated** has water molecules built into its crystal structure; for example copper(II) sulfate: $CuSO_4.5H_2O$

**hydrocarbon** a compound containing *only* carbon and hydrogen

**hydrogenation** adding hydrogen

**hydrogen fuel cell** it uses the reaction between hydrogen (from a tank), and oxygen (from the air), to give an electric current

# REFERENCE

**hydrolysis** the breaking down of a compound by reaction with water

**hypothesis** a statement you can test by doing an experiment and taking measurements

## I

**incomplete combustion** the burning of fuels in a limited supply of oxygen; it gives carbon monoxide instead of carbon dioxide

**indicator** a chemical that shows by its colour whether a substance is acidic or alkaline

**inert** does not react (except under extreme conditions)

**inert electrode** is not changed during electrolysis; all it does is conduct the current

**in excess** more than is needed for a reaction; some will be left at the end

**insoluble** does not dissolve in a solvent

**insulator** a poor conductor of heat or electricity

**intermolecular forces** forces between molecules

**ion** a charged atom or group of atoms formed by the gain or loss of electrons

**ionic bond** the bond formed between ions of opposite charge

**ionic compound** a compound made up of ions, joined by ionic bonds

**ionic equation** shows only the ions that actually take part in a reaction, and ignores any other ions present; the other ions are called spectator ions

**isomers** compounds that have the same formula, but a different arrangement of atoms

**isotopes** atoms of the same element, that have a different numbers of neutrons

## L

**lattice** a regular arrangement of particles

**lime** the common name for calcium oxide

**limewater** a solution of the slightly soluble compound calcium hydroxide, which is used to test for carbon dioxide

**locating agent** used to show up colourless substances, in chromatography; it reacts with them to give coloured substances

## M

**macromolecule** a very large molecule; for example a molecule in a polymer

**malleable** can be bent or hammered into shape

**mass spectrometer** an instrument used to find the masses of atoms and molecules

**melting point** the temperature at which a solid substance melts

**melting** the physical change from a solid to a liquid

**metal** an element that shows metallic properties (for example conducts electricity, and forms positive ions)

**metallic bond** the bond that holds the atoms together in a metal

**metalloid** an element that has properties of both a metal and a non-metal

**microbe** a microscopic (very tiny) living organism, such as a bacterium or virus

**minerals** compounds that occur naturally in the Earth; rocks contain different minerals

**mixture** contains two or more substances that are not chemically combined

**molar solution** contains one mole of a substance in 1 dm$^3$ (1 litre) of water

**mole** the amount of a substance that contains the same number of elementary units as the number of carbon atoms in 12g of carbon-12; you obtain it by weighing out the $A_r$ or $M_r$ of the substance, in grams

**molecular** made up of molecules

**molecule** a unit of two or more atoms held together by covalent bonds

**monatomic** made up of single atoms; for example neon is a monatomic element

**monomers** small molecules that join together to form polymers

## N

**native** describes a metal that is found in the Earth as the element

**negative electrode** another name for the cathode, in an electrolysis cell

**negative ion** an ion with a negative charge

**neutral (electrical)** has no charge

**neutral (oxide)** is neither acidic nor basic; carbon monoxide is a neutral oxide

**neutral (solutions)** neither acidic nor alkaline; neutral solutions have a pH of 7

**neutralisation** the chemical reaction between an acid and a base or a carbonate, giving a salt and water

**neutron** a particle with no charge and a mass of 1 unit, found in the nucleus of an atom

**nitrogenous fertiliser** it provides nitrogen for plants, in the form of nitrate ions or ammonium ions

**noble gases** the Group 0 elements of the Periodic Table; they are called 'noble' because they are so unreactive

**non-metal** an element that does not show metallic properties: the non-metals lie to the right of the zig-zag line in the Periodic Table, (except for hydrogen, which sits alone)

**non-renewable resource** a resource such as petroleum that we are using up, and which will run out one day

**non-toxic** not harmful health

**nucleon number** the number of protons plus neutrons in an atom of an element

**nuclear fuel** contains radioisotopes such as uranium-235; these are forced to break down, giving out energy

**nucleus** the centre part of the atom, made up of protons and neutrons

## O

**ore** rock containing a metal, or metal compounds, from which the metal is extracted

**organic chemistry** the study of organic compounds (there are millions of them!)

**organic compound** a compound containing carbon, and usually hydrogen; petroleum is a mixture of many organic compounds

**oxidation** a chemical reaction in which a substance gains oxygen, or loses electrons

**oxidation state** every atom in a formula can be given a number that describes its oxidation state; for example in NaCl, the oxidation states are +I for sodium, and −I for chlorine

**oxide** a compound formed between oxygen and an other element

**oxidising agent** a substance that brings about the oxidation of another substance

**ozone** a gas with the formula $O_3$

**ozone layer** the layer of ozone up in the atmosphere, which protects us from harmful UV radiation from the sun

## P

**paper chromatography** a way to separate the substances in a mixture, using a solvent and special paper; the substances separate because they travel over the paper at different speeds

**percentage composition** it tells you which elements are in a compound, and what % of each is present by mass

**period** a horizontal row of the Periodic Table; its number tells you how many electron shells there are

**periodicity** the pattern of repeating properties that shows up when elements are arranged in order of proton number; you can see it in the groups in the Periodic Table

**Periodic Table** the table showing the elements in order of increasing proton number; similar elements are arranged in columns called groups

**petroleum** a fossil fuel formed over millions of years from the remains of tiny sea plants and animals; it is also called *crude oil*

**pH scale** a scale that tells you how acidic or alkaline a solution is; it is numbered 0 to 14

**photochemical reaction** a reaction that depends on light energy; photosynthesis is an example

**photodegradeable** can be broken down by light

**photosynthesis** the process in which plants convert carbon dioxide and water to glucose and oxygen

**physical change** a change in which no new chemical substance forms; melting and boiling are physical changes

**physical properties** properties such as density and melting point (that are not about chemical behaviour)

**pipette** a piece of laboratory equipment used to deliver a known volume of liquid, accurately

**plastics** a term used for synthetic polymers (made in factories, rather than in nature)

**pollutant** a substance that causes harm if it gets into the air or water

**pollution** when harmful substances are released into the environment

**polymer** a compound containing very large molecules, formed by polymerisation

**polymerisation** a chemical reaction in which many small molecules join to form very large molecules; the product is called a polymer

**positive ion** an ion with a positive charge

**precipitate** an insoluble chemical produced during a chemical reaction

**precipitation reaction** a reaction in which a precipitate forms

**product** a chemical made in a chemical reaction

**protein** a polymer made up of many different amino acid units joined together

**proton number** the number of protons in the atoms of an element; it is sometimes called the atomic number

**proton** a particle with a charge of 1+ and a mass of 1 unit, found in the nucleus of an atom

**pure** there is only one substance in it

### Q
**quicklime** another name for calcium oxide

### R
**radioactive isotopes (radioisotopes)** unstable atoms that break down, giving out radiation

**random motion** the zig-zag path a particle follows as it collides with other particles and bounces away again

**rate of reaction** how fast a reaction is

**reactant** a starting chemical for a chemical reaction

**reactive** tends to react easily

**reactivity** how readily a substance reacts

**reactivity series** the metals listed in order of their reactivity

**recycling** reusing resources such as scrap metal, glass, paper and plastics

**redox reaction** any reaction in which electrons are transferred; one substance is oxidised (it loses electrons) and another is reduced (it gains electrons)

**reducing agent** a substance which brings about the reduction of another substance

**reduction** when a substance loses oxygen, or gains electrons

**refining (petroleum)** the process of separating petroleum (crude oil) into groups of compounds with molecules fairly close in size; it is carried out by fractional distillation

**refining (metals)** the process of purifying a metal; copper is refined using electrolysis

**relative atomic mass ($A_r$)** the average mass of the atoms of an element, relative to the mass of an atom of carbon-12

**relative formula mass ($M_r$)** the mass of one formula unit of an ionic compound; you find it by adding together the relative atomic masses of the atoms in the formula

**relative molecular mass** the mass of a molecule; you find it by adding the relative atomic masses of the atoms in it

**renewable resource** a resource that will not run out; for example water, air, sunlight

**residue** the solid you obtain when you separate a solid from a liquid by filtering

**respiration** the reaction between glucose and oxygen that takes place in the cells of all living things (including you) to provide energy

**reversible reaction** a reaction that can go both ways: a product can form, then break down again; the symbol ⇌ is used to show a reversible reaction

**rusting** the name given to the corrosion of iron; oxygen and water attack the iron, and rust forms

### S
**sacrificial protection** allowing one metal to corrode, in order to protect another metal

**salt** an ionic compound formed when an acid reacts with a metal, a base, or a carbonate

**saturated compound** an organic compound in which all the bonds between carbon atoms are single covalent bonds

**saturated solution** no more of the solute will dissolve in it, at that temperature

**single bond** the bond formed when two atoms share just one pair of electrons

**slaked lime** another name for calcium hydroxide

**solubility** the amount of solute that will dissolve in 100 grams of a solvent, at a given temperature

**soluble** will dissolve in a solvent

**solute** the substance you dissolve in the solvent, to make a solution

**solution** a mixture obtained when a solute is dissolved in a solvent

**solvent** the liquid in which a solute is dissolved, to make a solution

**sonorous** makes a ringing noise when struck

**spectator ions** ions that are present in a reaction mixture, but do not actually take part in the reaction

**stable** unreactive

**state symbols** these are added to an equation to show the physical states of the reactants and products ($g$ = gas, $l$ = liquid, $s$ = solid, $aq$ = aqueous)

**structural formula** the formula of a compound displayed to show the bonds between the atoms as lines; we often show organic compounds this way

### T
**thermal decomposition** the breaking down of a compound by heating it

**thermite process** the redox reaction between iron oxide and aluminium, which produces molten iron

**titration** a laboratory technique for finding the exact volume of an acid solution that will react with a given volume of alkaline solution, or vice versa

**toxic** poisonous

**transition elements** the elements in the wide middle block of the Periodic Table; they are all metals and include iron, tin, copper, and gold

**trend** a gradual change; the groups within the Periodic Table show trends in their properties; for example as you go down Group I, reactivity increases

**triple bond** the bond formed when two atoms share three pairs of electrons; a nitrogen molecule has a triple bond

### U
**universal indicator** a paper or liquid you can use to find the pH of a solution; it changes colour across the whole range of pH

**unreactive** does not react easily

**unsaturated compound** an organic compound with at least one double bond between carbon atoms

### V
**valency** a number that tells you how many electrons an atom gains, loses or shares, in forming a compound

**valency electrons** the electrons in the outer shell of an atom

**variable valency** an element shows variable valency if its atoms can lose different numbers of electrons, in forming compounds; for example copper forms $Cu^+$ and $Cu^{2+}$ ions

**viscosity** a measure of how runny a liquid is; the more runny it is, the lower its viscosity

**viscous** thick and sticky

**volatile** evaporates easily, to form a vapour

### W
**water of crystallisation** water molecules built into the crystal structure of a compound; for example in copper(II) sulfate, $CuSO_4.5H_2O$

**weak acids** acids in which only some of the molecules are dissociated, to form $H^+$ ions; ethanoic acid is a weak acid

### Y
**yield** the actual amount of a product obtained in a reaction; it is often given as a % of the theoretical yield (which you can work out from the equation)

# REFERENCE

## The Periodic Table of the Elements

| Group | | | | | | | | | | | | | | | | | |
|---|---|---|---|---|---|---|---|---|---|---|---|---|---|---|---|---|---|
| I | II | | | | | | | | | | | | | | III | IV | |
| | | | | | | 1<br>**H**<br>Hydrogen<br>1 | | | | | | | | | | | |
| 7<br>**Li**<br>Lithium<br>3 | 9<br>**Be**<br>Beryllium<br>4 | | | | | | | | | | | | | 11<br>**B**<br>Boron<br>5 | 12<br>**C**<br>Carbon<br>6 | | |
| 23<br>**Na**<br>Sodium<br>11 | 24<br>**Mg**<br>Magnesium<br>12 | | | | | | | | | | | | | 27<br>**Al**<br>Aluminium<br>13 | 28<br>**Si**<br>Silicon<br>14 | | |
| 39<br>**K**<br>Potassium<br>19 | 40<br>**Ca**<br>Calcium<br>20 | 45<br>**Sc**<br>Scandium<br>21 | 48<br>**Ti**<br>Titanium<br>22 | 51<br>**V**<br>Vanadium<br>23 | 52<br>**Cr**<br>Chromium<br>24 | 55<br>**Mn**<br>Manganese<br>25 | 56<br>**Fe**<br>Iron<br>26 | 59<br>**Co**<br>Cobalt<br>27 | 59<br>**Ni**<br>Nickel<br>28 | 64<br>**Cu**<br>Copper<br>29 | 65<br>**Zn**<br>Zinc<br>30 | 70<br>**Ga**<br>Gallium<br>31 | 73<br>**Ge**<br>Germanium<br>32 | | | | |
| 85<br>**Rb**<br>Rubidium<br>37 | 88<br>**Sr**<br>Strontium<br>38 | 89<br>**Y**<br>Yttrium<br>39 | 91<br>**Zr**<br>Zirconium<br>40 | 93<br>**Nb**<br>Niobium<br>41 | 96<br>**Mo**<br>Molybdenum<br>42 | 101<br>**Tc**<br>Technetium<br>43 | 101<br>**Ru**<br>Ruthenium<br>44 | 103<br>**Rh**<br>Rhodium<br>45 | 106<br>**Pd**<br>Palladium<br>46 | 108<br>**Ag**<br>Silver<br>47 | 112<br>**Cd**<br>Cadmium<br>48 | 115<br>**In**<br>Indium<br>49 | 119<br>**Sn**<br>Tin<br>50 | | | | |
| 133<br>**Cs**<br>Caesium<br>55 | 137<br>**Ba**<br>Barium<br>56 | 139<br>**La**<br>Lanthanum<br>57 * | 178<br>**Hf**<br>Hafnium<br>72 | 181<br>**Ta**<br>Tantalum<br>73 | 184<br>**W**<br>Tungsten<br>74 | 186<br>**Re**<br>Rhenium<br>75 | 190<br>**Os**<br>Osmium<br>76 | 192<br>**Ir**<br>Iridium<br>77 | 195<br>**Pt**<br>Platinum<br>78 | 197<br>**Au**<br>Gold<br>79 | 201<br>**Hg**<br>Mercury<br>80 | 204<br>**Tl**<br>Thallium<br>81 | 207<br>**Pb**<br>Lead<br>82 | | | | |
| **Fr**<br>Francium<br>87 | 226<br>**Ra**<br>Radium<br>88 | 227<br>**Ac**<br>Actinium<br>89 † | | | | | | | | | | | | | | | |

*58–71 Lanthanoid series
†90–103 Actinoid series

| | | | | | | | | | |
|---|---|---|---|---|---|---|---|---|---|
| 140<br>**Ce**<br>Cerium<br>58 | 141<br>**Pr**<br>Praseodymium<br>59 | 144<br>**Nd**<br>Neodymium<br>60 | **Pm**<br>Promethium<br>61 | 150<br>**Sm**<br>Samarium<br>62 | 152<br>**Eu**<br>Europium<br>63 | 157<br>**Gd**<br>Gadolinium<br>64 | 159<br>**Tb**<br>Terbium<br>65 | 163<br>**Dy**<br>Dysprosium<br>66 | 165<br>**Ho**<br>Holmium<br>67 |
| 232<br>**Th**<br>Thorium<br>90 | **Pa**<br>Protactinium<br>91 | 238<br>**U**<br>Uranium<br>92 | **Np**<br>Neptunium<br>93 | **Pu**<br>Plutonium<br>94 | **Am**<br>Americium<br>95 | **Cm**<br>Curium<br>96 | **Bk**<br>Berkelium<br>97 | **Cf**<br>Californium<br>98 | **Es**<br>Einsteinium<br>99 |

## The symbols and proton numbers of the elements

| Element | Symbol | Proton number | Element | Symbol | Proton number | Element | Symbol | Proton number | Element | Symbol | Proton number |
|---|---|---|---|---|---|---|---|---|---|---|---|
| actinium | Ac | 89 | calcium | Ca | 20 | francium | Fr | 87 | lawrencium | Lw | 103 |
| aluminium | Al | 13 | californium | Cf | 98 | gadolinium | Gd | 64 | lead | Pb | 82 |
| americium | Am | 95 | carbon | C | 6 | gallium | Ga | 31 | lithium | Li | 3 |
| antimony | Sb | 51 | cerium | Ce | 58 | germanium | Ge | 32 | lutetium | Lu | 71 |
| argon | Ar | 18 | chlorine | Cl | 17 | gold | Au | 79 | magnesium | Mg | 12 |
| arsenic | As | 33 | chromium | Cr | 24 | hafnium | Hf | 72 | manganese | Mn | 25 |
| astatine | At | 85 | cobalt | Co | 27 | helium | He | 2 | mendelevium | Md | 101 |
| barium | Ba | 56 | copper | Cu | 29 | holmium | Ho | 67 | mercury | Hg | 80 |
| berkelium | Bk | 97 | curium | Cm | 96 | hydrogen | H | 1 | molybdenum | Mo | 42 |
| beryllium | Be | 4 | dysprosium | Dy | 66 | indium | In | 49 | neodymium | Nd | 60 |
| bismuth | Bi | 83 | einsteinium | Es | 99 | iodine | I | 53 | neon | Ne | 10 |
| boron | B | 5 | erbium | Er | 68 | iridium | Ir | 77 | neptunium | Np | 93 |
| bromine | Br | 35 | europium | Eu | 63 | iron | Fe | 26 | nickel | Ni | 28 |
| cadmium | Cd | 48 | fermium | Fm | 100 | krypton | Kr | 36 | niobium | Nb | 41 |
| caesium | Cs | 55 | fluorine | F | 9 | lanthanum | La | 57 | nitrogen | N | 7 |

# REFERENCE

| V | VI | VII | 0 |
|---|---|---|---|
| | | | 4<br>**He**<br>Helium<br>2 |
| 14<br>**N**<br>Nitrogen<br>7 | 16<br>**O**<br>Oxygen<br>8 | 19<br>**F**<br>Fluorine<br>9 | 20<br>**Ne**<br>Neon<br>10 |
| 31<br>**P**<br>Phosphorus<br>15 | 32<br>**S**<br>Sulfur<br>16 | 35.5<br>**Cl**<br>Chlorine<br>17 | 40<br>**Ar**<br>Argon<br>18 |
| 75<br>**As**<br>Arsenic<br>33 | 79<br>**Se**<br>Selenium<br>34 | 80<br>**Br**<br>Bromine<br>35 | 84<br>**Kr**<br>Krypton<br>36 |
| 122<br>**Sb**<br>Antimony<br>51 | 128<br>**Te**<br>Tellurium<br>52 | 127<br>**I**<br>Iodine<br>53 | 131<br>**Xe**<br>Xenon<br>54 |
| 209<br>**Bi**<br>Bismuth<br>83 | **Po**<br>Polonium<br>84 | **At**<br>Astatine<br>85 | **Rn**<br>Radon<br>86 |

**Key**

| a |
|---|
| **X** |
| b |

a = relative atomic mass
**X** = atomic symbol
b = proton number

| 167<br>**Er**<br>Erbium<br>68 | 169<br>**Tm**<br>Thulium<br>69 | 173<br>**Yb**<br>Ytterbium<br>70 | 175<br>**Lu**<br>Lutetium<br>71 |
|---|---|---|---|
| **Fm**<br>Fermium<br>100 | **Md**<br>Mendelevium<br>101 | **No**<br>Nobelium<br>102 | **Lr**<br>Lawrencium<br>103 |

## Relative atomic masses ($A_r$) for calculations

| Element | Symbol | $A_r$ |
|---|---|---|
| aluminium | Al | 27 |
| bromine | Br | 80 |
| calcium | Ca | 40 |
| carbon | C | 12 |
| chlorine | Cl | 35.5 |
| copper | Cu | 64 |
| fluorine | F | 19 |
| helium | He | 4 |
| hydrogen | H | 1 |
| iodine | I | 127 |
| iron | Fe | 56 |
| lead | Pb | 207 |
| lithium | Li | 7 |
| magnesium | Mg | 24 |
| manganese | Mn | 55 |
| neon | Ne | 20 |
| nitrogen | N | 14 |
| oxygen | O | 16 |
| phosphorus | P | 31 |
| potassium | K | 39 |
| silver | Ag | 108 |
| sodium | Na | 23 |
| sulfur | S | 32 |
| zinc | Zn | 65 |

| Element | Symbol | Proton number | Element | Symbol | Proton number | Element | Symbol | Proton number |
|---|---|---|---|---|---|---|---|---|
| nobelium | No | 102 | rhodium | Rh | 45 | thallium | Tl | 81 |
| osmium | Os | 76 | rubidium | Rb | 37 | thorium | Th | 90 |
| oxygen | O | 8 | ruthenium | Ru | 44 | thulium | Tm | 69 |
| palladium | Pd | 46 | samarium | Sm | 62 | tin | Sn | 50 |
| phosphorus | P | 15 | scandium | Sc | 21 | titanium | Ti | 22 |
| platinum | Pt | 78 | selenium | Se | 34 | tungsten | W | 74 |
| plutonium | Pu | 94 | silicon | Si | 14 | uranium | U | 92 |
| polonium | Po | 84 | silver | Ag | 47 | vanadium | V | 23 |
| potassium | K | 19 | sodium | Na | 11 | xenon | Xe | 54 |
| praseodymium | Pr | 59 | strontium | Sr | 38 | ytterbium | Yb | 70 |
| promethium | Pm | 61 | sulfur | S | 16 | yttrium | Y | 39 |
| protactinium | Pa | 91 | tantalum | Ta | 73 | zinc | Zn | 30 |
| radium | Ra | 88 | technetium | Tc | 43 | zirconium | Zr | 40 |
| radon | Rn | 86 | tellurium | Te | 52 | | | |
| rhenium | Re | 75 | terbium | Tb | 65 | | | |

# Index

Where several page numbers are given and one is **bold**, look that one up first.

## A
acetone 17
acetylene 213
acidic oxides 157
acidity 150
  and pH number 149
acid rain **214**, 225
acids 148–155
  and pH 149
  as proton donors 154, **155**
  carboxylic 258–259
  reactions of 152–153
  strong and weak 148, 150
addition
  polymerisation 264–265
air 210–213
alanine 274
alcoholic drinks 256
alcohols 256–257
algae 229
alkali metals 166, **168–169**
alkaline earth metals 166, 182
alkalis 148–152
alkanes 250, 251, **252–253**
alkenes 254–255
allotropes (of carbon) 61, 234
alloys 203, 204
alumina (aluminium oxide) 200, 201
aluminium
  alloys of 203
  apparent unreactivity 191
  corrosion in 191, 201
  extraction of 200–201
  in the Earth's crust 194
  in the reactivity series 188
  ore 195, 197, **200**
  properties of 201
  uses of 200, **202**
aluminium ions, test for 286
aluminium oxide 195, 197, 201
amide linkage **266**, 274
amino acids 24, **274**, 276, 277
ammonia 225–227
  as an alkali 148, 151
  bonding in 57
  Haber process for 226
  in making fertilisers 229
  laboratory preparation 225
  manufacture 126–127, **226–227**
  test for 285
ammonium compounds
  chloride 225
  nitrate 225, 228

ammonium ion, test for 285
amphoteric oxides 157
anode 104
aqueous solution 17
$A_r$ 70
argon 37, 48, **172–173**
artificial elements 30, **167**
artificial fertilisers 228
atmosphere 210
atoms 30, **32**
Avogadro constant 76
Avogadro's Law 80
aquifer 218

## B
back reaction 124
bacteria 35, 141, 142, 218
baking soda 52, 153
balancing equations 68–69
barium sulfate 160, 287
bases 152–153
  alkalis (soluble bases) 148–152
  and acids 152–153
  as proton acceptors 155
  reactions of 152
basic oxides 156
batteries 122–123
bauxite (aluminium ore) 195, **200**
biological catalysts 140
biological detergents 141
bio-polymers 271
bitumen 247
blast furnace 198–199
bleaching
  by chlorine 285
  by hydrogen peroxide 87
  by sulfur dioxide 231
boiling (change of state) 11
boiling points 9
  and purity 19
  for ionic compounds 59
  for molecular substances 54, 59
bond breaking and making 116
bond energies 116
bonding
  between ions 50–51
  in giant covalent structures 60–61
  in metals 62
  in molecules 54–57
brass 203
breathalyser test 99
brine 108
  electrolysis of 108–109
bromide ion, test for 287
bronze 206
burning of fuels 114, 118–119
  energy from 118, 119

butane 252
butanoic acid 258
but-1-ene 254

## C
calamine lotion 153
calcium
  reaction with oxygen 93
  reactivity of 184, **188**
calcium compounds
  carbonate 236, 241
  hydroxide 16, 148, 241
  oxide 93, 199, 205, 240, 241
calcium ions, test for 286
calculations from equations 78–79
cancer treatment by radioisotopes 35
carbohydrates 272–273
  complex 273
carbon
  allotropes of 61
  as a reducing agent 196
  in steel 204, **205**
  isotopes of 34
  reactivity compared to metals 186, **188**
carbon-12 **34**, 70, 76
carbonates 236
carbon cycle 234–235
carbon dioxide 236
  and global warming 238–239
  and the carbon cycle 234–235
  bonding in 57
  in photosynthesis 235
  properties of 236
carbonic acid 148, 236
carbon monoxide 236
  as a reducing agent 196
carboxyl functional group 258
carboxylic acids 258–259
car exhausts 215
cast iron 199
catalysts 140–141
  biological (enzymes) **140**, 142–143
  in car exhausts 215
  transition elements as **141**, 175
cathode 104
cell (producing electricity) 120
cellulose 263
cement 241
chain length
  and boiling point 251
  and homologous series 251
  in polythene 264

changes of state 8–11
charcoal 234
charge (for blast furnace) 198
charge (on sub-atomic particles) 32
chemical change 46, **47**
chemical equations 68–69
chemical properties 42
chemical reaction 47
chlorine
  as a member of Group VII 170–171
  bonding in 55
  from electrolysis of brine 108
  reaction with sodium 48–50
  uses of 109
chlorine water 171
chloromethane 253
chlorophyll 144
chromatogram 23
chromatography (paper) 23–25
chromium 175, 204
  use in plating 111
citric acid 148
climate change 239
coal 214, **244**
cobalt chloride 124
coke 198
collision theory (for reaction rate) 138–139
combustion 93, 114
competition among metals 186–187
compound ions 53
compounds 46
  and mixtures 46
  covalent **56–57**, 58–59
  forming 48
  ionic **50–51**, 58–59
  names and formulae 66–67
  percentage composition of 73
compression of gases 12
concentration of a solution 82–83
  finding by titration 162–163
concrete 241
condensation 11
condensation polymerisation 263, **266–267**
condenser 22
conductivity of metals 42, 63, **103**
conductors 102–103
Contact process 232

316

# INDEX

copper 174
  bonding in 62
  in reactivity series 188
  refining of 110
  uses of 202
copper ion, test for 286
copper(II) oxide 92
copper(II) sulfate 158
  electrolysis of solution 110
  in testing for water 124
corrosion
  of aluminium 191
  of iron 216–217
covalent bond 54–55
covalent compounds 56–57
covalent giant structures 60–61
cracking of hydrocarbons 248–249
crude oil (petroleum) 244–245
crystallisation 20
cysteine 274

## D
decay (radioactivity) 34
denature (enzymes) 141
density 182
detergents (biological) 141
diamond 60
diaphragm cell 108
diatomic 54
dichloromethane 253
diesel (fraction from petroleum) 247
diffusion 7, 13
displacement reactions 171, 185, 187
dissolving 16
distillation 22
distilled water 22
double bonds 55
  in alkenes 254
ductile (of metals) 63, **182**
dynamic equilibrium 125

## E
Earth's crust 194
electricity 102
electrode 104
electrolysis 102–111
  in extraction of aluminium 201
  in extraction of metals 196–197
  of brine 108
  of copper(II) sulfate solution 110
  of molten compounds 104
  of solutions 105
electrolyte 103
electron distribution 36–37
electrons **32**, 36
electron shells 36
electron transfer (in redox reactions) 94
electroplating 111
elements 30–31
empirical formula 84–85
endothermic reactions 115
energy changes during reactions 114–115
energy from fuels 118–119
energy level diagram 114, 115
enzymes **140**, 141, 142–143
equations 68–69
  calculations using 78–79
  half-equations 94
  ionic equations **95**, 154
equilibrium (in reversible reactions) 125
ester linkage **259**, 267, 275
esters 259
ethane 252
ethanoates 259
ethanoic acid **258**, 259
ethanol **256**, 257
ethene **254**, 255
evaporation 11
exothermic reactions 114
explosions 137
extraction of metals 196–197

## F
fats 275
fatty acids 275
fermentation 142, **256**
fertilisers 228–229
filtering 20
filtrate 20
flue gas desulfurisation 241
fluoride (in water supply) 219
formulae **46**, 66
  empirical 84–85
  molecular 86–87
  of compounds 66–67
  of ionic compounds 52–53
formula mass 71
forward reaction 124
fossil fuels 244
fountain experiment 225
fractional distillation 22
fractions in fractional distillation 246
fractions from refining petroleum 247
freezing 8
freezing point 8
fuel cell 121
fuel oil 247
fuels 118–119
functional group **251**, 254, 256, 258

## G
galena (ore) 230
galvanising **191**, 217
gas chromatography 26
gases **8**, 10, 12–13
  and Avogadro's Law 80
  compression of 12
  diffusion in 13
  effect of temperature on 12
  pressure in 12
gasoline (petrol) 247
giant covalent structures 60–61
global warming 238–239
glucose 272, **273**
glycerol 275
glycine 274
gold 195
graphite 61
greenhouse gases 238
groundwater 218
groups in Periodic Table 31, **166**
  Group 0 (noble gases) 172–173
  Group I (alkali metals) 168–169
  Group II (alkaline earth metals) 166
  Group VII (halogens) 170–171
gypsum 241

## H
Haber process 226–227
haemoglobin **236**, 274
half-equations 94
halides 170
halogens 170–171
heating curve 9
helium 37, 172, 173
hematite (iron ore) 198
homologous series 251
hydrocarbons **251**, 252, 254
hydrochloric acid **148**, 150
  electrolysis of dilute 105
  electrolysis of concentrated 105
hydrogen 224
  as a fuel 119, **121**
  bonding in 54
  displacement of 185
  from electrolysis of brine 108
  in Periodic Table 167
  reactivity relative to metals 188
  test for 285
  uses of 109
hydrogen chloride 116, 150
  bonding in 56
hydrogen ions, in acids **150**, 154
hydrogen sulfide 67, 230
hydrolysis
  in digestion 276
  in the lab 277
hydroxide ions (and alkalinity) 151
hydroxides **152**, 184, 189

## I
impurities 18–17
incomplete combustion 252
indicators 149
indigestion 155
insoluble salts 160–161
insulators 102
iodide ion, test for 287
iodine 170
  displacement of 171
ionic bonds 50–51
ionic compounds 50, 51
  and electrolysis 104
  formulae of 52
  names of 52–53
  properties of 59
ions **49**, 52, 53
  compound ions 53
  mass of 71
iron **174**, 175
  alloys of 204–205
  corrosion (rusting) of 216–217
  extraction of 198–199
  in the Earth's crust 194
  in reactivity series 188
  ore 198
  properties of 174, 175, 204
iron ions, test for 286
iron(III) oxide 156, **198**
iron sulfide 46
isomers **253**, 255
isotopes **34**, 70

## K
kerosene 247
kilojoule (kJ) 114
kinetic particle theory 11
krypton 172, 173

## L
lattice **50**, 58
lead
  and extraction methods 196
  'lead' in pencils 61
  reactivity of 185, 186, 188
lead bromide, electrolysis of 104, 106
Le Chatelier's principle 126
light (and photochemical reactions) 144–145, 253
lime 153, 240
limestone 153, **240–241**
limewater **241**, 285
liquids 8–11
lithium 123, **168**, 169
litmus 148
locating agent 25
lubricating fraction 247

## M
macromolecules 262, 272–275
magnesium 51, 132, 185
magnesium chloride 51, 132, 185
magnesium oxide 51, 85, 186
magnesium sulfate (Epsom salts) 52

# INDEX

malleable (of metals) 63, **174**
manganese (and variable valency) 96
manganese(IV) oxide 122, **140**
mass of atoms, molecules and ions 70–71
melting 8, 9, **10**
melting points **8**, 9
  and purity 19
  of ionic compounds 59
  of metals 62
  of molecular covalent compounds 59
  of giant covalent structures 60
metallic bonds 62
metalloids 176
metal oxides 156, 157
metals 42, 43, **182–183**
  bonding in 62–63
  compared with non-metals 42–43
  extraction from ores 196–197
  in Earth's crust 194
  in Periodic Table 182
  in reactivity series 188
  ores 195
  properties of 182–183
  reactions with oxygen 156
  reactivity of 184–188
  stability of compounds 189
  transition elements 174–175
  uses of 175, 202–203
metal salts 152, 158–161
methane 252
  as fuel (natural gas) 244
  as greenhouse gas 238
  bonding in 56
methanoic acid 252, **258**
methyl orange indicator 149
mild steel 204
mining of metal ores 195
mixtures 16
  compared with compounds 46
mobile phase (in chromatography) 26
molar volume of a gas 80
molar solution 82
mole 76
molecular formulae 46, **66**
molecular substances **54–57**, 59
molecules 54
  masses of 71
monomer **262**, 264, 266
monosaccharides 273
$M_r$ 71

## N

names for compounds 52–53, 66–67
naphtha **247,** 249
natural gas 118, **244**
natural polymers 263
neon 48, **172–173**
neutralisation reactions 153, **154–155**
neutral liquids (and pH) 149
neutral oxides 157
neutrons **32,** 33
nickel 174
ninhydrin 25
nitrate ion, test for 287
nitrates **152,** 160, 189
nitric acid 148
nitrogen 224
  bonding in 55
  for plants 228
  production in industry 212
  uses 213
nitrogen oxides as pollutants 214
noble gases 172–173
  electron distribution in 48
  in Periodic Table 166
  uses of 173
non-conductor 102
non-metals 42, 43
  compared with metals 42
  in Periodic Table 31
  reactions with oxygen 157
non-renewable resource 245
nuclear fuels 119
nucleon number 33
nucleus of atoms 32
nylon **266,** 269

## O

oil (petroleum) 244–245
  and pollution 214
  refining of 246–247
  sulfur extraction from 230
ores **195,** 196
organic chemistry 237
organic compounds **237,** 250
oxidation 92
  and electron transfer 94–95
oxidation state 96–97
oxides 156–157
oxidising agents 98–99
oxy-acetylene torch 213
oxygen
  in air 210–211
  bonding in the molecule 55
  bonding in ionic compounds 51
  reaction with metals and non-metals 156–157
  separation from air 212
  uses 213

oxygen mask 213
oxygen tent 213
oxygen, test for 285

## P

painting (against rust) 217
palmitic acid 275
paper chromatography 23, 24–25
paraffin (kerosene) 247
particles in matter 6–7, **10–11**
  evidence for 6, 7
pentane 250
percentage composition of compounds 73
percentage purity 73, **88–89**
percentage yield 88
Periodic Table **166–167,** 178–179
petrol (gasoline) 87, **247**
petroleum (crude oil) 244–245
  and pollution 214
  refining of 246–247
  sulfur extraction from 230
pH scale 149
phenolphthalein 149
phosphorus
  combustion in oxygen 157
  impurity in steel-making 205
  for plants 228
phosphorus pentoxide 157
photographic film **145,** 161
photosynthesis **144,** 234
physical change 47
physical properties 42
phytoplankton 235
plastics 262, **268–269**
  and pollution 270–271
  made by polymerisation 264–267
platinum
  in catalytic converters 215
  in the Earth's crust 194
  for inert electrodes 104
pollution
  in air 214–215
  by plastic 270–271
polyamide 266
poly(chloroethene) or PVC 265
polyester 267
polymerisation 255, 263, **264–267**
polymers **262–263,** 264–267
poly(propene) 265
polysaccharide 273
polystyrene 265
polythene, or poly(ethene) 262, **264**
  and chain lengths 264
  and pollution 270

changing properties of 268
  uses of 269
potassium (as Group I metal) 168–169
  reactivity 184, 185, **188**
potassium compounds
  bromide, and displacement 171
  bromide, and electrolysis 105
  bromide, and photography 161
  dichromate(VI), as oxidising agent 99
  iodide, and displacement 171
  iodide, as reducing agent 99
  hydroxide **148,** 151, 184
  manganate(VII) and diffusion 7
  manganate(VII), as oxidising agent **98**
potassium for plants 228
power stations **118,** 119, 231, 241
precipitation 160–161
pressure in gases 12
propane 250, **252**
propane-1,2,3-triol (glycerol) 275
propanoic acid 258
propanone 17
propene 254
proteins 263, **272**
  from polymerisation of amino acids 274
  hydrolysis of 276, 277
proton donors and acceptors 155
proton number **32,** 33
protons 32
purity 18
  and melting and boiling points 19
  percentage 73, 88, 89
PVC (polychloroethene) **265,** 269

## Q

quartz 60
quicklime **240,** 241

## R

radiation **34,** 35
radioactivity 34–35
radioisotopes 34
random motion of particles 6
rate of reaction **131,** 130–145
  and catalysts 140–141
  and concentration **134,** 138

# INDEX

and photochemical
 reactions 140–141
and surface area **136–137,**
 139
and temperature **135,** 139
measuring **131,** 132–137
reactants 68
reactions, equations for
 68–69
 energy changes in 114–115
 reversible 124–125
reactivity
 of Group I metals 168–169
 of Group VII
  (halogens) 170–171
 of Group 0 (noble
  gases) 172
 of metals 184–187
reactivity series of
 metals 188
recycling iron and steel 205
redox reactions 92–97
 and electron transfer
  94–95
 and oxidation state 96–97
reducing agent **98,** 99
reduction 92–95
 in metal extraction 196
refining petroleum (oil) 22,
 **246–247**
relative atomic mass, $A_r$ 70
relative formula mass, $M_r$ 71
relative molecular mass,
 $M_r$ 71
residue from filtration 20
respiration 234, **235**
reversible reactions 124–125
$R_f$ value 25
rock salt 195
rotary kiln 240
rubber (vulcanizing of) 230
rusting and its prevention
 216–217

## S

sacrificial protection **191,**
 217
salt (sodium chloride) 50, **58**
salts 152
 making in laboratory
  158–161
 solubility of some 160

sand 60
saturated compounds
 (alkanes) 252
saturated solutions 17
scientific method 280
semi-conductors 177
separation methods 20–25
shells (electron) 36
silicon dioxide (silica) **60,**
 61, 66
silver (a transition
 element) 174
 and electroplating 111
 in film photography 145
 reactivity of 184, 185, **188**
silver bromide **145,** 161
single bond 54
slag 199, 205
slaked lime 153, **241**
smoke particles 6
soap 277
sodium
 as alkali metal 168–169
 extraction of 196
 properties of 168–169
sodium chloride 48
 bonding in 48, 50
 electrolysis of molten
  compound 105
 electrolysis of
  solutions **105,** 106–108
 structure and
  properties 58, 59
sodium hydroxide 184
 as alkali 148, 151
 from electrolysis of
  brine 108
 uses of 109
soil acidity **153,** 240
solidifying 11
solids **8,** 11
solutions 16–17
 concentration of 82–83
 electrolysis of **105,**
  106–107
solvents **16,** 17
sonorous 182
stability of metal
 compounds 189
stainless steel **204,** 205
standard solution 162

starch 263, 272
 from polymerisation of
  glucose 273
 hydrolysis of 276, 277
state symbols 68
steel 175
 electroplating of 111
 galvanising of 191
 manufacture of 204–205
 recycling of 205
styrene 265
substitution reaction 253
sulfates 152
 solubility of 160
sulfur 30, **230**
sulfur dioxide 157, **231**
 and flue gas
  desulfurisation 241
 as pollutant 214, **231**
sulfur trioxide 232
sulfuric acid 233
 manufacture of 232
 properties 233
sulfurous acid 231
 surface area of solid
  reactant
  and rates of
   reaction **136–137,** 139
 and risk of explosions 137
symbols for elements 30
synthetic polymers 262

## T

Teflon **265,** 269
Terylene **267,** 269
tests in the laboratory
 for anions and
  cations 286–287
 for gases 285
 for unsaturation 255
 for water 124
tetrachloromethane 253
thermal decomposition **189,**
 240, 248
thermite process 190
tin (for plating steel) 111
titanium 108, 206, 207, 264
titrations 159, **162–163**
tracers 35
transition elements 167,
 **174–175**
 as catalysts 141, 175
trichloromethane 253

## U

universal indicator 149
unsaturated (organic
 compounds) 254
 test for unsaturation 255
unsaturated fats 275

## V

valency **66,** 67, 176
valency electrons 166
variables (in experiments) 280
vinegar 258
vinyl chloride 265
viscosity 247, 251
volatile liquids **17,** 246
voltage (of cells) 190
vulcanizing 230

## W

water 224
 as solvent 16, 17
 bonding in 56
 changes of state for 8
 formula of 46
 heating curve for 9
 reaction with metals 184
 tests for 124
 uses of 218
water supply 218–219
water vapour 8

## X

xenon 172, 173

## Y

yeast 142, **256**
yield 88
 in making ammonia 126,
  **227**

## Z

zinc **174,** 182
 extraction of 197
 reactivity of 184–186, **188**
 uses of 202
zinc blende (ore) 197
zinc chloride 185
zinc oxide 157
zinc sulfate 131

**Acknowledgments:**

The publisher would like to thank the following for their kind permission to reproduce photographs:

P6t: RMG; P6bl: Shutterstock; P6br: Dan Gair Photographic/Photolibrary; P7: Philippe Plailly/Science Photo Library; P9t: Bowater Peter/Photo Researchers,Inc/Photo Library; P9b: Tepic/Shutterstock; P10t: Istockphoto; P10m: Istockphoto; P10b: Istockphoto; P12t: Jose AS Reyes/Shutterstock; P12m: Prestige; P12b: Cebas/Istockphoto; P13t: Ekaterina Pokrovskaga/Shutterstock; P13b: Bobby Yip/Reuters; P16t: OUP; P16b: OUP; P17: Rich Legg/Istockphoto; P18bl: Mariusz S.Jurgielewicz/Shutterstock; P18br: Sean Locke/Istockphoto; P19:OUP; P20: Maria Toutoudaki/Istockphoto; P21t: Wijnand Loven/Istockphoto; P21m: OUP; P21b: OUP; P22: Robert Francis/Photolibrary; P24: Yuri Arcurs/Shutterstock; P26: Loren Rodgers/Shutterstock; P27: Mara Radeva/Istockphoto; P27b: Phil Augustavo/Istockphoto; P30tl: OUP; P30tm: Shutterstock; P30tr: OUP; P30m: The Art Gallery Collection/Alamy; P30b: AK Photo/Shutterstock; P31: Hulton Archive/Getty Images; P32: Iofoto/Shutterstock; P34: Science Photo Library/Photolibrary; P35t: Hank Morgan/Science Photo Library; P35bl: JTB Photo/Photolibrary; P35br: Roca/Shutterstock; P36: Baron/Hulton Archive/Getty Images; P37: OUP; P38t: Georgios Kollidas/Shutterstock; P38m: Stephen Kiers/Shutterstock; P38b: Science Photo Library; P39t: Sheila Terry/Science Photo Library; P39b: Philippe Plailly/Science Photo Library; P40t: Science Photo Library; P40b: Mark Yuill/Shutterstock; P41bl: CERN; P41br: A. Barrington Brown/Science Photo Library; P42: Omar Farooque; P43tl: Alan Crawford/Istockphoto; P43tr: Omar Farooque; P43b: Kushch Dmitry/Shutterstock; P46bl: Journal of Chemical Education Softward, Chemistry Comes Alive! Vol 3 © 1999, Division of Chemical Education; P46bm: Journal of Chemical Education Softward, Chemistry Comes Alive! Vol 3 © 1999, Division of Chemical Education; P46br: Journal of Chemical Education Softward, Chemistry Comes Alive! Vol 3 © 1999, Division of Chemical Education; P47: Tlorna/Shutterstock; P48tl: Photocritical/Shutterstock; P48tr: Xtrekx/Shutterstock; P50: OUP; P52: Katarzyna Krawiec/Shutterstock; P54: Paul Ingram; P55t: OUP; P55m: OUP; P55b: OUP; P56t: OUP; P56m: OUP; P56b: OUP; P57(t-b): OUP; P57m: Paul Ingram; P58tl: OUP; P58tr: Daniel Taeger/Shutterstock; P58bl: Stefan Glebowski/Shutterstock; P58br: Dmitry Galanternik/Istockphoto; P59t: Maximilian Stock Ltd/Science Photo Library; P59b: Stefan Redel/Shutterstock; P60t: Dmitry Kalinovsky/Shutterstock; P60b: OUP; P61t: OUP; P61b: RMG; P62t: Andrew Lambert/University of Liverpool; P62bl: Andrew Lambert/University of Liverpool; P62br: Andrew Lambert/University of Liverpool; P63t: Crafts Council; P63b: Omar Farooque; P66: ImageState/Alamy; P67: Paul Rapson/Science Photo Library; P68: Johnson Space Center/NASA; P69t: Yoav Levy/Photolibrary; P69b: OUP; P72t: OUP; P72b: French School,(20th century)/Musee de la Ville de Paris,Musee Carnavalet,Paris, France/Archives Charmet / The Bridgeman Art Library; P73: Iain McGillivray/Shutterstock; P76bl: OUP; P76bm: OUP; P76br: OUP; P78: Journal of Chemical Education Softward, Chemistyr Comes Alive!Vol3©1999, Division of Chemical Education; P79t: OUP; P79b: OUP; P80: OUP; P81: Thomas J.Casadevall/U.S. Geological Survey; P84: OUP; P85: Ron Kloberdanz/Shutterstock; P86: Geoff Tompkinson/Science Photo Library/Photolibrary; P87t: Aj Photo/Hop Americain/Science Photo Library/Photolibrary; P87b: David H. Lewis/Istockphoto; P88t: Thor Jorgen Udvang/Shutterstock; P88b: Chepko Danil Vitalevich/Shutterstock; P89: Tina Lorien/Istockphoto; Page 92t: OUP; P92bl: Jordache/Shutterstock; P92br: Claudio Arnese/Istockphoto; P93t: Olga Utlyakova/Istockphoto; P93b: MorePixels/Istockphoto; P95: Charles D. Winters/Science Photo Library; P96t: OUP; P96b: OUP; P97t: OUP; P97b: OUP; P98: OUP; P99t: OUP; P99m: Jim Varney/Science Photo Library/Photolibrary; P99b: OUP; P102t: OUP; P102b: Muellek Josef/Shutterstock; P103l: OUP; P103m: OUP; P103r: OUP; P108t: Georg Gerster/Photo Researchers; P108b: OUP; P109l: Andrzej Drozdza/Istockphoto; P109r: Mona Makela/Shutterstock; P111l: Ullstein Ecopix/Still Pictures; P111r: OUP; P114l: OUP; P114m: OUP; P114r: OUP; P115l: OUP; P115m: Sciencephotos/Alamy; P115r: OUP; P116: Charles D. Winters/Science Photo Library; P118l: Jiri Jura/Shutterstock; P118m: Sagasan/Shutterstock; P118r: Chris G. Parkhurst/Shutterstock; P119t: Carolina K. Smith, M.D./Shutterstock; P119b: Postnikova Kristina/Shutterstock; P121: American Honda Motor Co., Inc.; P122t: Jon Schutte/Istockphoto; P122b: Hywit Dimyadi/Shutterstock; P123: Stepanov/Shutterstock, Design/Shutterstock, Agita Leimane/Shutterstock, Konstantin Chagin/Shutterstock; P124l: OUP; P124r: OUP; P125: Peter Albrektsen/Shutterstock; P130tl: OUP; P130tm: Jorgen Schytte/Photolibrary; P130tr: Charles Schug/Istockphoto; P130bl: Best Images/Shutterstock; P130bm: Valeria/Shutterstock; P130br: Seraficus/Istockphoto; P131l: OUP; P131m: OUP; P131r: OUP; P132: OUP; P134: OUP; 135t: Filipe Varela/Istockphoto; 135b: Christine Glade/Istockphoto; P136l: OUP; P136r: OUP; P137t: Tyler Stableford/Getty Images; P137b: STR New/Reuters; P141t: James L. Amos/Corbis; P141b: Gabi Moisa/Shutterstock; P142t: Dmitry Knorre/Istockphoto; P142m: Grasko/Shutterstock; P142bl: Volker Steger/Science Photo Library; P142br: OUP; P143t: Shannon Matteson/Shutterstock; P143m: Tomas Bercic/Istockphoto; P143b: NOAA Photo Library; P144t: Power And Syred/Science Photo Library; P144b: Nigel Cattlin/FLPA; P148l: OUP; P148tm: OUP; P148tr: OUP; P148m: OUP; P148b: OUP; P150t: Dino O./Shutterstock; P150b: Rob Cruse/Istockphoto; P151: RMG; P152t: OUP; P152m: OUP; P152b: OUP; P153t: INSADCO Photography/Alamy; P153b: Ivanschenko Roman/Shutterstock; P154: Andrew Lambert Photography/Science Photo Library; P155: RMG; P157t: OUP; P157b: Jodi Jacobson/Istockphoto; P158: Canis Maior/Shutterstock; P159: Laurence Gough/Istockphoto; P160: OUP; 161t: Brenda Carson/Shutterstock; 161bl: Marianne Rosenstiehl/Corbis; 161br: Niall McDiarmid/Alamy; P163tl: Laurence Gough/Shutterstock; P163tr: Andrew Lambert Photography/Science Photo Library; P163br: Nell Redmond/Istockphoto; P167t: Photodisc/OUP; P167b: Omar Farooque; P168: OUP; P169: OUP; P170: David H. Lewis/Istockphoto; P171: Andrew Lambert Photography /Science Photo Library; P173tl: Amy Nichole Harris/Shutterstock; P173tr: OUP; P173br: One clear vision/Istockphoto; P174tl: OUP; P174tm: OUP; P174tr: OUP; P174br: Collection of 'Bizarre' series ceramics (ceramic), Cliff, Clarice (1899-1972)/Private Collection/Photo © Bonhams, London, UK/The Bridgeman Art Library; P175: Mike Clarke/Istockphoto; P177t: Maria Skaldina/Shutterstock; P177b: John Keith/Shutterstock; P178t: March Cattle/Shutterstock; P178b: Skyhawk/Shutterstock; P179tl: Mary Evans Picture Library/Photolibrary; P179tr: A. Zschau, GSI; P179mr: Dmitry Yashkin/Shutterstock; P179br: Graca Victoria/Istockphoto; P183tl: Chris Schmidt/Istockphoto; P183tr: Pgiam/Istockphoto, P183bl: OUP; P183bm: OUP; P183br: Elianet Ortiz/Istockphoto; P185: OUP; P188t: Martin Anderson/Shutterstock; P188b: Sergey Peterman/Shutterstock; P189: Dean Conger/Corbis; P190: The Metropolitan Council; P191t: Leslie Garland Picture Library/Alamy; P191b: George Peters/Istockphoto; P194t: STS-114 Crew/NASA; P194m: Thorsten Rust/Shutterstock; P194b: Zoran Karapancev/Shutterstock; P195tl: OUP; P195tm: OUP; P195tr: OUP; P195br: Lee Prince/Shutterstock; P196t: Paul Fleet/Shutterstock; P197t: Maximilian Stock Ltd/Science Photo Library; P197b: Tamara Kulikova/Istockphoto; P198t: Crown Copyright/Health & Safety Laboratory/Science Photo Library; P198b: SergioZ/Shutterstock; P199: David Reilly/Shutterstock; P200tl: Andreas Reh/Istockphoto; P200tm: Iuoman/Istockphoto; P200tr: Robert Weber/Istockphoto; P200bl: OUP; P200bm: OUP; P200br: Colin Young/Istockphoto; P201t: Maximilian Stock Ltd/Science Photo Library; P201b: Nolimitpictures/Istockphoto; P202l: Christian Darkin/Science Photo Library/Photolibrary; P202m: RMG; P202r: Vasily Smirnov/Shutterstock; P203: Garloon/Shutterstock; P204tl: OUP; P204tm: Anton Foltin/Istockphoto; P204tr: Svetlana Lukienko/Shutterstock; P204bl: Miguel Malo/Istockphoto; P204br: Fancy/Photolibrary; P205l: Rihardzz/Shutterstock; P205r: Oleg-F/Shutterstock; P207tl: Erhan Dayi/Shutterstock; P207tr: Jasmin Awad/Istockphoto; P207mr: Eoghan McNally/Shutterstock; P207br: Georgios Kollidas/Istockphoto; P210l: NASA; P210br: Elisei Shafer/Shutterstock; P211t: Sebastian Kaulitzki/Istockphoto; P211b Grzegorz Choinski/Istockphoto; P212: Pasquale Sorrentino/Science Photo Library; P213tl: Medical-on-Line/Alamy; P213tr: Jordache/Shutterstock; P213b: Jonathan Feinstein/Shutterstock; P214t: OUP; P214b: Ian Bracegirdle/Shutterstock; P215: Barnaby Chambers/Shutterstock; 216tl: Shutterstock; 216tr: Piotr Tomicki/Shutterstock; 216b: Kolbjorn Hoseth Larssen/Istockphoto; P217tl: Shutterstock, P217tr: Robert Buchanan Taylor/Shutterstock; P217b: Ricardo Azoury/Istockphoto; P218t: OUP; P218r: Kris Jacobs/Shutterstock; P220: NASA; P221: NASA; P224: Mohammed Yousuf/The Hindu; P225: OUP; P226: ICI; P228tl: P Wei/Shutterstock; P228tm: Shutterstock; P228tr: Adam Hart-Davis/Science Photo Library; P228m: Jean-Michel Groult/Photolibrary; P228b: Ron Giling/Lineair/Still Pictures; P229t: Yara International ASA; P229b: P229: Heike Kampe/Istockphoto; P230m: OUP; P230b: OUP; P231t: OUP; P231bl: Kuttelvaserova/Shutterstock; P231br: Alessandro Minguzzi/Istockphoto; P232: Manamana/Shutterstock; P233t: Martyn F. Chillmaid/Science Photo Library; P233b: JinYoung Lee/Shutterstock; P234: Jurie Maree/Shutterstock; P235t: Sportgraphic/Shutterstock; P235b: Elwynn1130/Istockphoto; P236t: Jam-design.cz/Shutterstock; P236b: Don Hammond/Design Pics/Corbis; P237t: Scott Lomenzo/Shutterstock; 237bl: Ranplett/Istockphoto; P237br: Ghaint/Shutterstock; P238: Javarman/Shutterstock; P239t: Marc van Vuren/Shutterstock; P239b: Pauline S Mills/Istockphoto; P240: Christopher Jones/Istockphoto; P241t: Emmanuel R Lacoste/Shutterstock; P241b: Ruslan Gilmanshin/Istockphoto; P244tl: Paul Rapson/Science Photo Library; P244tm: Oleksandr Kalinichenko/Shutterstock; P244tl: Hywit Dimyadi/Istockphoto; P245tl: Lya Cattel/Istockphoto; P245tm: OUP; P245tr: OUP; P245b: Eyeidea/Istockphoto; P247: Mike Clarke/Istockphoto; P248t: Zybr/Shutterstock; P248b: Christian Lagerek/Shutterstock; P249t: Tony Tremblay/Istockphoto; P249b: MirAgareb/Istockphoto; P250: Rtimages/Istockphoto; P252: Sciencephotos/Alamy; P253: Mark Evans/Istockphoto; P245: NFSphoto/Shutterstock; P255: Andre Blais/Shutterstock; P256t: Originalpunkt/Shutterstock; P256b: Tish1/Shutterstock; P257t: Celso Pupo/Shutterstock; P257bl: Carroteater/Istockphoto; P257br: Tomas Bercic/Istockphoto; P258: Gordana Sermek/Shutterstock; P259: OUP; P262tl: OUP; P262tm: OUP; P262tr: Gelpi/Shutterstock; P262b: RMG; P263tl: Highviews/Shutterstock; P263tm:Shutterstock; P263tr: Gabi Moisa/Shutterstock; P263b: Chris Hellyar/Istockphoto; P264t: Miguel Malo/Istockphoto; P264b: Andriscam/Shutterstock; P265: Istockphoto; P266t: Charles D.Winters/Science Photo Library; P266b: Dreamstime; P267t: OUP; P267b: Shutterstock; P268t: P268t: Tomas Bercic/iStockphoto; P268b: OUP; P269t:  Alexey U/Shutterstock; P269bl: Christina Richards/Istockphoto; P269br: Uyen Le/Istockphoto; P270t: Izabela Habur/Istockphoto; P270bl: Andreas Weiss/Istockphoto; P270br: Andy Davies/Science Photo Library; P271t: Maxine Adcock/Photolibrary; P271b: Peter Yates/Science Photo Library; P273t: Michał Krakowiak/Istockphoto; P273b: Tania A/Shutterstock; P275: Heldur Netocny/Photolibrary; P276t: Anita Patterson Peppers/Big Stock; P276m: Martin Turzak/Shutterstock; P276b: Shout/Rex Features; P277: Alandj/Istockphoto.